DESIGN OF EXPERIMENTS

STATISTICS: Textbooks and Monographs

A SERIES EDITED BY

D. B. OWEN, Coordinating Editor

Department of Statistics
Southern Methodist University
Dallas, Texas

Vol. 1: The Generalized Jacknife Statistic, *H. L. Gray and W. R. Schucany*
Vol. 2: Multivariate Analysis, *Anant M. Kshirsagar*
Vol. 3: Statistics and Society, *Walter T. Federer*
Vol. 4: Multivariate Analysis: A Selected and Abstracted Bibliography, 1957-1972, *Kocherlakota Subrahmaniam and Kathleen Subrahmaniam* (out of print)
Vol. 5: Design of Experiments: A Realistic Approach, *Virgil L. Anderson and Robert A. McLean*
Vol. 6: Statistical and Mathematical Aspects of Pollution Problems, *John W. Pratt*
Vol. 7: Introduction to Probability and Statistics (in two parts), Part I: Probability; Part II: Statistics, *Narayan C. Giri*
Vol. 8: Statistical Theory of the Analysis of Experimental Designs, *J. Ogawa*
Vol. 9: Statistical Techniques in Simulation (in two parts), *Jack P. C. Kleijnen*
Vol. 10: Data Quality Control and Editing, *Joseph I. Naus* (out of print)
Vol. 11: Cost of Living Index Numbers: Practice, Precision, and Theory, *Kali S. Banerjee*
Vol. 12: Weighing Designs: For Chemistry, Medicine, Economics, Operations Research, Statistics, *Kali S. Banerjee*
Vol. 13: The Search for Oil: Some Statistical Methods and Techniques, *edited by D. B. Owen*
Vol. 14: Sample Size Choice: Charts for Experiments with Linear Models, *Robert E. Odeh and Martin Fox*
Vol. 15: Statistical Methods for Engineers and Scientists, *Robert M. Bethea, Benjamin S. Duran, and Thomas L. Boullion*
Vol. 16: Statistical Quality Control Methods, *Irving W. Burr*
Vol. 17: On the History of Statistics and Probability, *edited by D. B. Owen*
Vol. 18: Econometrics, *Peter Schmidt*
Vol. 19: Sufficient Statistics: Selected Contributions, *Vasant S. Huzurbazar (edited by Anant M. Kshirsagar)*
Vol. 20: Handbook of Statistical Distributions, *Jagdish K. Patel, C. H. Kapadia, and D. B. Owen*
Vol. 21: Case Studies in Sample Design, *A. C. Rosander*
Vol. 22: Pocket Book of Statistical Tables, *compiled by R. E. Odeh, D. B. Owen, Z. W. Birnbaum, and L. Fisher*
Vol. 23: The Information in Contingency Tables, *D. V. Gokhale and Solomon Kullback*

Vol. 24: Statistical Analysis of Reliability and Life-Testing Models: Theory and Methods, *Lee J. Bain*
Vol. 25: Elementary Statistical Quality Control, *Irving W. Burr*
Vol. 26: An Introduction to Probability and Statistics Using BASIC, *Richard A. Groeneveld*
Vol. 27: Basic Applied Statistics, *B. L. Raktoe and J. J. Hubert*
Vol. 28: A Primer in Probability, *Kathleen Subrahmaniam*
Vol. 29: Random Processes: A First Look, *R. Syski*
Vol. 30: Regression Methods: A Tool for Data Analysis, *Rudolf J. Freund and Paul D. Minton*
Vol. 31: Randomization Tests, *Eugene S. Edgington*
Vol. 32: Tables for Normal Tolerance Limits, Sampling Plans, and Screening, *Robert E. Odeh and D. B. Owen*
Vol. 33: Statistical Computing, *William J. Kennedy, Jr. and James E. Gentle*
Vol. 34: Regression Analysis and Its Application: A Data-Oriented Approach, *Richard F. Gunst and Robert L. Mason*
Vol. 35: Scientific Strategies to Save Your Life, *I. D. J. Bross*
Vol. 36: Statistics in the Pharmaceutical Industry, *edited by C. Ralph Buncher and Jia-Yeong Tsay*
Vol. 37: Sampling from a Finite Population, *J. Hajek*
Vol. 38: Statistical Modeling Techniques, *S. S. Shapiro*
Vol. 39: Statistical Theory and Inference in Research, *T. A. Bancroft and C.-P. Han*
Vol. 40: Handbook of the Normal Distribution, *Jagdish K. Patel and Campbell B. Read*
Vol. 41: Recent Advances in Regression Methods, *Hrishikesh D. Vinod and Aman Ullah*
Vol. 42: Acceptance Sampling in Quality Control, *Edward G. Schilling*
Vol. 43: The Randomized Clinical Trial and Therapeutic Decisions, *edited by Niels Tygstrup, John M. Lachin, and Erik Juhl*
Vol. 44: Regression Analysis of Survival Data in Cancer Chemotherapy, *Walter H. Carter, Jr., Galen L. Wampler, and Donald M. Stablein*
Vol. 45: A Course in Linear Models, *Anant M. Kshirsagar*
Vol. 46: Clinical Trials: Issues and Approaches, *edited by Stanley H. Shapiro and Thomas H. Louis*
Vol. 47: Statistical Analysis of DNA Sequence Data, *edited by B. S. Weir*
Vol. 48: Nonlinear Regression Modeling: A Unified Practical Approach, *David A. Ratkowsky*
Vol. 49: Attribute Sampling Plans, Tables of Tests and Confidence Limits for Proportions, *Robert E. Odeh and D. B. Owen*
Vol. 50: Experimental Design, Statistical Models, and Genetic Statistics, *edited by Klaus Hinkelmann*
Vol. 51: Statistical Methods for Cancer Studies, *edited by Richard G. Cornell*
Vol. 52: Practical Statistical Sampling for Auditors, *Arthur J. Wilburn*
Vol. 53: Statistical Signal Processing, *edited by Edward J. Wegman and James G. Smith*
Vol. 54: Self-Organizing Methods in Modeling: GMDH Type Algorithms, *edited by Stanley J. Farlow*
Vol. 55: Applied Factorial and Fractional Designs, *Robert A. McLean and Virgil L. Anderson*
Vol. 56: Design of Experiments: Ranking and Selection, *edited by Thomas J. Santner and Ajit C. Tamhane*

OTHER VOLUMES IN PREPARATION

DESIGN OF EXPERIMENTS
RANKING AND SELECTION

Essays in Honor of Robert E. Bechhofer

Edited by

Thomas J. Santner
School of Operations Research
and Industrial Engineering
Cornell University
Ithaca, New York

Ajit C. Tamhane
Department of Industrial Engineering
and Management Sciences
Northwestern University
Evanston, Illinois

MARCEL DEKKER, INC. New York and Basel

Library of Congress Cataloging in Publication Data

Main entry under title:

Design of experiments.

(Statistics, textbooks and monographs ; v. 56)
Includes index.
1. Ranking and selection (Statistics) 2. Experimental design. I. Santner, Thomas J.
II. Tamhane, Ajit C. III. Series.
QA278.75.D47 1984 519.5 84-12121
ISBN 0-8247-7274-1

MARCEL DEKKER, INC.
270 Madison Avenue, New York, New York 10016

Current printing (last digit):
10 9 8 7 6 5 4 3 2 1

PRINTED IN THE UNITED STATES OF AMERICA

Dedicated to Robert E. Bechhofer
On the Occasion of His 65th Birthday

PREFACE

This volume of research papers in statistics is presented as a tribute to Bob Bechhofer on the occasion of his sixty-fifth birthday. Many of the papers included in it study problems of designing experiments for selection. Coincidentally, the year 1984 also marks the 30th anniversary of the publication of Bob's pioneering paper "A single-sample multiple decision procedure for selecting the best of several normal populations with unknown variances" in the Annals of Mathematical Statistics which introduced a radical change in the classical way of formulating a class of problems that had been treated previously by analysis of variance techniques. In the simplest case, it proposed the problem of designing a single-factor experiment to select the "best" level of the factor. Since 1954 the subject of ranking and selection has developed into a field of its own with well over 700 published papers and several books. In addition to his work on ranking and selection, Bob Bechhofer has also been instrumental in the formulation and solution of other problems of designing experiments, in particular, when the goal of the experiment is to make multiple comparison statements and for clinical trials. During the over thirty years of his Cornell career, he has not only personally contributed to these subjects but has also interested numerous colleagues and students in working on them. This volume is a token of appreciation for the inspiration, encouragement and warm friendship that Bob has provided to all of us.

There is a long list of individuals who gladly gave of their time to help bring this volume to a successful completion. We would like to identify them in the chronological order in which the project evolved. Initially an advisory committee consisting of Saul Blumenthal, Charles Dunnett, and Shanti Gupta was formed to help us identify potential contributors. We are grateful for their help with this important task which was made difficult because of page and cost restrictions for the book. The invitees included Bob's former students, present and former colleagues, and other researchers who have been influenced by Bob's work.

They were asked to write articles on research topics of their choice, not necessarily on ranking and selection, although this ultimately was the main focus. To maintain the high standards which Bob has always insisted in his own and others' work, all articles were refereed anonymously as well as by one of us. Most of the authors offered their services as referees and in addition we would like to acknowledge the following individuals who also acted in this capacity: Sid Dalal, Bill Fairweather, Fred Faltin, Leon Gleser, Tony Hayter, John Healy, Yosef Hochberg, Jiunn Hwang, Iain Johnstone, Harji Patel, John Ramberg, Harold Sackrowitz, and Leonard Stefanski. We sincerely thank all the authors and referees for their contributions.

Saul Blumenthal, Charlie Dunnett, Bill Fairweather, Lionel Weiss, and especially Joan Bechhofer provided details about Bob's family history and academic career that were invaluable in the preparation of the dedication article. The articles were typed by Kathy King, Anne Kline and Elaina Jeddry at Cornell and by June Wayne at Northwestern. We thank them for their patience and an excellent job. We would also like to acknowledge the support of our individual department Chairmen (George Nemhauser and Les Trotter at Cornell and Art Hurter at Northwestern) in this work. Finally, we express our appreciation to the staff at Marcel Dekker for their help in the production of this volume.

Thomas J. Santner

Ajit C. Tamhane

CONTENTS

CONTRIBUTORS

ROGER L. BERGER Department of Statistics, North Carolina State University, Raleigh, North Carolina

SAUL BLUMENTHAL Department of Statistics, Ohio State University, Columbus, Ohio

LAWRENCE D. BROWN Department of Mathematics, Cornell University, Ithaca, New York and Department of Statistics, Hebrew University, Jerusalem, Israel

RAYMOND J. CARROLL Department of Statistics, University of North Carolina, Chapel Hill, North Carolina

C. L. CHANG* Department of Mathematics, National Taiwan Normal University, Taipei, Republic of China

EDWARD J. DUDEWICZ Department of Mathematics, Syracuse University, Syracuse, New York

CHARLES W. DUNNETT Department of Clinical Epidemiology and Biostatistics, McMaster University, Hamilton, Ontario, Canada

WALTER T. FEDERER Biometrics Unit, Cornell University, Ithaca, New York

PAUL P. GALLO Statistical Design and Analysis Department, Lederle Laboratories, Pearl River, New York

SHANTI S. GUPTA Department of Statistics, Purdue University, West Lafayette, Indiana

IRWIN GUTTMAN Department of Statistics, University of Toronto, Toronto, Canada

JASON C. HSU Department of Statistics, The Ohio State University, Columbus, Ohio

D. Y. HUANG Department of Mathematics, National Taiwan Normal University, Taipei, Republic of China

WEN-TAO HUANG Institute of Mathematics, Academia Sinica, Taipei, Republic of China

C. JENNISON Department of Mathematical Sciences, University of Durham, Durham, England

*Current affiliation: Examination Yuan, Taipei, Republic of China.

S. JEYARATNAM Department of Mathematics, Southern Illinois University, Carbondale, Illinois

WOO-CHUL KIM Department of Computer Science and Statistics, Seoul National University, Seoul, Korea

R.V. KULKARNI SAS Institute, Cary, North Carolina

T. L. LAI Department of Mathematical Statistics, Columbia University, New York, New York

THOMAS J. LORENZEN Mathematics Department, General Motors Research Laboratories, Warren, Michigan

CHARLES E. McCULLOCH Biometrics Unit, Cornell University, Ithaca, New York

GARY C. McDONALD Mathematics Department, General Motors Research Laboratories, Warren, Michigan

KLAUS J. MIESCKE Department of Mathematics, Statistics, and Computer Science, University of Illinois at Chicago, Chicago, Illinois

SATYA N. MISHRA Department of Mathematics and Statistics, University of South Alabama, Mobile, Alabama

INGRAM OLKIN Department of Statistics, Stanford University, Stanford, California

S. PANCHAPAKESAN Department of Mathematics, Southern Illinois University, Carbondale, Illinois

HERBERT ROBBINS Department of Mathematical Statistics, Columbia University, New York, New York

THOMAS J. SANTNER School of Operations Research and Industrial Engineering, Cornell University, Ithaca, New York

MILTON SOBEL Department of Mathematics, University of California at Santa Barbara, Santa Barbara, California

AJIT C. TAMHANE Department of Industrial Engineering and Management Sciences, Northwestern University, Evanston, Illinois

Y.L. TONG Department of Mathematics and Statistics, University of Nebraska, Lincoln, Nebraska

BRUCE W. TURNBULL School of Operations Research and Industrial Engineering, Cornell University, Ithaca, New York

LIONEL I. WEISS School of Operations Research and Industrial Engineering, Cornell University, Ithaca, New York

DAVID E. WETZELL* Department of Mathematics and Statistics, University of Nebraska, Lincoln, Nebraska

HSIU-FEN WU Department of Mathematics, University of California at Santa Barbara, Santa Barbara, California

*Current affiliation: Texas Instruments, Inc.

ROBERT E. BECHHOFER'S CONTRIBUTIONS TO STATISTICS

1. FAMILY HISTORY

Robert Eric Bechhofer was born on March 11, 1919 in New York City. He was the first of two children of Julius and Lillian Bechhofer. His family moved to Teaneck, New Jersey when he was quite young and he spent most of his youth there.

Many of Bob's salient characteristics can be traced to this period. His diligent work habits were learned by example from his parents while his sense of social justice and involvement in human rights traces to the 1930's when his family housed relatives escaping from Nazi Germany. Bob's interest in collecting original prints and Eskimo carvings have replaced the coin, stamp and tropical fish collections of his youth. His early interest in chess with its requirements of logical reasoning and critical thinking has been replaced by mathematics and statistics.

Bob married Joan Lebrecht on October 4, 1952. Over the past 30 years they have shared a deep interest in art, music, and politics. Family has been a central focus for Bob and Joan; their four children are Robin (b. 1954), David (b. 1957), Ellen (b. 1962) and Laurie (b. 1967).

2. EDUCATION AND EARLY CAREER

Bob attended Teaneck's public school system, then entered Columbia College in 1937 and earned the A.B. degree in 1941 with majors in mathematics and statistics. He took 7 courses in statistics while at Columbia, three from Harold Hotelling, and became increasingly interested in the subject. Upon graduation he was offered a position as a statistician with the Arms and Ammunition Division at Aberdeen Proving Ground which he held until nearly the end of the war. During this period he took evening classes in statistics at the Department of Agriculture; Girshick and Deming were among his teachers. While attending these classes he met his future collaborator Milton Sobel who was stationed in Washington, D.C. At the same time a group of biostatisticians including Sam Greenhouse, Max Halpern and Jerome Cornfield were beginning their careers in the Washington area; later in his career their work would become one subject area of interest to Bob.

While at Aberdeen, Bob had his first encounter with real-world applied statistics through the design and analysis of ballistics experiments. This experience established a theme that has permeated much of his subsequent research work, namely designing experiments to produce the maximum amount of information.

As the war in Europe drew to a close, he was encouraged by a close high school friend to apply for an opening at the Carbide and Carbon Chemicals Corporation in Oak Ridge, Tennessee. Bob was offered the job and by good fortune was assigned to work for Cuthbert Daniel. Daniel's habit of critically examining the formulation of the scientific questions to be answered by experiments made a lasting impression on Bob.

As the war in the Pacific ended, Bob enrolled for graduate study in the Department of Mathematical Statistics at Columbia University. During the summer of 1946 he attended the famous session in Raleigh, North Carolina, taught by Sir Ronald Fisher, William Cochran, Gertrúde Cox, and Chester Bliss, among others. It was here that he met his future Cornell colleague, Lionel Weiss, as well as Charles Stein and George Nicholson.

Bob was a graduate student at Columbia from 1946 to 1951; the environment was exceptionally stimulating. The faculty included A. Wald, J. Wolfowitz, T. Anderson, H. Levine, H. Scheffe, with visits by R.C. Bose, J. Doob, M. Loeve, J. Neyman, E.J.G. Pitman, and S.N. Roy. Bob's fellow students at Columbia were equally illustrious; they later were to form a who's who of the statistical world. Among them were future colleagues and collaborators Charles Dunnett, Jack Kiefer, Milton Sobel and Lionel Weiss plus Alan Birnbaum, Al Bowker, Herman Chernoff, Leon Herback, William Kruskal, Gottfreid Noether, Ingram Olkin and Charles Stein.

Bob's Ph.D. thesis reflected his background in experimental design. It is entitled <u>The effect of preliminary tests of significance on the size and power of certain tests of univariate linear hypotheses</u> ([1])* and was written under the guidance of T.W. Anderson.

During the academic year 1951-1952, immediately following the completion of his Ph.D., Bob served as Assistant Professor in the Department of Industrial Engineering at Columbia. During his Columbia years, Bob occasionally consulted with his former supervisor, Cuthbert Daniel, who had moved to New York City from Oak Ridge. At one of these consulting sessions, Cuthbert asked Bob about the probability that a set of sample means all line up in the same order as the population means. This question led Bob to initiate his study of designing experiments when the goal is ranking and selection. His initial investigations culminated in the pioneering paper "A single sample multiple decision procedure for ranking means of normal populations with known variances" ([2]). This paper was his formalization of Cuthbert Daniel's question regarding the correct ordering of sample means. In it, Bob introduced the so-called indifference zone formulation for selection problems (which are a particular type of multiple decision problem). In the simplest case the question is to determine the level of a (qualitative) factor yielding the largest mean response in a single factor experiment. He considered the intuitive procedure which takes n observations on each level of the factor and claims that the largest population mean is associated with the largest sample mean. The

* The bracketed numbers refer to Bechhofer's publications listed in Section 4.

problem is to design the experiment on such a scale (i.e. to choose the minimum sample size n) so that the intuitive procedure succeeds with a prespecified probability P^*. If the population means are allowed to be arbitrarily close, then it is mathematically impossible to achieve this goal. Bob introduced the idea of requiring the level P^* of correct decision (or selection) to be met only so long as the largest and second largest population means differed by a prespecified amount $\delta^* > 0$. If the largest and second largest means were closer than δ^*, then the experimenter is indifferent as to whether the best level of the factor was selected, hence the name of the formulation.

By 1952, numerous changes had occurred at Columbia including the tragic death of Wald in 1950 and, in 1951, the departure of Wolfowitz whom Bob respected enormously. These events set the stage for the next phase of Bob's career.

3. CORNELL CAREER

A. Research Activities

In 1951, Jacob Wolfowitz joined the Department of Mathematics at Cornell University with his student Jack Kiefer and invited Bob to also consider moving. After securing an Air Force contract to continue his work on multiple decision procedures for selection and ranking, Bob made the move to Cornell in 1952. Bob invited Milton Sobel and Charles Dunnett to join him in working on the project. Milton stayed for two years until he joined Bell Laboratories and Charlie for one year until he returned to his regular position with the Canadian Food and Drug Laboratories. In addition, Lionel Weiss also spent the year 1952-53 at Cornell teaching in the Mathematics Department.

During the years that followed, an amazing amount of research was produced in ranking and selection and related topics such as multiple comparisons. Multiple decision procedures were developed for the selection of a variety of parameters in single factor experiments based on the normal and other sampling models. Papers in this class included ones on selection of (i) normal means with common unknown variance ([4,7,11,19]), (ii) normal variances ([5]), and (iii) multinomial success probabilities ([8,14]). A second direction the research took was the study of factorial experiments. Here, for example, selection rules were developed for means ([2,27,30]) and variances ([9,14,16]) under various structural models. In other papers, the research explored the use of multi-stage and sequential sampling rules. Such procedures often save on the expected total sample size in comparison to competing single-stage procedures when both guarantee the same specified probability requirement ([28,32]). Furthermore in some cases, for intuitively reasonable indifference zones, single-stage procedures do not exist which guarantee P^* probability levels of correct selection and two-stage or even fully sequential rules must be used. For example, a two-stage procedure is required to select normal means in the common unknown variance case when the indifference zone is defined by differences of the means ([4]) and an open sequential procedure is required to select the Bernoulli population with the largest success probability when the indifference zone is defined by the odds ratio ([15]). Bob, Jack Kiefer and Milton Sobel extended Wald's sequential hypothesis test to a sequential selection procedure for exponential families in terms of their natural parameters; the Bernoulli problem described above is a special case of their general problem. This joint work, spread over a period of nearly 15 years resulted in the monograph Sequential Identification and Ranking Procedures published in 1968 ([15]).

In the 1970's, under the influence of Jerry Cornfield, Sam Greenhouse and others, Bob became interested in encouraging the application of selection procedures to biomedical research, especially controlled clinical trials. One set of papers partially motivated by this interest provided a modification of Milton Sobel and Marylin Huyett's 1957 single-stage Bernoulli selection procedure to allow early stopping while retaining the feature that only a fixed maximum number of observations (set prior to experimentation) need be taken from each population ([38,43,44]). Bob and his student, Radhika Kulkarni, identified various adaptive sampling rules, in the class of sequential procedures with the above two properties, which minimized (i) the mean total number of patients (or experimental units) used in the trial and (ii) the mean number of patients (or experimental units) on the worst treatment.

Bob and his coworkers have also studied the construction of optimal experimental designs for multiple comparisons problems. This branch of research was undoubtedly influenced by his long association with Charles Dunnett who shared Bob's view that ANOVA was often an inappropriate statistical tool for the practical analysis of data. Two series of papers in the multiple comparisons area will be mentioned. The first consists of four papers on the problem of optimally allocating a fixed total number of observations among a control population and a set of treatments to minimize the length of simultaneous confidence intervals for comparison of each treatment mean to the control mean ([18,21,22,41]). The second set is written with his former student, Ajit Tamhane. It introduces a theory of optimal designs for comparing treatments with a control in incomplete blocks ([35,36,42]).

Bob is noted for the clarity and organization of his articles, class lectures, and research seminars. His classes and seminars always pose unanswered questions. Being in an Operations Research Department, he moves with infectious zeal to insure that students make informed thesis area decisions. Not all of the students he has recruited into the field have written theses with Bob, but his concern for his students both intellectually and personally has influenced the lives of all of them. Bob has formally guided the Ph.D. studies of Saul Blumenthal (joint with L. Weiss), Edward Dudewicz (joint with L. Weiss), John Ramberg, Vijay Bawa (joint with L. Weiss), Ricardo Frishtak, Ajit Tamhane and Radhika Kulkarni and the M.S. studies of Norman Morse, George Gershefski, Donald Cohen, William Fairweather, Frederick Faltin, and Thomas Frisardi. In addition, he has informally worked with a large number of graduate students who wrote theses in other areas. This list includes Salah Elmaghraby, Dominique Nocturne, Bruce Turnbull, Richard Smith, Chuck McCulloch, Iain Johnstone, Chris Jennison, and David Goldsman. In a similar way, many visiting faculty members and colleagues have been interested in problems he has posed. Lastly, a host of researchers outside Cornell have been motivated by Bob's writing and invited talks. In 1979 he co-organized (with Jean Gibbons, Shanti Gupta, and Ingram Olkin) the annual American Statistical Association Meeting short course on "Selecting and Ordering Populations".

In sum, Bob's research career has spread the message of optimality and experimental design in selection and multiple comparison problems. His accomplishments have been recognized by his peers who have elected him a Fellow of (i) the Institute of Mathematical Statistics (1962), (ii) the American Statistical Association (1963), (iii) the American Association for the Advancement of Science (1980) and as a member of the International Statistical Institute (1975). He is also a Fellow of the Royal Statistical Society, a senior member of the American Society for Quality Control (elected 1982) and a member of Phi Kappa Phi, Sigma Xi, and Tau Beta Pi. In 1981 he was awarded the first Samuel S. Wilks Award for Contributions to Statistical Methodologies in Army Research.

B. Leadership Roles at Cornell and in the Statistical Community

Bob became an Associate Professor in the Department of Industrial Engineering at Cornell in 1953 and a Professor in the Department of Industrial Engineering and Operations Research in 1957. In 1967 he became Chairman of the Department of Operations Research and devoted enormous energy to develop the Department into one of the prominent OR groups in the country. Bob continued his duties as chairman until 1977 when he returned to full-time teaching and research. In addition to his Cornell duties Bob has served on various committees of the American Statistical Association and other societies; he is currently a member of the Electorate Nominating Committee, Section U (statistics) of the American Association for the Advancement of Science. In the summer of 1983 he was Chairman of the Jack Kiefer-Jacob Wolfowitz Memorial Research Conference held in honor of his late colleagues.

Bob's inquisitive attitude, willingness to work, honesty and open manner have been the hallmarks of a highly successful career. His colleagues and friends everywhere rank him a first-rate scientist and compassionate human being.

4. PUBLICATIONS OF ROBERT E. BECHHOFER

[1] The effect of preliminary tests of significance on the size and power of certain tests of univariate linear hypotheses, unpublished Ph.D. dissertation, Columbia University, NY, (1951). Abstract, Annals of Mathematical Statistics, 22 (1951), 143.

[2] A single sample multiple decision procedure for ranking means of normal populations with known variances, Annals of Mathematical Statistics, 25 (1954), 16-39.

[3] On designing experiments for ranking means, Proceedings of Tenth Annual Clinic, Rochester Society for Quality Control, Rochester, NY, (1954), 6-10.

[4] A two-sample multiple decision procedure for ranking means of normal populations with a common unknown variance, (with C.W. Dunnett and M. Sobel), Biometrika, 41 (1954), 170-176.

[5] A single-sample multiple decision procedure for ranking variances of normal populations, (with M. Sobel), Annals of Mathematical Statistics, 25 (1954), 273-289.

[6] Multiple decision procedures for ranking means, Transactions of the National Convention of the American Society for Quality Control, NY, (1955), 513-519.

[7] A sequential multiple decision procedure for selecting the best one of several normal populations with a common unknown variance, and its use with various experimental designs," Biometrics, 14 (1958), 408-429.

[8] A single-sample multiple decision procedure for selecting the multinomial event which has the highest probability, (with S. Elmaghraby and N. Morse), Annals of Mathematical Statistics, 30 (1959), 102-119.

[9] A multiplicative model for analyzing variances which are affected by several factors, Journal of the American Statistical Association, 55 (1960), 245-264.

[10] A note on the limiting relative efficiency of the Wald sequential probability ratio test, Journal of the American Statistical Association 55 (1960), 660-663.

[11] A sequential multiple-decision procedure for selecting the best one of several normal populations with a common unknown variance, II: Monte Carlo sampling results and new computing formulae, (with S. Blumenthal), Biometrics, 18 (1962), 52-67.

[12] A fixed-sample size procedure for ranking means of finite populations with an application to bulk sampling problems, Report on Seminar on Sampling of Bulk Materials, (1965), Tokyo, Japan sponsored by the National Science Foundation and the Japan Society for Promotion of Science, 39-49.

[13] A two-stage subsampling procedure for ranking means of finite populations to bulk sampling problems, Technometrics, 9 (1967), 355-364.

[14] Designing factorial experiments to rank variances, Transactions of the Twenty-Second Annual Technical Conference of the American Society for Quality Control, Philadelphia, PA, (1968), 69-73.

[15] Sequential Identification and Ranking Procedures, (with J. Kiefer and M. Sobel), The University of Chicago Press, (1968).

[16] Single-stage procedures for ranking multiply-classified variances of normal populations, Technometrics, 10 (1968), 693-714.

[17] Multiple comparisons with a control for multiply-classified variances of normal populations, Technometrics, 10 (1968), 715-718.

[18] Optimal allocation of observations when comparing several treatments with a control, Multivariate Analysis, II (ed. by P.R. Krishnaiah), Academic Press, (1969), 463-473.

[19] An undesirable feature of a sequential multiple-decision procedure for selecting the best one of several normal populations with a common unknown variance, Correction Note, Biometrics, 26 (1970), 347-349.

[20] On ranking the players in a 3-player tournament, Nonparametric Techniques in Statistical Inference (ed. by M.L. Puri), Cambridge University Press, (1970), 545-549.

[21] Optimal allocation of observations when comparing several treatments with a control (II): 2-sided comparisons, (with D.J. -M. Nocturne), Technometrics, 14 (1972), 423-436.

[22] Optimal allocation of observations when comparing several treatments with a control (III): globally best one-sided intervals for unequal variances, (with B. Turnbull), Statistical Decision Theory and Related Topics (ed. by S.S. Gupta and J. Yakel), Academic Press, (1971), 41-78.

[23] An iterated integral representation for a multivariate normal integral having block covariance structure, (with A.C. Tamhane), Biometrika, 61 (1974), 615-619.

[24] A two-sample procedure for selecting the population with the largest mean from several normal populations with unknown variances: some comments on Ofosu's paper, Department of Operations Research Technical Report 233 (1974), Cornell University.

[25] Chebyshev type lower bounds for the probability of correct selection, I: the location problem with one observation from each of two populations, (with B.W. Turnbull). (Preliminary Report) Department of Operations Research Technical Report 236 (1974), Cornell University.

[26] Ranking and selection procedures, Proceedings of the Twentieth Conference on the Design of Experiments in Army Research Development and Testing, ARO Report 75-2, Part 2, (1975), 929-949.

[27] Selecting the largest interaction in a two-factor experiment, (with T.J. Santner and B.W. Turnbull), Statistical Decision Theory and Related Topics II (ed. by S.S. Gupta and D.S. Moore), Academic Press, (1977), 1-18.

[28] A two-stage minimax procedure with screening for selecting the largest normal mean, (with A.C. Tamhane), Communications in Statistics - Theory and Methods, A6 (11) (1977), 1003-1033.

[29] On selecting the process with the highest fraction of conforming product, (with B.W. Turnbull), Proceedings of the 31st Technical Conference of the American Society for Quality Control, Philadelphia, PA, (1977), 568-573.

[30] Selection in factorial experiments, Proceedings of 1977 Winter Simulation Conference, Gaithersburg, MD, (1977), I, 65-70.

[31] Two (k+1)-decision selection procedures for comparing k normal means with a fixed known standard, (with B.W. Turnbull), Journal of the American Statistical Association, Theory and Methods Section, 73 (1978), 385-392.

[32] A two-stage minimax procedure with screening for selecting the largest normal mean (II): an improved PCS lower bound and associated tables, (with A.C. Tamhane), Communications in Statistics - Theory and Methods, A8 (4) (1979), 337-358.

[33] Sampling plans for testing combination drugs. Abstract of paper read at the Eastern and Western North American Regions of the Biometric Society, Chicago, Ill., (1977), Biometrics, 34 (1978), 153-154.

[34] A note on the lower bound for the P{CS} of Gupta's subset selection procedure, (with T.J. Santner), School of Operations Research and Industrial Engineering Technical Report 401 (1979), Cornell University.

[35] Review of Selecting and Ordering Populations: A New Statistical Methodology, (with J.D. Gibbons, I. Olkin, and M. Sobel) (John Wiley 1977), Journal of the American Statistical Association, 75 (1980), 751-756.

[36] Incomplete block designs for comparing treatments with a control: general theory, (with A.C. Tamhane), Technometrics, 23 (1981), 45-57. Corrigendum, Technometrics, 24 (1982), 171.

[37] Closed adaptive sequential procedures for selecting the best of $k \geq 2$ Bernoulli populations, (with R.V. Kulkarni), Proceedings of the Third Purdue Symposium on Statistical Decision Theory and Related Topics (ed. by S.S. Gupta and J. Berger), Academic Press, 1, (1982), 61-108.

[38] Multiple comparisons for orthogonal contrasts: tables and applications, (with C.W. Dunnett), Technometrics, 24 (1982), 213-222.

[39] In memoriam, Jack Carl Kiefer, 1924-1981, The American Statistician, 36 (1982), 356-357.

[40] On the performance characteristics of a closed adaptive sequential procedure for selecting the best Bernoulli population, (with R.V. Kulkarni), Sequential Analysis, Communications in Statistics (C), 1 (1982), 315-354.

[41] Incomplete block designs for comparing treatments with a control (II): optimal designs for $p = 2(1)6$, $k = 2$ and $p = 3$, $k = 3$, (with A.C. Tamhane), Sankhyā (B), 45 (1983), 193-224.

[42] Design of experiment for comparing treatments with a control: tables of optimal allocations of observations, (with A.C. Tamhane), Technometrics, 25 (1983), 87-95.

[43] A Monte Carlo study of the performance of a closed adaptive sequential procedure for selecting the best Bernoulli population, (with T. Frisardi), Journal of Statistical Computation and Simulation, 18, (1983), 179-213.

[44] Discussion of "Experimental design: review and comment" by D.M. Steinberg and W.G. Hunter, Technometrics, 6, (1984), 125-127.

[45] Tables of admissible and optimal balanced treatment incomplete block (BTIB) designs for comparing treatments with a control, (with A.C. Tamhane), to appear in Selected Tables in Mathematical Statistics, 8.

ACKNOWLEDGMENT

We would like to express our deep appreciation to Joan Bechhofer, Saul Blumenthal, Charles Dunnett, William Fairweather, and Lionel Weiss whose insights and criticisms led to substantial improvements in this paper.

A NOTE ON THE TUKEY-KRAMER PROCEDURE FOR PAIRWISE COMPARISONS OF CORRELATED MEANS

Lawrence D. Brown†

Department of Statistics
Hebrew University, Jerusalem, Israel,
and Department of Mathematics
Cornell University, Ithaca, New York

1. MAIN RESULT

Let X have a multivariate normal distribution on R^k with mean μ and covariance matrix $\sigma^2\Sigma$. Suppose μ and σ^2 are unknown and Σ is known. Let (Z^2/σ^2) be an independent chi-square random variable with v degrees of freedom.

Introduce $q_{k,v}^{(\alpha)}$, the upper α^{th} quantile of the Studentized range. By definition $q_{k,v}^{(\alpha)}$ is the value for which

$$P\{\max_{i,j}|X_i-X_j| \geq q_{k,v}^{(\alpha)}Z\} = \alpha \tag{1}$$

when $\mu = 0$, $\Sigma = I$.

The purpose of this note is to prove

<u>Theorem</u>: Let k = 3. Let $q = q_{k,v}^{(\alpha)}$ and

$$\omega_{ij}^2 = (\sigma_{ii} - 2\sigma_{ij} + \sigma_{jj})/2 . \tag{2}$$

Then

$$\Pr\{|X_i - X_j - (\mu_i-\mu_j)| \geq \omega_{ij}qZ \text{ for some } i,j\} \leq \alpha . \tag{3}$$

The following trivial corollary contains the statistical motivation for (3).

---

† Research supported in part by N.S.F. grant 8200031.

<u>Corollary</u>: Let $k = 3$. The simultaneous confidence intervals

$$\mu_i - \mu_j \in [X_i - X_j \pm \omega_{ij} qZ] \qquad (1 \leqq i \leqq j \leqq k) \tag{4}$$

have confidence level at least $1 - \alpha$.

2. DISCUSSION

One class of problems covered by the theorem is one-way analysis of variance. Here one observes

$$Y_{in} = \mu + \tau_i + \varepsilon_{in}, \quad n = 1,\ldots,N_i,\ i = 1,\ldots,k \tag{5}$$

with ε_{in} independent normal $(0,\sigma^2)$ variables and one sets $X_i = Y_{i.} = N_i^{-1} \sum_{n=1}^{N_i} Y_{in.}$. In the balanced case one has $N_i \equiv N$. For this case Tukey (1953) proposed what has come to be called the T-procedure for simultaneously setting confidence intervals for all the contrasts $\tau_i - \tau_j = \mu_i - \mu_j$. This is the special procedure defined by the intervals in (4) in the special case $\omega_{ij} \equiv \omega = N^{-1/2}$. The constant $q_{k,v}^{(\alpha)}$ defined by (1) is devised precisely for this classic case, and has been extensively tabled.

Tukey (1953) and Kramer (1956) proposed the confidence intervals in (4) as the natural extension of the T-procedure for cases where the N_i are not necessarily equal. Here Σ is diagonal with diagonal elements N_i^{-1} so that

$$\omega_{ij}^2 = (N_i^{-1} + N_j^{-1})/2 . \tag{6}$$

These proposals were based on a conjecture that (3) is valid, or nearly valid. When $k = 3$ Kurtz (1956) proved that (3) is valid in this situation. Kurtz' proof is the foundation for the proof of our main theorem, which of course includes also the case of non-diagonal Σ.

Dunnett (1980) provided extensive numerical evidence of the validity of (4) when the ω_{ij} are given by (6). Brown (1979) then proved the validity of (4) in this case when $k = 4,5$. Recently Hayter (1984) has proved the validity of (4) for all values of k when the ω_{ij} are given by (6).

In analysis of covariance one observes

$$Y_{in} = \mu + \tau_i + b_i\beta + \varepsilon_{in} \tag{7}$$

with μ, τ, ε_{in} as before, b_i known constants and β an unknown constant. Then the least squares estimates $X_i = \hat{\mu} + \hat{\tau}_i$ of $\mu + \tau_i$ have covariance matrix $\sigma^2\Sigma$. Σ is not, in general, a diagonal matrix. For such situations Tukey (1953) and Kramer (1957) again proposed the multiple comparison procedure described by (4) for comparisons involving all contrasts $\tau_i - \tau_j$. Our main theorem validates this proposal when $k = 3$.

There are linear models other than the analysis of covariance, (7), which lead to the above canonical model with Σ non-diagonal. One such model is that of an unbalanced two-way design mixed model (blocks random, treatments fixed) in which simultaneous confidence intervals are desired for all treatment contrasts.

3. PROOF OF THE THEOREM

Let $\beta_1 = (1,-1,0)'$, $\beta_2 = (1,0,-1)'$, $\beta_3 = (0,1,-1)'$. Assume, without loss of generality, that $\mu = 0$. Then the left side of (3) can be rewritten as

$$\Pr\{|\beta_i'X| \geq (\beta_i' \Sigma \beta_i/2)^{1/2} qZ, \text{ some } i = 1,2,3\}. \tag{8}$$

Let $\Sigma = C'C$, $Y = C'^{-1}X$, and $\gamma_i = C\beta_i$. Then Y is multivariate normal, mean 0, covariance $\sigma^2 I$. The expression (8) equals

$$\Pr\{|\gamma_i'Y| \geq (\gamma_i'\gamma_i/2)^{1/2} qZ, \text{ some } i = 1,2,3\}. \tag{9}$$

Note that $\{\gamma_1,\gamma_2,\gamma_3\}$ spans a two-dimensional subspace of R^3, since $\beta_2 - \beta_1 = \beta_3$ so that $\gamma_2 - \gamma_1 = \gamma_3$. We may thus assume, without loss of generality, that C has been chosen to yield

$$(\gamma_i)_3 = 0, \quad i = 1,2,3\ . \tag{10}$$

Then Y_3 plays no role in the expression (9) (i.e., $\gamma_i'Y$ depends only on Y_1,Y_2).

The conditional distribution of (Y_1,Y_2) given $Y_1^2 + Y_2^2 = r^2$ is uniform on the circle of radius r^2. As Kurtz (1956) noted, a simple geometric argument then shows that

$$\Pr\{|\gamma_i'Y| \geq t\,\|\gamma_i\|\ ,\ \text{some } i = 1,2,3 | Y_1^2 + Y_2^2 = r^2\} \tag{11}$$

is maximized (for every t,r) by choosing $\gamma_1,\gamma_2,\gamma_3$ to form equal angles (i.e., $\gamma_1 = (1,0,0)'$, $\gamma_2 = (+1/2,-\sqrt{3}/2,0)'$, $\gamma_3 = (-1/2,-\sqrt{3}/2,0)'$, or any rotation of these preserving the condition (10)). In fact for this choice of $\{\gamma_i\}$ the probability in (11) equals $\min(6 \cos^{-1} t/2\pi, 1)$ since the sets $\{\gamma_i' Y \geqq t \| \gamma_i \|\}$, $i = 1,2,3$ overlap only when $t \leqq 1/2$, and then the probability in (11) is one.

The expression (9) results from setting $\Sigma = qZ/\sqrt{2}$ and taking the expectation of (11) over the distribution of Z and $R^2 = Y_1^2 + Y_2^2$. It is thus maximized by taking $\gamma_1,\gamma_2,\gamma_3$ to form equal angles in a two-dimensional plane. This is equivalent to choosing $\Sigma = I$ in (8). When $\Sigma = I$ the probability in (8) is exactly α, by definition of q. □

4. GENERALIZATIONS: REMARKS ON THE GEOMETRY OF THE PROOF

A look at the proof shows that the manipulations leading to (11) reduce the theorem to a geometric question on the unit circle. (Take r = 1 without loss of generality.) When $k \geq 4$ the problem can be similarly rephrased as a geometric question on the k-1 dimensional unit sphere. For example, when k = 4 the theorem can be reduced as follows to a geometric question:

Begin by defining $\beta_1 = (1,-1,0,0)'$, $\beta_2 = (1,0,-1,0)'$, $\beta_3 = (1,0,0,-1)'$, ..., $\beta_6 = (0,0,1,-1)'$, $\Sigma = C'C$, and $\gamma_i = C\beta_i$, as in the preceding proof. To simplify the picture, as before, note that C can be chosen to yield $(\gamma_i)_4 = 0$, $i = 1,\dots,6$; and then define $\gamma_i^* = (\gamma_{i1},\gamma_{i2},\gamma_{i3})'$ and define the unit vectors in R^3, $\rho_i = \gamma_i^*/\| \gamma_i^* \|$, $i = 1,\dots,6$. Note that the set of vectors $\{\rho_i\}$ is constrained by three linear relations, e.g.

$$\| \gamma_3 \| \rho_3 - \| \gamma_2 \| \rho_2 = \| \gamma_6 \| \rho_6 \tag{12}$$

(since $\beta_3 - \beta_2 = \beta_6$), and by the angular relations

$$\rho_i'\rho_j = (\beta_i' \Sigma \beta_j)/(\beta_i' \Sigma \beta_i)^{1/2}(\beta_j' \Sigma \beta_j)^{1/2}, \quad 1 \leq i,j \leq 6. \tag{13}$$

These relations will be examined later from a geometric perspective. Note that Σ plays an important role in (13) and also in (12). (In fact, $\{\rho_i\}$ is uniquely determined, up to rotation, by Σ through (13).)

Let $A_3(S)$ denote the area of a set S on the unit sphere in R^3. Then the theorem for k = 4 would be valid if

$$A_3(\{x: \| x \| = 1, |\rho_i' x| \geq t, \text{ some } i = 1,\dots,6\}) \tag{14}$$

is maximized by the choice $\Sigma = I$ for each $0 \le t \le 1$. Note that each of the sets $\{x: \|x\| = 1, \rho_i'x \ge t\}$ or $\{x: \|x\| = 1, -\rho_i'x \ge t\}$ appearing in (14) is the interior of a circle on the unit sphere.

The conjecture that (14) is maximized for all k by the choice $\Sigma = I$ is sufficient to prove the theorem for k = 4, but it is not necessary, since the expression in (9) corresponds to an expectation of the area in (14) with t^2 a rescaled F random variable $q^2Z^2/(2\sum_{i=1}^{3} Y_i^2)$. Indeed, the proofs of Brown (1979) and Hayter (1984) for the case Σ diagonal, as in (6), proceed without establishing the truth of the conjecture.

The conditions (12), (13) can be rephrased geometrically as follows:

(i) There are 4 2-dimensional planes containing the origin.
(ii) Each plane contains 3 of the 6 vectors $\{\rho_i\}$.
(iii) Each ρ_i lies in 2 of these planes. (15)

Thus, the set $\{\pm\rho_i\}$ is uniquely determined as the set of unit vectors determined by the intersections of four planes through the origin. Different choices of Σ lead to all possible configurations for these planes.

The assumption that Σ is diagonal, as in (6), corresponds to an additional geometric condition as follows: To each ρ_i there corresponds a unique vector $\rho_{j(i)}$ not lying in either of the two planes in (15) which contain ρ_i. Σ is diagonal if and only if

$$\rho_{j(i)}'\rho_i = 0, \qquad i = 1,\ldots,6 . \tag{16}$$

When $\Sigma = I$ the vectors $\{\rho_i\}$ of course satisfy the conditions in (15) and (16). Furthermore, the three lines $\pm\rho_i$ lying in each plane in (15) cut the unit circle of that plane into six equal segments. The convex body with vertices $\pm\rho_i$, $i = 1,\ldots,6$ is a well-known semi-regular convex body with 14 faces--six squares and eight equilateral triangles, all with edges of length 1; each square is bordered by four triangles and each triangle by three squares.

After drawing pictures and examining cardboard constructions, I (and others I have talked with) believe that, subject to (15) and (16) the area in (14) is indeed maximized by the choice $\Sigma = I$. It also seems reasonable that (14) is maximized by $\Sigma = I$ subject to (15). However, it appears to be quite difficult to establish the truth (or falsity) of these geometric conjectures.

SUMMARY

This note contains a proof that the Tukey-Kramer multiple comparison procedure for three correlated means (with known correlations) is conservative. The geometry of the proof is then discussed in the hope that a proof using it can be found for more than three means.

REFERENCES

Brown, L. D. (1979). A proof that the Tukey-Kramer multiple comparison procedure for differences between treatment means is level-α for 3, 4, or 5 treatments. Statistics Center Technical Report, Cornell University, Ithaca, N.Y.

Dunnett, C. W. (1980). Pairwise multiple comparisons in the homogeneous variance, unequal sample size case. J. Amer. Statist. Assoc., 75, 789-795.

Hayter, A. J. (1984). A proof of the conjecture that the Tukey-Kramer multiple comparisons procedure is conservative. Ann. Statist., 12, 61-75.

Kramer, C. Y. (1956). Extensions of multiple range tests to group means with unequal numbers of replications. Biometrics, 12, 307-310.

Kramer, C. Y. (1957). Extensions of multiple range tests to group correlated adjusted means. Biometrics, 13, 13-18.

Kurtz, T. E. (1956). An extension of a multiple comparison procedure. Ph.D. thesis, Princeton University, Princeton, N.J.

Tukey, J. W. (1953). The problem of multiple comparisons. Unpublished manuscript. Princeton University, Princeton, N.J.

MULTIPLE COMPARISONS PROCEDURES FOR SOME SPLIT PLOT AND SPLIT BLOCK DESIGNS

Walter T. Federer

Biometrics Unit
Cornell University
Ithaca, New York

Charles E. McCulloch

Biometrics Unit
Cornell University
Ithaca, New York

1. INTRODUCTION

The subject of multiple comparisons for a set of v means has received considerable attention in scientific and textbook literature. Some of the earlier references are Duncan (1947, 1955), Keuls (1952), Tukey (1953), Federer (1955), Dunnett (1955), and Miller (1966). More recent discussions are given by Miller (1976), and Chew (1977). These latter two references contain bibliographies.

A closely related area is that of ranking and selection. Historically, the area has been divided into two branches, the indifference zone approach and the subset selection approach (Gibbons, Olkin and Sobel, 1977). Some early references are Bechhofer (1954), Gupta and Sobel (1958), Paulson (1964), Gupta (1965) and Bechhofer, Kiefer and Sobel (1968). Recently, however, Hsu (1981, 1982) has demonstrated a relationship between the two branches and also between them and a special kind of multiple comparisons, called simultaneous confidence intervals with the best.

All of these procedures relate to means of treatments for which there is no structure among the treatments. When there is structure, it should be taken into account. Failure to do this has resulted in misuses of multiple comparisons procedures. Examples of structure that should be utilized include sets of orthogonal contrasts or quantitative levels of a factor. Some of these are discussed by Chew (1977), Peterson (1977), Little (1978),

Brian-Jones and Finney (1983) and Carmer and Walker (1982). Also, thought should be given to the goal of setting simultaneous confidence intervals and the goal of significance testing. In many experimental situations, it is known (almost certainly) that the overall null hypothesis of equality for a set of v treatments is false. Hence there is no point in significance testing since finding a significant difference among the v means is mainly a function of sample size. In this paper, we will concentrate on simultaneous confidence intervals for the set of contrasts under consideration.

Some of the more widely known and used multiple comparisons procedures are the following:

(i) lsd: The comparisonwise confidence interval on pairs of means is computed as $\bar{Y}_{i.} - \bar{Y}_{i'.} \pm t_{\alpha,f}\sqrt{2s^2/r}$, where s^2 is an estimate of the experimental error with f degrees of freedom, $\bar{Y}_{i.}$ is the sample mean of treatment $i \neq i' = 1,2,\ldots,v$, r is the number of replicates for the ith treatment mean, and $t_{\alpha,f}$ is the tabular value of the two-tailed Students' t at the α level and for f degrees of freedom. This is the least significant difference procedure.

(ii) hsd: The experimentwise confidence interval on pairs of means is computed as $\bar{Y}_{i.} - \bar{Y}_{i'.} \pm q_{\alpha,v,f}\sqrt{s^2/r}$, where $q_{\alpha,v,f}$ is the tabulated value of the Studentized-range statistic at the α level for v treatments and for f degrees of freedom, and the other values are as defined in (i). This is commonly called the honestly significant difference or Tukey's range procedure.

(iii) esd: The per experiment confidence interval on m-pairs of means is computed as $\bar{Y}_{i.} - \bar{Y}_{i'.} \pm t_{\alpha/m,f}\sqrt{2s^2/r}$. This is sometimes called the Bonferroni procedure.

(iv) ssd: The experimentwise confidence interval for contrasts of the form $\sum_{i=1}^{v} c_i\mu_i$, where $\sum_{i=1}^{v} c_i = 0$, is computed as

$$\sum_{i=1}^{v} c_i \bar{Y}_{i.} \pm \sqrt{(v-1)F_\alpha(v-1,f)s^2 \sum_{i=1}^{v} c_i^2/r}$$

, where $F_\alpha(v-1,f)$ is the tabulated value of F at the α level for v-1 degrees of freedom in the numerator and f in the denominator. This is known as Scheffe's procedure.

(v) csd: The experimentwise confidence interval for comparing a control or standard treatment mean with p other treatment means is computed as $\bar{Y}_{c.} - \bar{Y}_{i.} \pm d_{v,f,\alpha/2}\sqrt{2s^2/r}$, where $i = 1,2,\ldots,p$ and $d_{v,f,\alpha/2}$, is the two-sided α point of a p-variate t-distribution (p = v-1) with f degrees of freedom and common correlation $\rho = 1/2$. These values have been tabulated by Dunnett (1964) and are also given by others, e.g., Chew (1977) and Miller (1966).

Ranking and selection methods are more recent in origin than the other multiple comparison procedures and are considerably less known and used. The recently proposed confidence intervals by Hsu (1981) give a subset of the treatments that is guaranteed (with probability $1-\alpha$) to contain the population which has the largest mean and, at the same time, a set of simultaneous confidence intervals for the differences between the population means and the largest mean. Specifically, if we let $\mu_{[v]}$ denote the largest population mean then the procedure is:

(vi) Choose the ith population to be in the selected subset if

$$\overline{Y}_{i\cdot} \geq \max_{i' \neq i} \{\overline{Y}_{i'\cdot}\} - d_{v,f,\alpha}\sqrt{\frac{2s^2}{r}} .$$

$d_{v,f,\alpha}$ is the one-sided α point of a $(v-1)$-variate t-distribution with f degrees of freedom and common correlation coefficient $\rho = 1/2$. These values have been tabulated by Dunnett (1955) and are also given by others, e.g., Miller (1966), Krishnaiah and Armitage (1966) or Gupta and Sobel (1957). (To use the tables in Dunnett or Miller, enter with $k = v-1$. The entries in Gupta and Sobel need to be divided by $\sqrt{2}$ before use. The tables in Gupta and Sobel or Krishnaiah and Armitage need to be entered with $\rho = 1/2$.) The simultaneous confidence intervals for $\mu_{[v]} - \mu_i$ are given by $[0, D_i]$, where

$$D_i = \max\{0, \max_{i' \neq i}\left(\overline{Y}_{i'\cdot}\right) - \overline{Y}_{i\cdot} + d_{v,f,\alpha}\sqrt{\frac{2s^2}{r}}\}.$$

Thus, Dunnett's one-sided comparisons with a control are related to simultaneous confidence intervals with the best in the following way. If the control treatment has the largest sample mean, the upper confidence intervals for the difference between the control mean and the other treatment means will be the same as those for $\mu_{[v]} - \mu_i$.

In the following we show how to apply these procedures to more complex designs of the split plot and split block nature.

2. SPLIT PLOT DESIGNS

There are many split plot designs (e.g. see Federer, 1975) but we shall confine our attention to one which has a whole plot treatments laid out in a randomized complete block design with r blocks. Then, each whole plot is divided into b split plots to which the b split plot treatments are randomly allocated within each whole plot. Thus, there are r randomizations for the whole plot treatments and ra randomizations for the split plot treatments.

Table 2.1a

Standard ANOVA for the Split Plot Design

Source of variation	Degrees of freedom	Sums of squares	Mean square
Total	rab	$\sum_{h=1}^{r}\sum_{i=1}^{a}\sum_{j=1}^{b} Y_{hij}^2$	-
Correction for mean	1	$\frac{Y_{\cdot\cdot\cdot}^2}{rab}$	-
Blocks = R	r-1	$\sum_{h=1}^{r}\frac{Y_{h\cdot\cdot}^2}{ab} - \frac{Y_{\cdot\cdot\cdot}^2}{rab}$	-
Whole plot treatments = A	a-1	$\sum_{i=1}^{a}\frac{Y_{\cdot i\cdot}^2}{rb} - \frac{Y_{\cdot\cdot\cdot}^2}{rab}$	MSA
R × A = error (a)	(r-1)(a-1)	$\sum_{h=1}^{r}\sum_{i=1}^{a}\frac{Y_{hi\cdot}^2}{b} - \sum_{i=1}^{a}\frac{Y_{\cdot i\cdot}^2}{rb} - \sum_{h=1}^{r}\frac{Y_{h\cdot\cdot}^2}{ab} + \frac{Y_{\cdot\cdot\cdot}^2}{rab}$	MSE(a)
Split plot treatments = B	b-1	$\sum_{j=1}^{b}\frac{Y_{\cdot\cdot j}^2}{ra} - \frac{Y_{\cdot\cdot\cdot}^2}{rab}$	MSB
A × B	(a-1)(b-1)	$\sum_{i=1}^{a}\sum_{j=1}^{b}\frac{Y_{\cdot ij}^2}{r} - \sum_{i=1}^{a}\frac{Y_{\cdot i\cdot}^2}{rb} - \sum_{j=1}^{b}\frac{Y_{\cdot\cdot j}^2}{ra} + \frac{Y_{\cdot\cdot\cdot}^2}{rab}$	MSAB
B × R within A = error (b)	a(r-1)(b-1)	by subtraction	MSE(b)

Table 2.1b

Alternative ANOVA

Source of variation	Degrees of freedom	Whole plot (sums of squares) 1	2	.... a	Sum
Total	rb	T_1	T_2	T_a	T.
Correction for mean	1	C_1	C_2	C_a	C.
Blocks = R	r-1	R_1	R_2	R_a	R.
Split plot treatments = B	b-1	B_1	B_2	B_a	B.
R × B	(r-1)(b-1)	E_1	E_2	E_a	E.

This is the design considered in most textbooks. Let Y_{hij} be the observation in block h (h = 1,2,...,r), on whole plot treatment i (i = 1,2,...,a), and on split plot treatment j (j = 1,2,...,b) and μ_{hij} be the expected value of Y_{hij}. Two analyses of variance (ANOVA) tables for this design are presented in Table 2.1. The first one is the standard one given in most textbooks that discuss an analysis of variance for a split plot design. The second one given in Table 2.1b is a useful one in many situations, especially where possibly different split plot treatments are to be recommended for each whole plot treatment. It is also appropriate when the split plot treatments differ from whole plot to whole plot; it should be used as a prior and/or supplementary ANOVA to the first one. Some of the sums of squares in the second ANOVA may be used to obtain those in the first. As examples, $T_.$ is the total sum of squares with rab degrees of freedom, $E_.$ is the error (b) sum of squares with a(r-1)(b-1) degrees of freedom, and $B_.$ is the sum of the B and A × B sums of squares.

If the error rate base is comparisonwise, then the lsd procedure, (i), is appropriate. The procedure is straightforward using error (a), error (b), or possibly the individual $E_i/(r-1)(b-1)$ mean squares to compute the lsd. If, however, one wishes to use some sort of family error rate (see Tukey, 1953, and Miller, 1966), several procedures come to mind. These are:

a) Make comparisons among the a whole plot treatment means $\{\bar{\mu}_{.i.}-\bar{\mu}_{.i'.}, i \neq i'\}$, and among the b split plot treatment means $\{\bar{\mu}_{..j}-\bar{\mu}_{..j'}, j \neq j'\}$. These comparisons would be pertinent if there were no A × B interaction; the error terms are error (a) and error (b) from 2.1a.

b) Make comparisons among the a whole plot treatment means $\{\bar{\mu}_{.i.}-\bar{\mu}_{.i'.}, i \neq i'\}$, and among the split treatment means separately for each whole plot treatment $\{\bar{\mu}_{.ij}-\bar{\mu}_{.ij'}, j \neq j', i = 1,2,...,a\}$. The error terms to be used are error (a) and error (b) for the whole plot treatments and split plot treatments respectively.

c) Make comparisons as in b) above, but use $E_i/(b-1)(r-1)$ within each whole plot treatment to compare the split plot treatment means. This would be appropriate if the variances of observations under different whole plot treatments are unequal. If possibly different split plot treatments are to be recommended for each whole plot treatment, the experimenter might wish to control the family error rate for each whole plot treatment.

d) Make comparisons among the b split plot treatment means $\{\bar{\mu}_{\cdot\cdot j} - \bar{\mu}_{\cdot\cdot j'}, j \neq j'\}$, and among the a whole plot treatment means separately for each split plot treatment, $\{\bar{\mu}_{\cdot ij} - \bar{\mu}_{\cdot i'j}, i \neq i', j = 1,2,\ldots,b\}$.

e) Make comparisons among the b split plot treatment means $\{\bar{\mu}_{\cdot ij} - \bar{\mu}_{\cdot ij'}, j \neq j', i = 1,2,\ldots,a\}$ separately for each of the whole plot treatments. Use error (b) as an error estimate.

To see how the standard procedures can be adapted, consider the following model for a split plot design:

$$Y_{hij} = \mu + \alpha_i + \rho_h + \delta_{hi} + \beta_j + (\alpha\beta)_{ij} + \epsilon_{hij}, \qquad (2.1)$$

where

Y_{hij} = observation in block h with whole plot treatment i and split plot treatment j,

μ = overall mean,

α_i = effect of whole plot treatment i,

ρ_h = effect of block h,

δ_{hi} = whole plot error term,

β_j = effect of split plot treatment j,

$(\alpha\beta)_{ij}$ = interaction of whole plot treatment i and split plot treatment j,

and

ϵ_{hij} = split plot error term.

We will assume $\delta_{hi} \sim \text{iid } N(0,\sigma_\delta^2)$ independently of $\epsilon_{hij} \sim \text{iid } N(0,\sigma_\epsilon^2)$ and that $\sum_i \alpha_i = \sum_j \beta_j = \sum_i (\alpha\beta)_{ij} = \sum_j (\alpha\beta)_{ij} = 0$. For situation c) we allow the possibility that the ϵ_{hij} have different variances, $\sigma_{\varepsilon_i}^2$, for each whole plot treatment. Under model (2.1), comparisons among the whole plot means are based on

$$\hat{\alpha}_i - \hat{\alpha}_{i'} - (\alpha_i - \alpha_{i'}) = \overline{Y}_{\cdot i\cdot} - \overline{Y}_{\cdot i'\cdot} - (\alpha_i - \alpha_{i'})$$
$$= \overline{\delta}_{\cdot i} - \overline{\delta}_{\cdot i'} + (\overline{\epsilon}_{\cdot i\cdot} - \overline{\epsilon}_{\cdot i'\cdot}).$$

Denoting $\mu + \alpha_i + \beta_j + (\alpha\beta)_{ij}$ by μ_{ij}, comparisons among the subplot means within whole plot i'' are based on

$$\hat{\mu}_{i''j} - \hat{\mu}_{i''j'} - (\mu_{i''j} - \mu_{i''j'}) = \overline{Y}_{\cdot i''j} - \overline{Y}_{\cdot i''j'} - (\mu_{i''j} - \mu_{i''j'})$$
$$= \overline{\epsilon}_{\cdot i''j} - \overline{\epsilon}_{\cdot i''j'} \, .$$

The covariance between these is

$$\mathrm{Cov}(\overline{\epsilon}_{\cdot i\cdot} - \overline{\epsilon}_{\cdot i'\cdot}, \overline{\epsilon}_{\cdot i''j} - \overline{\epsilon}_{\cdot i''j'}) \tag{2.2}$$

since δ_{hi} and ϵ_{hij} are independent. (2.2) equals

$$\mathrm{Cov}(\overline{\epsilon}_{\cdot i\cdot}, \overline{\epsilon}_{\cdot i''j} - \overline{\epsilon}_{\cdot i''j'}) - \mathrm{Cov}(\overline{\epsilon}_{\cdot i'\cdot}, \overline{\epsilon}_{\cdot i''j} - \overline{\epsilon}_{\cdot i''j'})$$

which is zero since $\mathrm{Cov}(\overline{\epsilon}_{\cdot i\cdot}, \overline{\epsilon}_{\cdot i''j})$ is independent of j. Thus, inferences about the whole plot means and the split plot means are independent. The same is true for error estimates based on them (MSE(a) and MSE(b)). So, for example, considering situation (a)

$$P\{\left|\hat{\alpha}_i - \hat{\alpha}_{i'} - (\alpha_i - \alpha_{i'})\right| \le q_{[\alpha_1, a, (r-1)(a-1)]}\sqrt{\frac{\mathrm{MSE(a)}}{rb}} \quad i,i'$$

and

$$\left|\hat{\beta}_j - \hat{\beta}_{j'} - (\beta_j - \beta_{j'})\right| \le q_{[\alpha_2, b, a(r-1)(b-1)]}\sqrt{\frac{\mathrm{MSE(b)}}{ra}} \quad j,j'\}$$

$$= (1 - \alpha_1)(1 - \alpha_2).$$

Also, inferences among subplots within a particular whole plot are independent of those in a different whole plot.

We illustrate some of these situations with an example from Federer (1955). The means and two ANOVA's are presented in Table 2.2. The data are number of seeds germinating out of 100. Several transformations of the data were attempted but none affected the analysis appreciably so the analysis on the original scale is presented. In Federer (1955), single degree of freedom contrasts were presented for comparisons among the split plot treatment means, seed treatments, and for partitioning the seed treatment x variety (whole plot) interaction. If any three contrasts among seed treatment means were the only ones of interest, one could use the esd procedure (iii) described above. For the three contrasts one could use $t_{\alpha/3,f}\sqrt{\frac{2\ \text{error (b)}}{ra}} = t_{.05/3,48}\sqrt{2(24.23)/8(3)}$ for $\alpha = .05$. If one wished to make these three

Table 2.2a

Variety and Seed Treatment Means

Seed treatment	Variety								Means
	a_1	a_2	a_3	a_4	a_5	a_6	a_7	a_8	
b_1	66	63	65	50	49	58	46	48	56
b_2	12	18	13	10	16	8	15	20	14
b_3	12	14	26	14	10	30	22	31	20
b_4	8	11	11	10	12	17	10	12	11
Means	25	27	29	21	22	28	23	28	25

Table 2.2b

Classical Textbook ANOVA

Source of variation	Degrees of freedom	Sum of squares	Mean square
Total	96	98,195.00	-
Correction for mean	1	61,458.76	-
Blocks	2	38.58	19.29
Varieties	7	763.16	109.02
Error (a)	14	1,377.25	98.38
Seed treatments	3	30,774.28	10,258.09
Seed treatments x varieties	21	2,620.13	124.77
Error (b)	48	1,162.84	24.23

Table 2.2c

ANOVA as Suggested in Table 2.1b. Whole Plot Treatment (variety)

Source of variation	Degrees of freedom	Sum of squares								
		a_1	a_2	a_3	a_4	a_5	a_6	a_7	a_8	Sum
Total	12	14326	14956	16184	9112	8889	13972	9107	11649	98195
Correction for mean	1	7301.3	8640.3	9976.3	5292.0	5764.1	9520.3	6486.8	9240.8	62221.9
Blocks = R	2	2.2	763.2	338.2	8.0	4.2	83.2	120.5	96.5	1416
Treatments = B	3	6972.0	5403.0	5645.7	3476.7	3066.2	4271.0	2350.9	2208.9	33394.4
R × B	6	50.5	149.5	223.8	335.3	54.5	97.5	148.8	102.8	1162.7

contrasts within each whole plot treatment, there would be 3a = 3(8) = 24 = m contrasts. In this case use $t_{\alpha/24,48}\sqrt{2(24.23)/8(3)}$. If p contrasts among the varietal means were desired, one could add these to the preceding to obtain m + p contrasts or one could consider m and p separately.

We will illustrate the hsd and esd procedures and Hsu's simultaneous confidence intervals with the best for situations a) and c). For situation a) we take the error rate for whole plot comparisons and the error rate for split plot comparisons to be .05 so that the overall error is $.95^2 \doteq .90$. The hsd procedure uses $q_{.05,8,14}\sqrt{\frac{MSE(a)}{3(4)}} = 4.99\sqrt{\frac{98.38}{12}} = 14.29$ for whole plot comparisons and $q_{.05,4,48}\sqrt{\frac{MSE(b)}{3(8)}} = 3.77\sqrt{\frac{24.23}{24}} = 3.79$ for the split plot comparisons. The esd procedure uses $t_{.05/28,14}\sqrt{\frac{2(MSE(a))}{3(4)}} = 3.85\sqrt{\frac{2(98.38)}{12}} = 15.57$ for whole plot comparisons and $t_{.05/6,48}\sqrt{\frac{2(MSE(b))}{3(8)}} = 2.76\sqrt{\frac{2(24.23)}{24}} = 3.92$ for split plot comparisons Since the hsd always gives smaller intervals when comparing all pairs of means, it is to be preferred in this situation. Tables 2.3a and 2.3b list the means, differences between the means and the hsd length to be used for confidence intervals:

$$\bar{Y}_{\cdot i\cdot} - \bar{Y}_{\cdot i'\cdot} \pm 14.3$$
$$(\text{for } \alpha_i - \alpha_{i'})$$

and

$$\bar{Y}_{\cdot\cdot j} - \bar{Y}_{\cdot\cdot j'} \pm 3.8$$
$$(\text{for } \beta_j - \beta_{j'}).$$

Calculating simultaneous confidence intervals with the best, we need $d_{v,f,\alpha}\sqrt{\frac{2(MSE)}{r}}$. For the whole plot treatments, we use $d_{8,14,.05}=2.59$ and for split plots we use $d_{4,48,.05}=2.11$. Thus the selected subsets which contain the population with the largest mean (with confidence coefficient .90) are

$$\{a_1,a_2,a_3,a_4,a_5,a_6,a_7,a_8\} \quad \text{for the whole plots}$$

and

$$\{b_1\} \quad \text{for the split plots.}$$

Table 2.4 lists the simultaneous confidence intervals from the best calculated as $[0,D_i^{(\alpha)}]$ for the whole plots, where

Table 2.3a

Whole Plot hsd Comparisons for the Data of Table 2.2 for Situation (a). The Body of the Table Gives Pairwise Mean Differences. (hsd = 14.3 for Confidence Coefficient .90)

		Variety						
		a_3	a_6	a_8	a_2	a_1	a_7	a_5
Variety	Mean	29	28	28	27	25	23	22
a_4	21	8	7	7	6	4	2	1
a_5	22	7	6	6	5	3	1	
a_7	23	6	5	5	4	2		
a_1	25	4	3	3	2			
a_2	27	2	1	1				
a_8	28	1	0					
a_6	28	1						

Table 2.3b

Split Plot hsd Comparisons for the Data of Table 2.3b for Situation (a). The Body of the Table Gives Pairwise Mean Differences. (hsd = 3.8 for Confidence Coefficient .90)

		Seed Treatment		
		b_1	b_3	b_2
Seed Treatment	Mean	56	20	14
b_4	11	45	9	3
b_2	14	42	6	
b_3	20	36		

Table 2.4

Simultaneous Confidence Intervals from the Best for the Data of Table 2.2 Using Situation (a). (Confidence Coefficient .90)

Confidence interval	Parameter
[0,14.5]	$\alpha_{[8]} - \alpha_1$
[0,12.5]	$\alpha_{[8]} - \alpha_2$
[0, 9.5]	$\alpha_{[8]} - \alpha_3$
[0,18.5]	$\alpha_{[8]} - \alpha_4$
[0,17.5]	$\alpha_{[8]} - \alpha_5$
[0,11.5]	$\alpha_{[8]} - \alpha_6$
[0,16.5]	$\alpha_{[8]} - \alpha_7$
[0,11.5]	$\alpha_{[8]} - \alpha_8$
[0,0]	$\beta_{[4]} - \beta_1$
[0,45]	$\beta_{[4]} - \beta_2$
[0,39]	$\beta_{[4]} - \beta_3$
[0,48]	$\beta_{[4]} - \beta_4$

$$D_i^{(\alpha)} = \max\{0, \max_{i' \neq i} \bar{Y}_{\cdot i' \cdot} - \bar{Y}_{\cdot i \cdot} + 10.5\}$$

and as $[0, D_j^{(\beta)}]$ for the split plots, where

$$D_j^{(\beta)} = \max\{0, \max_{j' \neq j} \bar{Y}_{\cdot\cdot j'} - \bar{Y}_{\cdot\cdot j} + 3.0\}.$$

Notice that the "yardstick" used for confidence intervals from the best are much shorter than the hsd (10.5 compared to 14.3 and 3.0 compared to 3.8), reflecting the smaller number of comparisons made.
Thus confidence intervals from the best is clearly advantageous when the goal is to make statements about which treatments have the largest means.

Next we consider situation c). We will choose an error rate of .05 for the whole plot comparisons and each of the within-whole-plot comparisons. This will give an overall error rate of $.95^9 \doteq .63$. The whole plot comparisons are the same as above. We will illustrate the calculations for the comparisons within whole plot treatment a_6. The hsd procedure uses $q_{.05,4,6}\sqrt{\frac{E_6/6}{3}} = 4.90\sqrt{\frac{97.5/6}{3}} = 11.39$ and the confidence intervals from the best uses $d_{4,6,.05}\sqrt{\frac{2(E_6/6)}{3}} = 2.56\sqrt{\frac{2(97.5/6)}{3}} = 8.43.$ Thus, for the hsd, the intervals are

$$\bar{Y}_{\cdot 6j} - \bar{Y}_{\cdot 6j'} \pm 11.39$$

and the intervals from the best are $[0,\max\{0, \max_{j'\neq j} \bar{Y}_{\cdot 6j'} - \bar{Y}_{\cdot 6j} + 8.43\}]$ which are [0,0], [0,58.4], [0,36.4], and [0,49.4] for b_1, b_2, b_3 and b_4 respectively. Doing the calculations for the other whole plot treatments shows that the split plot treatment b_1 is the best within each of the whole plot treatments. This, as well as other statements, e.g., that all whole plot treatments are within 18.5 of the best, can be made with confidence coefficient .63.

3. EXTENSIONS

In a split block design, there are two sets of whole plot treatments with each set having a different error variance. For our purpose, let us consider the split block design for which each set of whole plots is in a randomized complete block design, i.e. there would be r randomizations for each set of whole plots (e.g. see Federer, 1955 and 1975). Thus, within a block, each plot in each of the two sets of whole plots is divided into subplots by the plots of the other set. An analysis of variance table for this situation is given in Table 3.1. There are three separate error mean squares, MSE(a), MSE(b), and MSE(ab), for this design. Resulting F ratios are MSA/MSE(a), MSB/MSE(b), and MSAB/MSE(ab). In setting up an "experimentwise" error rate, one could consider several situations, one of which is

f) Consider comparisons among each of the two sets of whole plot treatments. These will be independent (with different error terms) and could be handled as in Section 2.

For a split-split plot design let us consider the design in Section 2 for which the split plot experimental units are split into c split-split plot experimental units (sspeu) and the c split-split plot treatments are randomly allocated to these c sspeus within each of the rab split plot experimental units resulting in rab randomizations for split-split plot treatments. An analysis of variance table for this design may be found in several places, one of which is Table X-8 of Federer (1955). There are nine separate variances for contrasts listed as formulae (X-11) through

Table 3.1

Analysis of Variance Table for a Split Block Design

Source of variation	Degrees of freedom	Sum of squares	Mean square
Total	rab	$\sum_{h=1}^{r} \sum_{i=1}^{a} \sum_{j=1}^{b} Y_{hi}^2$	-
Correction for mean	1	$\frac{Y_{\cdots}^2}{rab}$	-
Blocks	r-1	$\sum \frac{Y_{h\cdot\cdot}^2}{ab} - \frac{Y_{\cdots}^2}{rab}$	-
One set of whole plots = A	a-1	$\sum \frac{Y_{\cdot i\cdot}^2}{rb} - \frac{Y_{\cdots}^2}{rab}$	MSA
Blocks × A	(r-1)(a-1)	subtraction	MSE(a)
Second set of whole plots = B	b-1	$\sum \frac{Y_{\cdot\cdot j}^2}{ra} - \frac{Y_{\cdots}^2}{rab}$	MSB
Block × B	(r-1)(b-1)	subtraction	MSE(b)
A × B	(a-1)(b-1)	$\sum\sum \frac{Y_{\cdot ij}^2}{r} \sum \frac{Y_{\cdot i\cdot}^2}{rb} - \sum \frac{Y_{\cdot\cdot j}^2}{ra} + \frac{Y_{\cdots}^2}{rab}$	MSAB
Blocks × A × B	(r-1)(a-1)(b-1)	subtraction	MSE(ab)

(X-19) in Federer (1955). These involve three error mean squares E_a, E_b and E_c. One could

g) Consider comparisons among whole plot treatments, among split plot treatments and among split-split plot treatments. Again, these are independent (with different error variances) and can be treated as in Section 2.

A multitude of other possibilities are conceivable.

4. DISCUSSION

We have illustrated three techniques for multiple comparisons, each with its advantages. When desiring statements about the best treatment, an investigator should use the procedure of Hsu (1981, 1982) which yields shorter intervals for the comparisons of interest. The hsd is a reasonable procedure for making comparisons among all means and is applicable for some situations in the multiple-error-term designs we considered here. The esd always gives intervals which are wider than the hsd when making comparisons among all means. However, it is extremely versatile and can be used easily even in the most complex situations.

SUMMARY

Three multiple comparisons procedures, least significant difference, Studentized range, and Bonferroni, and simultaneous confidence intervals containing the best population are demonstrated for comparing means from split plot designs. A numerical example is used to illustrate two of the five situations for a split plot design.

REFERENCES

Bechhofer, R.E, (1954). A single-sample multiple decision procedure for ranking means of normal populations with known variances. Ann. Math. Statist. 25, 16-39.

Bechhofer, R.E., Kiefer, J. and Sobel, M. (1968). Sequential identification and ranking procedures. University Chicago Press, Chicago.

Bryan-Jones, J., and Finney, D.J. (1983). On an error in "Instructions to Authors." HortScience 18, 279-282.

Carmer, S.G. and Walker, W.M. (1982). Baby bear's dilemma: a statistical tale. Agron. J. 74(1), 122-124.

Chew, V. (1977). Comparisons among treatment means in an analysis of variance. USDA, ARS/H/6.

Duncan, D.B. (1947). Significance tests for differences between ranked variates drawn from normal populations. Ph.D. thesis, Iowa State University.

Duncan, D.B. (1955). Multiple range and multiple F tests. Biometrics 11, 1-42.

Dunnett, C.W. (1955). A multiple comparisons procedure for comparing several treatments with a control. J. Amer. Statist. Assoc. 50, 1096-1121.

Dunnett, C.W. (1964). New tables for multiple comparisons with a control. Biometrics 20, 482-491.

Federer, W.T. (1955). Experimental Design-Theory and Application. MacMillan, NY (Republished by Oxford and IBH Company, New Delhi 1967, 1974), Chapter II.

Federer, W.T. (1975). The misunderstood split plot. In Applied Statistics (Ed. R.P. Gupta) North Holland Publishing Co. Amsterdam.

Gibbons, J.D., Olkin, I. and Sobel, M. (1977). Selecting and Ordering Populations: a new statistical methodology. Wiley: New York.

Gupta, S.S. (1965). On some multiple decision (selection and ranking) rules. Technometrics 6, 225-245.

Gupta, S.S. and Sobel, M. (1957). On a statistic which arises in selection and ranking problems. Ann. Math. Statist. 28, 957-967.

Gupta, S.S. and Sobel, M. (1958). On Selecting a subset which contains all populations better than a standard. Ann. Math Statist. 29, 235-244.

Hsu, J.C. (1981). Simultaneous confidence intervals for all distances from the "best". Ann. Statist. 9, 1026-1034.

Hsu, J.C. (1982). Simultaneous inference with respect to the best in block designs. J. Amer. Statist. Assoc. 77, 461-467.

Keuls, M. (1952). The use of the "studentized range" in connection with the analysis of variance. Euphytica 1, 112-122.

Krishnaiah, P.R. and Armitage, J.V. (1966). Tables for the multivariate t distribution. Sankhya, Ser.B. 28, 31-56.

Little, T.M. (1978). If Galileo published in HortScience. HortScience 13, 504-506.

Miller, R.G. Jr. (1966). Simultaneous Statistical Inference. McGraw-Hill, New York.

Miller, R.G. Jr. (1977). Developments in multiple comparisons 1966-1976. J. Amer. Statist. Assoc., 72, 779-788.

Peterson, R.G. (1977). Use and misuse of multiple comparisons procedures. Agron. J. 69, 205-208.

Paulson, E. (1964). A sequential procedure for selecting a population with the largest mean from k normal populations. Ann. Math. Statist. 35, 174-180.

Scott, A.J. and M. Knott (1974). A cluster analysis method for grouping means in an analysis of variance. Biometrics 30, 507-512.

Steel, R.G.D. and Federer, W.J. (1955). Yield-stand analyses. J. Indian Soc. Agri. Statist. VII (1&2), 27-45.

Tukey, J.W. (1953). The problem of multiple comparisons. Unpublished report in private circulation.

Waller, R.A. and D.B. Duncan (1969). A Bayes rule for the symmetric multiple comparisons problem. J. Amer. Statist. Assoc. 64, 1484-1503.

RANKING AND SELECTION AND MULTIPLE COMPARISONS WITH THE BEST*

Jason C. Hsu

Department of Statistics
The Ohio State University
Columbus, Ohio

1. INTRODUCTION

Beginning with the pioneering works of Bechhofer (1954), Gupta (1956, 1965) and others, a substantial theory has developed which has come to be known as Ranking and Selection. However, this theory is not extensively used in practice. Some cite the lack of computer software for Ranking and Selection as the reason for its nonusage. Some question the significance of the theory as a whole. We believe the lack of computer software is a manifestation of some deeper perceived obstacles to practical usage. By taking a new approach, it has recently been found that these obstacles are somewhat illusory. We first describe what we believe to be the perceived obstacles to practical usage, and then explain how these obstacles have been overcome by recent developments.

Traditional Indifference Zone Selection calls for a multi-stage experimentation in the usual case of unknown variability. Since most practitioners carry out their experiments in a single stage rather than follow specific multi-stage designs, there had been little incentive to write computer programs for Indifference Zone Selection. Subset Selection of course is applicable to single-stage designs. But a practical theory should, as a whole, be applicable to single-stage designs as well as multi-

---

* Research supported in part by Grant 1 R01 GM31287-01, awarded by the National Institute of General Medical Sciences, NIH.

stage designs. This, we believe, was the first perceived obstacle to practical usage.

In Hsu (1981), a multiple comparisons approach was taken for the problem of assessing the relative merits of k (≥ 2) treatments. The results show that Indifference Zone Selection inference is available for single-stage designs except, in contrast to multi-stage designs, the precision of the inference cannot be guaranteed prior to the experiment. The results further show that Indifference Zone Selection inference and Subset Selection inference can be given simultaneously without decreasing the minimum confidence level of either. So the first obstacle has been overcome by this new method of Hsu (1981), now called Multiple Comparisons with the Best (MCB). We summarize and give new insights into this result in Section 2.

A second obstacle to practical usage was a perceived difference between the equal and unequal sample size case and the resultant lack of uniformity in the proposed methodologies for the latter case. The simultaneous applicability of Indifference Zone Selection inference and Subset Selection inference was shown for the equal sample size case, where each school has a standard method of inference. But for unequal sample sizes, there is no general agreement on methodology even within a single school. Since many experiments in real life have unequal sample sizes, this lack of agreement has hampered practical usage. In Section 3 we document the fact that for MCB there is no difference in methodology between the equal and unequal sample size case with suitable computer software. In fact, sharper inferences are achieved by these computer implementable procedures than by the traditional procedures based on tabulable critical values.

2. MULTIPLE COMPARISONS WITH THE BEST FOR BALANCED DESIGNS

Consider the general one-way model

$$Y_{ia} = \theta_i + \varepsilon_{ia}, \quad i = 1,\ldots,k, \quad a = 1,\ldots,n_i,$$

where the ε_{ia} are i.i.d. $N(0,\sigma^2)$. We use the usual notations

$$\bar{Y}_i = \sum_{a=1}^{n_i} Y_{ia}/n_i,$$

$$s^2 = \sum_{i=1}^{k} \sum_{a=1}^{n_i} (Y_{ia}-\bar{Y}_i)^2 / \sum_{i=1}^{k} (n_i-1).$$

Also, for later use, let

$$\bar{\varepsilon}_i = \sum_{a=1}^{n_i} \varepsilon_{ia}/n_i \quad \text{for } i = 1,\ldots,k.$$

In this section, we restrict ourselves to the case $n_1 = \ldots = n_k = n$.

Let $(1),(2),\ldots,(k)$ denote the (unknown) indices for which

$$\theta_{(1)} \leq \theta_{(2)} \leq \cdots \leq \theta_{(k)}. \tag{2.1}$$

(In case of ties in the θ-values, any arrangement is permissible so long as (2.1) is satisfied.) Now suppose that a treatment with a larger θ-value is considered better than another with a smaller θ-value. Then $\pi_{(k)}$ is the best treatment and $\theta_{(k)}-\theta_i$ represents the amount by which π_i is inferior to the best treatment. MCB is appropriate when differences $\{\theta_{(k)}-\theta_i: i = 1,\ldots,k\}$ are of primary interest. Let

$$C = \{\pi_i: \min_{j \neq i} (\bar{Y}_i-\bar{Y}_j + ds(2/n)^{1/2}) \geq 0\} \quad \text{and}$$

$$D_i = \max_{j \neq i} (\bar{Y}_j-\bar{Y}_i + ds(2/n)^{1/2})^+$$

for $i = 1,\ldots,k$. Here + denotes the positive part and d satisfies

$$\int_0^\infty \int_{-\infty}^\infty \Phi^{k-1}(z + \sqrt{2}dv)\, d\Phi(z)\, dG(v) = 1 - \alpha,$$

where Φ is the standard normal distribution function, G is the distribution of s/σ, $1-\alpha$ is the desired confidence level.

If $\underset{\sim}{\theta} = (\theta_1,\ldots,\theta_k)$ then Hsu (1981) showed that

$$P_{\underset{\sim}{\theta}}[\pi_{(k)} \in C \text{ and } \theta_{(k)} - \theta_i \leq D_i \text{ for all } i] \geq 1-\alpha, \text{ for all } \underset{\sim}{\theta}. \tag{2.2}$$

Gupta's (1956) Subset Selection confidence statement is

$$P_{\underset{\sim}{\theta}}[\pi_{(k)} \in C] \geq 1 - \alpha \text{ for all } \underset{\sim}{\theta}. \tag{2.3}$$

Fabian's (1963) confidence statement (extended to the unknown variance case) for Bechhofer's Indifference Zone Selection procedure is

$$P_{\underset{\sim}{\theta}}[\theta_{(k)} - \theta_{[k]} \leq D_{[k]}] \geq 1 - \alpha \text{ for all } \underset{\sim}{\theta}, \tag{2.4}$$

where [k] is the random index such that $\bar{Y}_{[k]} = \max_i \bar{Y}_i$. Here $\pi_{[k]}$ is the treatment that appears to be the best. Desu's (1970) confidence statement (extended to the unknown variance case) for his good-bad populations selection procedure is implied by

$$P_{\underset{\sim}{\theta}}[\theta_{(k)} - \theta_i \leq D_i \text{ for all } i] \geq 1 - \alpha \text{ for all } \underset{\sim}{\theta}. \tag{2.5}$$

The connection between (2.4), (2.5) and Indifference Zone Selection is as follows. Suppose that treatments within δ^* of the best are considered good enough. Then by (2.5), if the experimenter selects the treatments $\Pi = \{\pi_i: D_i \leq \delta^*\}$, the probability that all those selected are good enough is at least $1-\alpha$. For single-stage designs, Π may be empty for a given data set. The multi-stage designs discussed in traditional Indifference Zone Selection in essence guarantee $D_{[k]} \leq \delta^*$ for specified $\delta^* > 0$, so that Π is nonempty always.

The confidence statement (2.2) shows that if all the inferences are given simultaneously, each at a minimum confidence level of $1-\alpha$, the joint confidence level is still at least $1-\alpha$. Some find this puzzling, since it is well known that the confidence level of each inference attains the $1-\alpha$ minimum. But the fact is that each of the events in (2.3), (2.4), and (2.5) contains the same $1-\alpha$ probability event

$$E = \{\bar{Y}_{(k)} - \theta_{(k)} \geq \bar{Y}_i - \theta_i - ds(2/n)^{1/2} \text{ for all } i, i \neq (k)\}.$$

Additional insight can be gained by examining the true joint and individual coverage probabilities of (2.2) - (2.5) as functions of $\underset{\sim}{\theta} = (\theta_1,\ldots,\theta_k)$. A simulation study was conducted for the case of $k = 5$, $\alpha = .10$, and $\sigma = 1$. Table 1 shows the true coverage probabilities of (2.2) - (2.5) for selected $\underset{\sim}{\theta}$, estimated from 10,000 replications of the experiment. Note that when $\theta_i = \theta_{i+1}$, θ_i is taken to be θ_{i+1}^- so that π_5 is the best treatment always.

It can be seen that the true coverage probability of Gupta's confidence set for the true best treatment $\pi_{(k)}$ (Eq. 2.3), increases rapidly from $1-\alpha$ to one as $\theta_{(k)} - \theta_{(k-1)}$ increases. On the other hand, the true coverage probability of the $\underset{\sim}{D} = (D_1,\ldots,D_k)$, upper bounds for $(\theta_{(k)}-\theta_1,\ldots,\theta_{(k)}-\theta_k)$ (Eq. 2.5) in Desu's good-bad populations selection inference, decreases rapidly from one to $1-\alpha$ as $\theta_{(k)} - \theta_{(k-1)}$ increases. Table 1 also shows that the true coverage probability of Fabian's $D_{[k]}$ upper bound on $\theta_{(k)} - \theta_{[k]}$ (Eq. 2.4) which implies Bechhofer's Indifference Zone Selection inference, generally far exceeds the nominal confidence level $1-\alpha$. Hsu's

Table 1

Parameters					Coverage probability			
(θ_1,	θ_2,	θ_3,	θ_4,	θ_5)	Eq. 2.2	Eq. 2.3	Eq. 2.4	Eq. 2.5
(0,	0,	0,	0,	0)	.892	.892	1.000	1.000
(-.2,	-.2,	0,	0,	0)	.927	.929	.995	.995
(-.4,	-.4,	0,	0,	0)	.931	.936	.994	.994
(-.6,	-.6,	0,	0,	0)	.931	.936	.996	.994
(-.8,	-.8,	0,	0,	0)	.931	.936	1.000	.994
(-1.0,	-1.0,	0,	0,	0)	.931	.936	1.000	.994
(-.2,	-.2,	-.2,	-.2,	0)	.957	.976	.972	.972
(-.4,	-.4,	-.2,	-.2,	0)	.949	.985	.955	.955
(-.6,	-.6,	-.2	-.2,	0)	.942	.986	.954	.947
(-.8,	-.8,	-.2,	-.2,	0)	.941	.986	.963	.946
(-1.0,	-1.0,	-.2,	-.2,	0)	.941	.986	.965	.946
(-.4,	-.4,	-.4,	-.4,	0)	.926	.997	.927	.927
(-.6,	-.6,	-.4,	-.4,	0)	.910	.999	.924	.910
(-.8,	-.8,	-.4,	-.4,	0)	.908	.999	.943	.908
(-1.0,	-1.0,	-.4,	-.4,	0)	.908	.999	.946	.908
(-.6,	-.6,	-.6,	-.6,	0)	.892	1.000	.922	.892
(-.8,	-.8,	-.6,	-.6,	0)	.892	1.000	.947	.892
(-1.0,	-1.0,	-.6,	-.6,	0)	.892	1.000	.952	.892
(-.8,	-.8,	-.8,	-.8,	0)	.892	1.000	.984	.892
(-1.0,	-1.0,	-.8,	-.8,	0)	.892	1.000	.991	.892
(-1.0,	-1.0,	-1.0,	-1.0,	0)	.892	1.000	.998	.892

simultaneous coverage probabilities of C and $\underset{\sim}{D}$ (Eq. 2.2) are on the whole more nearly equal to .90 reflecting the fact that the true coverage probabilities of C and D vary with $\theta_{(k)} - \theta_i$ in somewhat opposite ways.

On the one hand, it can be said that, with the presently available inference procedures, there is no reason not to give the inferences simultaneously. On the other hand, as can be seen from Table 1, if only one particular aspect of the full MCB inference is of interest, then there is potential for improvement, especially if the only inference of interest is on $\pi_{[k]}$, the treatment that appears to be the best.

3. MULTIPLE COMPARISONS WITH THE BEST FOR UNBALANCED DESIGNS

In traditional Indifference Zone Selection, the design aspect is an integral part of the formulation. Therefore, the literature has concentrated more on the allocation of samples, rather than the analysis of samples of unequal size obtained in an arbitrary fashion.

In Subset Selection, procedures for unequal sample sizes have been proposed by Chen, Dudewicz, and Lee (1976), Gupta and Huang (1974), Gupta and Huang (1976), Gupta and Wong (1976). All these procedures employ tabulable critical values. Berger and Gupta (1977) showed that the procedure of Gupta and Huang (1976) is the only minimax rule among those proposed according to the criterion they considered. Below we describe MCB for unequal sample sizes which employs a vector of critical values. The Subset Selection aspect of MCB is always at least as good as and sometimes better than the Gupta and Huang (1976) procedure in the unequal sample size case.

For each i, define d_i to be the constant such that

$$P[\overline{\varepsilon}_i \geq \overline{\varepsilon}_j - d_i\ s(n_i^{-1} + n_j^{-1})^{1/2} \quad \text{for all } j,\ j \neq i]$$

$$= \int_0^\infty \int_{-\infty}^\infty \prod_{j\neq i} \Phi((n_j/n_i)^{1/2} z + d_i v(1+n_j/n_i)^{1/2}) d\Phi(z)\ dG(v) = 1 - \alpha \quad (3.1)$$

where Φ and G are as before. We assume $\alpha \leq .5$, which implies $d_i \geq 0$. Define

$$C = \{\pi_i : \min_{j\neq i} (\overline{Y}_i - \overline{Y}_j + d_i\ s(n_i^{-1} + n_j^{-1})^{1/2}) \geq 0\}$$

$$D_i = \max\ \{\overline{Y}_j - \overline{Y}_i + d_j\ s(n_j^{-1} + n_i^{-1})^{1/2},\ j \in C,\ j \neq i\}$$

for i = 1,...,k. Here it is understood that if $C = \{\pi_i\}$ then $D_i = 0$. The MCB inference for unequal sample sizes is given by the following confidence statement.

Theorem 3.1. For all $\underset{\sim}{\theta}$,

$$P_{\underset{\sim}{\theta}}[\pi_{(k)} \in C \text{ and } \theta_{(k)} - \theta_i \leq D_i \text{ for } i = 1,\ldots,k] \geq 1 - \alpha. \quad (3.2)$$

Proof. We proceed by enlarging the event E below.

$$E = \{\bar{Y}_{(k)}-\theta_{(k)} \geq \bar{Y}_i-\theta_i - d_{(k)}\, s(n_{(k)}^{-1} + n_i^{-1})^{1/2} \text{ for all } i \neq (k)\}$$

$$= \{\bar{Y}_{(k)}-\bar{Y}_i + d_{(k)}\, s(n_{(k)}^{-1} + n_i^{-1})^{1/2} \geq 0 \text{ for all } i \neq (k) \text{ and}$$

$$\theta_{(k)}-\theta_i \leq \bar{Y}_{(k)} - \bar{Y}_i + d_{(k)}\, s(n_{(k)}^{-1} + n_i^{-1})^{1/2} \text{ for all } i \neq (k)\}$$

$$\subseteq \{\pi_{(k)} \in C \text{ and}$$

$$\theta_{(k)} - \theta_i \leq \max\, \{\bar{Y}_j - \bar{Y}_i + d_j\, s(n_j^{-1} + n_i^{-1})^{1/2},\ j \in C,\ j \neq i\}$$

$$\text{for } i \neq (k)\}$$

$$= \{\pi_{(k)} \in C \text{ and } \theta_{(k)} - \theta_i \leq D_i \text{ for } i = 1,\ldots,k\}.$$

Since $P[E] = 1-\alpha$, the proof follows.

In practice, Theorem 3.1 can be implemented as follows. Given $\underset{\sim}{n} = (n_1,\ldots,n_k)$ and α, each d_i can be obtained by solving (3.1) numerically. Thus, the computer implementation of MCB consists of

1. input data and α,
2. compute d_i, $i = 1,\ldots,k$, for the specified $\underset{\sim}{n}$ and α,
3. output inference according to (3.2).

Note that in the equal sample size case, (3.2) reduces to (2.2). Thus, computer MCB analysis for equal sample sizes requires the same Steps 1, 2, and 3 as computer MCB analysis for unequal sample sizes (except only one d needs to be computed in Step 2 in the equal sample size case). Thus, with the availability of the computers, the same methodology is used in the equal and unequal sample size case.

Remark 3.1. Note that in the computation of the upper bound D_i, the maximum is taken over $j \in C$ rather than $j \in \{1,\ldots,k\}$. While it makes no difference in the equal sample size parametric case, the distinction is important in the unequal sample size parametric case and the nonparametric case (equal sample size or not). Taking the maximum over $j \in C$ not only gives sharper bounds, but also guarantees that if only one treatment remains then its upper bound on distance from the best is 0. The nonparametric bounds in Hsu (1981) can be improved using this modification.

Remark 3.2. The Gupta and Huang (1976) confidence set for $\pi_{(k)}$ at confidence level $1-\alpha$ is

$$C' = \{\pi_i : \min_{j \neq i} (\overline{Y}_i - \overline{Y}_j + d's(n_i^{-1} + n_j^{-1})^{1/2}) \geq 0\}$$

where $d' = \max_{1 \leq i \leq k} d_i$. Clearly $C \subseteq C'$ with strict inequality in some cases. So sharper Subset Selection inference is achieved by MCB in the unequal sample size case.

R- and S-values

Given the data, one can calculate, for each treatment π_i, the smallest α value for which π_i can be rejected as the best treatment (i.e. $\pi_i \not\in C$). This value for π_i, which we call the R-value for π_i and denote by R_i, can be computed as follows.

$$R_i = 1 - P[\overline{\varepsilon}_i \geq \overline{\varepsilon}_j - t_i s(n_j^{-1} + n_i^{-1})^{1/2} \text{ for all } j \neq i]$$

$$= 1 - \int_0^\infty \int_{-\infty}^\infty \prod_{j \neq i} \Phi((n_j/n_i)^{1/2} z + t_i v(1+n_j/n_i)^{1/2}) d\Phi(z)\, dG(v)$$

where

$$t_i = \max_{m \neq i} \{(\overline{Y}_m - \overline{Y}_i)/[s(n_m^{-1} + n_i^{-1})^{1/2}]\}$$

is a realized value.

Of course it is senseless to report the R-value for the treatment π_h (say) that has the largest R-value. For that treatment, it is useful to report the smallest α value for which it can be selected as the best treatment (i.e. $D_h = 0$). We call this value the S-value for π_h and denote by S_h. Clearly $S_h = \max_{j \neq h} R_j$.

R- and S-values would only be reported for convenience. If the experimenter rejects those treatments whose R-values are smaller than his α as the best treatment and selects the treatment π_h as the best treatment if its S-value is smaller than his α, then the probability that all his inferences are correct is at least $1-\alpha$ because these inferences are implied by the full MCB inference. The only reason for reporting the R- and S-values is if another experimenter, having a different α, is interested in the rejection and selection inferences described above only, then he can deduce his inferences from the R- and S-values without having to perform the full MCB inference at his α.

4. AN EXAMPLE

In Lorenz, Hsu, and Tuovinen (1982), seven brands of membrane filters were compared for their ability to reveal microorganisms in river water.

As these filters are used for monitoring water quality, a better brand is one that is more sensitive in revealing the target organisms (fecal coliforms in this case) in sample water. Water samples were collected from different rivers around Columbus, Ohio, at different times of the year. Each river-time combination was considered a separate experiment. For each experiment, three filters of each brand were randomly selected and 100 ml of sample water poured through each filter. After 24 hours of incubation in a fecal coliform-selective medium, the number of colonies on each filter (representing the number of fecal coliforms revealed by the filter) was recorded. The result from one of the experiments is as follows.

Brand						
π_1	π_2	π_3	π_4	π_5	π_6	π_7
69	118	171	122	204	140	170
122	154	132	119	225	130	165
95	102	182	-	190	127	-

As it often happens in real life, there were missing data, caused by colony blurring in this case. Application of MCB with unequal sample sizes at α = .05 yielded the following results.

Brand	Sample Mean	Sample Size	Upper Bound on $\theta_{(k)} - \theta_i$
π_1	95.3	3	153.9
π_2	124.7	3	124.6
π_3	161.7	3	87.6
π_4	120.5	2	133.8
π_5	206.3	3	8.0
π_6	132.3	3	116.9
π_7	167.5	2	86.8

Brand	R-Value	Reject	S-Value	Select
π_1	.0001	*		
π_2	.0009	*		
π_3	.0418	*		
π_4	.0013	*		
π_5			.1007	
π_6	.0019	*		
π_7	.1007			

For this experiment, even though we fail to assert a particular brand is best at $\alpha = .05$, the data does indicate that π_5 is at least close to being the best, judging from its confidence bound. Interestingly, the brand currently specified by EPA for monitoring water quality is π_2, one of the brands rejected as the best by the analysis.

SUMMARY

In this article, we describe the implementation of Multiple Comparisons with the Best (MCB) inference for possibly unbalanced designs with suitable computer software. Implications toward Ranking and Selection inferences are discussed. An example with real data is given to illustrate the MCB inference for unbalanced designs.

REFERENCES

Bechhofer, R. E. (1954). A single-sample multiple decision procedure for ranking means of normal populations with known variance. Ann. Math. Statist. 25, 16-39.

Berger, R. L. and Gupta, S. S. (1977). Minimax subset selection rules with applications to unequal variance (unequal sample size) problems. Mimeo Series No. 495, Dept. of Statist., Purdue Univ.

Chen, H. J., Dudewicz, E. J., and Lee, Y. J. (1976). Subset selection procedures for normal means under unequal sample sizes. Sankhyā Ser. B 38, 249-255.

Desu, M. M. (1970). A selection problem. Ann. Math. Statist. 41, 1596-1603.

Fabian, V. (1962). On multiple decision methods for ranking population means. Ann. Math. Statist. 33, 248-254.

Gupta, S. S. (1956). On a decision rule for a problem in ranking means. Mimeo Series No. 150, Inst. of Statist., Univ. of North Carolina.

Gupta, S. S. (1965). On some multiple decision (selection and ranking) rules. Technometrics 7, 225-245.

Gupta, S. S. and Huang, W. T. (1974). A note on selecting a subset of normal populations with unequal sample sizes. Sankhyā Ser. A.36, 389-396.

Gupta, S. S. and Huang, D. Y. (1976). Subset selection for the means and variances of normal populations: Unequal sample size case. Sankhyā Ser. B 38, 112-128.

Gupta, S. S. and Wong, W. Y. (1976). Subset selection procedures for the means of normal populations with unequal variances: Unequal sample size case. Tech. Report No. 473, Dept. of Statist., Purdue Univ.

Hsu, J. C. (1981). Simultaneous confidence intervals for all distances from the "best". _Ann. Statist._ _9_, 1026-1034.

Lorenz, R. C., Hsu, J. C., and Tuovinen, O. H. (1982). Performance variability, ranking and selection analysis of membrane filters for enumerating coliform bacteria in river water. _J. Am. Waterworks Assoc._ _74_, 429-437.

THE ROBUSTNESS OF BECHHOFER'S NORMAL MEANS SELECTION PROCEDURE*

Edward J. Dudewicz

Department of Mathematics
Syracuse University
Syracuse, New York

Satya N. Mishra

Department of Mathematics
and Statistics
University of South Alabama
Mobile, Alabama

1. INTRODUCTION

Let us consider the following <u>selection problem</u>:

Suppose that $\pi_1,\ldots,\pi_k$ are k populations the observations from which are known to be identically distributed except for a location parameter (i.e., an observation X_i from population π_i has c.d.f. $F(x-\nu_i)$, where $F(\cdot)$ may be either known or unknown). Select any one of the populations associated with $\nu_{[k]}$, where $\nu_{[1]} \leq \ldots \leq \nu_{[k]}$ denote the ranked ν_i.

If the form of $F(\cdot)$ is completely unknown, then a nonparametric method such as the one suggested by Bechhofer and Sobel (1958) is one possible recourse. However, in practice one will usually have some information about $F(\cdot)$; for example, it may be known completely except for the location parameter, or the information may imply that means of independent observations with such a c.d.f. obey the Law of Large Numbers (LLN). If it is completely known except for the location parameter, one can usually find a specific parametric procedure which requires fewer observations than a nonparametric procedure to guarantee a specified requirement on the probability of correctly selecting a population associ-

*This research was supported by PHS Grant Number F5 R01 CA26254-03, 7 R01 CA35186-01 awarded by the National Cancer Institute, DHHS.

ated with $\nu_{[k]}$. If means of independent observations with such a c.d.f. obey the LLN, one can base the decision on means and be assured of large-sample consistency (i.e., of having the probability of selecting at least one of the populations associated with $\nu_{[k]}$ approach one as the sample size increases). Dudewicz (1966) determined how much one pays (in terms of increased sample sizes) for the certainty provided by the Bechhofer and Sobel (1958) nonparametric method of guaranteeing a specified requirement on the probability of a correct selection (P(CS)) under specific parametric alternatives, for the given selection problem.

In practice, if one has some conjecture about the form of the true $F(\cdot)$, one will have to weigh the increased sample sizes needed by a nonparametric procedure (e.g., that of Bechhofer and Sobel (1958)) against the possibility that the deviations of the true $F(\cdot)$ from the conjectured c.d.f. may be so large that the requirement on the P(CS), will not be satisfied if a parametric procedure appropriate for the conjectured c.d.f. is used. For example, Dudewicz (1966) showed that the nonparametric procedure of Bechhofer and Sobel (1958) has a low efficiency relative to the specific parametric alternatives he considered and would therefore be useful only if real doubt existed as to the actual distribution and if a wrong decision would be costly.

Bechhofer (1954) proposed a selection procedure based on sample means (the Means Procedure) for a problem more general than the one under consideration here, but in which an observation from population π_i was assumed to be normally distributed with an unknown mean ν_i and a known variance σ_i^2 $(i = 1,\ldots,k)$. If the σ_i^2 are equal to a common value σ^2 then the k populations differ only in their location parameters ν_i, and $F(\cdot)$ is then the $N(0,\sigma^2)$ distribution function. In this paper we consider the robustness (sensitivity to deviations of the true c.d.f. from the conjectured c.d.f. $F(\cdot)$) of the Means Procedure when the conjectured $F(\cdot)$ is $N(0,\sigma^2)$. The issue of the robustness of the Means Procedure is one of practical importance since Hall (1959) has shown it to be the most economical, and hence it would be the recommended choice if normality holds.

2. THE MEANS PROCEDURE

The Means Procedure is as follows:

Take N independent vectors $\underline{X}_j = (X_{1j},\ldots,X_{kj})$, $j=1,\ldots,N$, where X_{ij} denotes the jth observation from the ith population π_i. Choose the population yielding the largest $\overline{X}_i = (1/N)\sum_{j=1}^{N} X_{ij}$ $(i = 1,\ldots,k)$ as being associated with $\nu_{[k]}$.

Adopting the approach of Bechhofer (1954), we let $\{\delta^*, P^*\}$ be two specified constants $(0 < \delta^* < \infty,\ 1/k < P^* < 1)$ and ask that our procedure guarantee the following probability requirement:

$$P(CS) \geq P^* \quad \text{whenever } \nu_{[k]} - \nu_{[k-1]} \geq \delta^*. \tag{1}$$

For the Means Procedure and any c.d.f. $F(\cdot)$, if we assume (without loss) that $\nu_{[k]}$ is associated with π_k then we have

$$\begin{aligned} P(CS) &= P[\overline{X}_k > \overline{X}_i \quad (i = 1,\ldots,k-1)] \\ &= \int_{-\infty}^{\infty} \left[\prod_{i=1}^{k-1} F_{\overline{X}_i}(x)\right] dF_{\overline{X}_k}(x) \end{aligned} \tag{2}$$

where $F_{\overline{X}_i}(\cdot)$ is the c.d.f. of $\overline{X}_i (i=1,\ldots,k)$. The configuration of the $\nu_{[i]}$'s which, for fixed $\{\delta^*, P^*\}$, minimizes P(CS) for fixed N subject to $\nu_{[k]} - \nu_{[k-1]} \geq \delta^*$ is called the least-favorable configuration (LFC). Although in general the LFC will depend on the c.d.f. $F(\cdot)$ as well as the selection procedure, it is clear from (2) that for the Means Procedure the LFC is

$$\nu_{[1]} = \cdots = \nu_{[k-1]},\ \nu_{[k]} - \nu_{[k-1]} = \delta^*. \tag{3}$$

In order to eliminate the dependence on σ^2 from our study of robustness, in our work below we use not (1) but a modified probability requirement with specified constants $\{\lambda^*, P^*\}$ $(0 < \lambda^* < \infty,\ 1/k < P^* < 1)$:

$$P(CS) \geq P^* \quad \text{whenever } \nu_{[k]} - \nu_{[k-1]} \geq \delta^* = \lambda^*\sigma. \tag{4}$$

Note that if σ is unknown, the selection problem can in some cases (e.g., the normal case) still be solved if one is willing to specify the "distance" between $\nu_{[k]}$ and $\nu_{[k-1]}$ in terms of a multiple λ^* of the unknown standard deviation σ by using probability requirement (4). However, if one knows σ then it is immaterial whether (1) or (4) is used.

To study the robustness of the Means Procedure for this problem, we will compute the P(CS) actually achieved in the LFC, say $P_{LFC}(CS)$, for various λ^* values and various sample sizes N, when $F(\cdot)$, the true c.d.f., is of various forms.

3. SELECTION PROBABILITIES: NORMAL DISTRIBUTION

In this section, we describe the calculation of the $P_{LFC}(CS)$ for the Means Procedure when probability requirement (4) is used and the $F(\cdot)$ is $N(0,\sigma^2)$. (Thus, we are considering the problem of selecting the normal population with the largest mean out of k normal populations $N(\nu_i,\sigma^2)$, where $\nu_1,\ldots,\nu_k$ are all unknown and the common variance σ^2 may be either known or unknown as noted in Section 2.)

The $P_{LFC}(CS)$ values for this case have been obtained by numerical quadrature and checked by linear interpolation in the tables of Teichroew (1955) where, for $k = 2(1)10$, $P_{LFC}(CS)$ is tabulated for various $\sqrt{N}\lambda^*$ values. Our values will be found in Tables 1, 2, and 3 at the end of Section 4. (Note that Table I of Bechhofer (1954) gives, for $k = 2(1)10$ (and for our problem $t = 1$), $\sqrt{N}\lambda^*$ for various $P_{LFC}(CS)$ values and is not well-suited to our needs here.)

4. SELECTION PROBABILITIES: NON-NORMAL DISTRIBUTIONS

In this section, we describe the calculation of the $P_{LFC}(CS)$ for the Means Procedure when probability requirement (4) is used and the true $F(\cdot)$ is either uniform with spread b (short-tailed) or Student's t with 3 degrees of freedom denoted by t_3 (long-tailed). (A t_1 is Cauchy so the P(CS) would not change as N changes, and a t_2 has no variance, so these alternatives were not usable.)

For the uniform distribution with spread b it is well-known (e.g., Freeman (1963, p. 133)) that the variance is $b^2/12$. Thus, in (4) $\delta^* = \lambda^*\sigma = \lambda^* b/\sqrt{12}$. In the LFC we may suppose that $X_{i1},\ldots,X_{iN}$ $(i=1,\ldots,k-1)$ are uniformly distributed on $[\mu - \frac{b}{2} - \delta^*, \mu + \frac{b}{2} - \delta^*]$ and that $X_{k1},\ldots,X_{kN}$ are uniformly distributed on $[\mu - \frac{b}{2}, \mu + \frac{b}{2}]$ for some μ. However, since a linear function of a uniformly distributed random variable is also uniformly distributed, it follows that

$$P_{LFC}(CS) = P[Y_{i1} + \ldots + Y_{iN} < Y_{k1} + \ldots + Y_{kN} \quad (i=1,\ldots,k-1)] \tag{5}$$

where $Y_{i1},\ldots,Y_{iN}$ $(i=1,\ldots,k-1)$ are uniformly distributed on $[-\delta^*/b, 1-\delta^*/b]$ and $Y_{k1},\ldots,Y_{kN}$ are uniformly distributed on $[0,1]$. Thus, with the probability requirement (4), the $P_{LFC}(CS)$ is independent of b and μ

and may be calculated using uniform populations of spread 1 such that π_k is uniform on [0,1] and $\pi_1,\ldots,\pi_{k-1}$ are uniform on $[-\lambda^*/\sqrt{12},\ 1-\lambda^*/\sqrt{12}]$.

Note that if $\frac{\lambda^*}{\sqrt{12}} \geq 1$, the $P_{LFC}(CS) = 1$ for all $N = 1,2,3,\ldots$. Thus, in this case probability requirement (4) will be trivially satisfied. We will use only values $\lambda^* \leq \sqrt{12} = 3.464102$.

Parzen (1969, p. 407) shows that if $X_1,\ldots,X_n$ are independent random variables, each uniform on [0,1], then the density function of $S_n = X_1 + \ldots + X_n$ is

$$f_{S_n}(x) = \begin{cases} \dfrac{1}{(n-1)!} \sum_{j=0}^{[x]} \binom{n}{j} (-1)^j (x-j)^{n-1}, & 0 \leq x \leq n \\ 0, \text{ otherwise,} & \end{cases} \tag{6}$$

where [x] denotes the greatest integer less than or equal to x. Thus, from (5),

$$P_{LFC}(CS) = \int_0^{N(1-\delta^*)} \left[\int_0^{x+N\delta^*} \frac{1}{(N-1)!} \sum_{j=0}^{[y]} \binom{N}{j} (-1)^j (y-j)^{N-1} dy \right]^{k-1} \cdot \frac{1}{(N-1)!} \sum_{j=0}^{[x]} \binom{N}{j} (-1)^j (x-j)^{N-1} dx \tag{7}$$

$$+ \int_{N(1-\delta^*)}^{N} \frac{1}{(N-1)!} \sum_{j=0}^{[x]} \binom{N}{j} (-1)^j (x-j)^{N-1} dx,$$

where δ^* will be taken to be $\lambda^*/\sqrt{12}$ for our computations.

The following finite series expansion was obtained for (7) and was used for computing purposes:

$$P_{LFC}(CS) = \frac{1}{N!} \sum_{s=[N(1-\delta^*)]}^{N-1} \sum_{\nu=0}^{s} (-1)^\nu \binom{N+1}{\nu} (s+1-\nu)^N$$

$$- \frac{1}{N!} \sum_{j=0}^{[N(1-\delta^*)]} \binom{N}{j} (-1)^j \{(N(1-\delta^*)-j)^N - ([N(1-\delta^*)]-j)^N\}$$

$$+ \sum_{r_0=0}^{k-1} \sum_{r_1=0}^{k-1-r_0} \cdots \sum_{r_{N-1}=0}^{k-1-r_0-\cdots-r_{N-2}} c_{k,N}(\underset{\sim}{r}) \sum_{v=0}^{[N(1-\delta^*)]} \left\{ \sum_{s=0}^{v+[N\delta^*]-1} \sum_{\nu=0}^{s} \right.$$

$$(-1)^\nu \binom{N+1}{\nu} (s+1-\nu)^N$$

$$\left. + \sum_{j=0}^{v+[N\delta^*]} \binom{N}{j} (-1)^j \{(N\delta^*-j)^N - (v+[N\delta^*]-j)^N\} \right\}^{r_0}$$

$$\times \left\{ \prod_{i=1}^{N-1} \left(\sum_{j=0}^{v+[N\delta^*]} \binom{N}{j} (-1)^j \binom{N}{i} (N\delta^*-j)^{N-i} \right)^{r_i} \right\} \left\{ \sum_{j=0}^{v+[N\delta^*]} \right.$$

$$\left. \binom{N}{j} (-1)^j \right\}^{k-1-\sum_{i=0}^{N-1} r_i}$$

$$\times \left\{ \sum_{j=0}^{v} \sum_{w=0}^{N-1} \binom{N}{j} (-1)^j \binom{N-1}{w} (-j)^{N-1-w} \right\} \times \left\{ \{v+1-(N\delta^*-[N\delta^*])\}^{d_{k,N}(\underset{\sim}{r})+w} \right.$$

$$- D_j(0,v)\{v-(N\delta^*-[N\delta^*])\}^{d_{k,N}(\underset{\sim}{r})+w} \times \frac{N}{\{d_{k,N}(\underset{\sim}{r})+w\}(N!)^k} .$$

Here

$$c_{k,N}(\underset{\sim}{r}) = \frac{(k-1)!}{r_0! r_1! \cdots r_{N-1}! (k-1-r_0-\cdots-r_{N-1})!} ,$$

$$d_{k,N}(\underset{\sim}{r}) = 1 + 1\cdot r_1 + 2\cdot r_2 + \cdots + (N-1) r_{N-1} + N(k-1-r_0-\cdots-r_{N-1}),$$

and

$$D_j(0,v) = \begin{cases} 0 & \text{if} \quad v = 0 \\ 1 & \text{if} \quad v \neq 0,\ j \neq v \\ (v/\{v-(N\delta^* - [N\delta^*])\})^{d_{k,N}(\underset{\sim}{r}) + w} & \text{if } j=v \neq 0. \end{cases}$$

Single-precision arithmetic was used to compute all the $P_{LFC}(CS)$ values. Double-precision arithmetic was used only to check the tables presented. In order to minimize the computer time needed for such checking, at first only the values most likely to be in error were checked, namely, the values for the highest N in each column of Tables 1, 2, and 3.

Table 1

$P_{LFC}(CS)$ for k=2 Populations, Means Procedure[a]

N \ λ^*	0.1	1.0	3.0
1	.528 .528 .547	.760 .747 .827	.983 .991 .982
2	.540 .538 .557	.841 .835 .877	.999 1.000 .995
3	.549 .548 .562	.890 .887 .920	1.000 1.000 .999
4	.556 .555 .567	.921 .920 .940	1.000 1.000 .999
5	.563 .562 .572	.943 .943 .957	1.000 1.000 .999
6	.569 .568 .577	.958 .958 .965	1.000 1.000 1.000
7	.574 .573 .585	.969 .970 .976	1.000 1.000 1.000
8	.579 .579 .588	.977 .978 .978	1.000 1.000 1.000
9	.583 .583 .600	.983 .983 .981	1.000 1.000 1.000
10	.588 .585 .599	.987 .988 .987	1.000 1.000 1.000

[a]The upper, middle and the lower entries are the $P_{LFC}(CS)$ values for the normal, uniform and Student's t_3 distribution, respectively.

Table 2

$P_{LFC}(CS)$ for k=3 Populations, Means Procedure[a]

N \ λ^*	0.1	1.0	3.0
1	.362	.634	.969
	.362	.614	.983
	.378	.724	.969
2	.374	.745	.997
	.373	.734	1.000
	.377	.819	.993
3	.383	.817	1.000
	.382	.811	1.000
	.394	.872	.996
4	.391	.866	1.000
	.390	.863	1.000
	.400	.894	.999
5	.398	.901	1.000
	.397	.900	1.000
	.415	.925	.999
6	.405	.926	1.000
	.404	.926	1.000
	.412	.942	.999
7	.411	.945	1.000
	.409	.945	1.000
	.422	.950	.999
8	.416	.959	1.000
	.418	.959	1.000
	.433	.961	1.000
9	.421	.969	1.000
	.421	.968	1.000
	.428	.970	1.000
10	.426	.976	1.000
	.426	.979	1.000
	.443	.978	1.000

[a]The upper, middle and the lower entries are the $P_{LFC}(CS)$ values for the normal, uniform and Student's t_3 distribution, respectively.

Table 3

$P_{LFC}(CS)$ for k=6 Populations, Means Procedure[a]

N \ λ^*	0.1	1.0	3.0
1	.189 .196 .193	.449 .455 .549	.935 .962 .945
2	.198 .198 .213	.586 .577 .671	.994 .999 .984
3	.206 .206 .216	.685 .679 .744	.999 1.000 .993
4	.212 .212 .223	.758 .755 .808	1.000 1.000 .996
5	.218 .218 .224	.815 .813 .849	1.000 1.000 .998
6	.223 .223 .229	.858 .857 .878	1.000 1.000 .999
7	.228 .228 .231	.891 .891 .901	1.000 1.000 .999
8	.233 .244 .240	.916 .917 .922	1.000 1.000 1.000
9	.238 — .240	.935 — .937	1.000 1.000 1.000
10	.242 — .256	.950 — .952	1.000 1.000 1.000

[a]The upper, middle and the lower entries are the $P_{LFC}(CS)$ values for the normal, uniform and Student's t_3 distribution, respectively.

When the underlying distribution F(·) is Student's t with 3 d.f. (for which the variance $\sigma^2 = 3$), a simulation study with 10,000 replications was used to estimate the $P_{LFC}(CS)$.

We now present tables of the $P_{LFC}(CS)$, computed with F(·) normal, uniform, and t_3, for $k = 2, 3, 6$ populations and for $\lambda^* = 0.1, 1.0, 3.0$. These afford a measure of the robustness of the Means Procedure for the given selection problem when probability requirement (4) is used. Note that we have concentrated on small-sample results since, as N grows large, the uniform distribution case will tend (for $\lambda^* < \sqrt{12}$) to produce the same $P_{LFC}(CS)$ as the normal distribution case due to the fact that means of independent uniform random variables (which tend in distribution to normal variables with the same mean and variance as $N \to \infty$) enter into the $P_{LFC}(CS)$. However, this does not mean that the ratio of the sample sizes needed to guarantee the probability requirement tends to 1 as $P^* = P_{LFC}(CS) \to 1$. (See Dudewicz (1966, p. 13).)

In interpreting the tables, bear in mind that the $P_{LFC}(CS)$ values calculated under normality and uniformity are exact, while those from simulation are based on 10,000 replications (and hence have a standard error of $2\sqrt{p(1-p)/10{,}000} \leq .01$). From the tables we see that the $P_{LFC}(CS)$ values under uniform and t_3 distributions are never more than .013 units below the value under normality over a broad range of k,N, and λ^*. This indicates that if normality is conjectured (and treated as being the true state of affairs) the resulting selection procedure will be very robust both to short-tailed and long-tailed deviations. We also note that for $\lambda^* = 0.1$ and 1.0, generally the $P_{LFC}(CS)$ values for the uniform distribution are slightly lower than those for the normal while those for the t_3 distribution are higher. It is unclear whether this behavior extends to other short-tailed and long-tailed distributions; more work is needed to investigate this point further.

SUMMARY

In many selection problems arising in practice one is interested not only in finding procedures which guarantee a specified probability requirement but also in the robustness of such procedures. This paper examines the robustness of Bechhofer's 1954 means procedure, which assumes normality. We show by using a combination of exact numerical methods and Monte Carlo simulations that Bechhofer's procedure is robust to both the short-tailed and long-tailed alternatives considered, over a broad range of possible specifications of number of populations, preference zone, and sample size.

REFERENCES

Bechhofer, R. E. (1954), "A single-sample multiple decision procedure for ranking means of normal populations with known variances," Annals of Mathematical Statistics, Vol. 25, pp. 16-39.

Bechhofer, R. E. and Sobel, M. (1958), "Non-parametric multiple-decision procedures for selecting that one of k populations which has the highest probability of yielding the largest observation (preliminary report)," Abstract, Annals of Mathematical Statistics, Vol. 29, p. 325.

Dudewicz, E. J. (1966), The Efficiency of a Nonparametric Selection Procedure: Largest Location Parameter Case, M. S. Thesis, Cornell University, Ithaca, New York, February 1966. (Reprinted as Technical Report No. 14, Department of Operations Research, Cornell University, Ithaca, New York.)

Freeman, H. (1963), Introduction to Statistical Inference, Addison-Wesley Publishing Co., Inc., Reading, Massachusetts.

Hall, W. J. (1959), "The most-economical character of some Bechhofer and Sobel decision rules," Annals of Mathematical Statistics, Vol. 30, pp. 964-969.

Parzen, E. (1960), Modern Probability Theory and Its Applications, John Wiley & Sons, Inc., New York.

Teichroew, D. (1955), Probabilities Associated with Order Statistics in Samples from Two Normal Populations with Equal Variance, ENASR No. ES-3, Chemical Corps Engineering Agency, Engineering Statistics Unit, Army Chemical Center, Maryland.

SELECTION OF THE BEST TREATMENT IN COMPARISON TO A CONTROL WITH AN APPLICATION TO A MEDICAL TRIAL

Charles W. Dunnett

Department of Clinical
Epidemiology and Biostatistics
McMaster University
Hamilton, Ontario, Canada

1. INTRODUCTION

The determination of sample sizes is an important aspect of the design of experiments in which different treatment groups are to be compared. The problem is particularly complex in a medical trial, where there are usually multiple outcome measurements to be considered requiring more than one type of analysis of the data. The usual procedure is to choose one particular outcome measurement (such as the proportion of successes) and base sample sizes on a test of a null hypothesis to achieve a specified power for detecting some suitably defined difference, Δ. Alternatively, the problem might be posed in terms of obtaining a confidence interval estimate of Δ of specified length. These methods are straightforward if there are two treatments to be compared, but complications arise if there are more than two due to the multiplicity of treatment comparisons that are of interest. Moreover, there may be a question whether testing a null hypothesis and estimating a parameter are appropriate formulations of the problem. In many medical trials, the most important problem is to correctly identify the best treatment or treatments, hence a formulation as a selection problem ought to be more realistic in these cases. Formulating the statistical problem in terms of selection would seem to be particularly pertinent if the choice of a therapy for a group of patients is to be made on the basis of the results of the trial, or if the treatments to be studied in a later trial depend upon which treatment turns out to be superior to the others in the current trial.

In this paper, we consider a particular type of treatment comparison trial in which one of the treatments is a control or standard treatment. This is often the case in medical studies, where the superiority of any proposed new treatment over a standard treatment must be shown before it will gain acceptance by medical practitioners. Alternatively, there may be no standard treatment and the main problem may be to establish whether the new treatment has any beneficial effect, in which case a placebo control treatment may be included in the trial as a standard for comparison.

Thus, we consider the situation where there is a standard or control treatment along with k test treatments ($k \geq 1$) in the trial. The problem is to determine which of the test treatments are better than the control and, if there is more than one, to specify which one is best. The problem is posed as a selection problem, in which either none of the k test treatments is selected (equivalent to selecting the control as best) or one or more of the test treatments are chosen, one of which is designated as best. Since the control treatment is either an established treatment or standard method of therapy, or an inactive placebo, it would be natural to require special protection against the possibility of discarding it in favor of one of the test treatments if, in reality, none of the test treatments are superior to it. This is achieved by specifying a suitably high value P_0^* for the probability of selecting the control treatment in this situation. On the other hand, if more than one of the test treatments are sufficiently superior to the control treatment to warrant the rejection of the latter, we would then want some assurance, expressed in terms of a suitably high probability value P_1^*, that the best of the test treatments is selected, for otherwise the control treatment could be correctly discarded, but replaced by a treatment that was not the best one possible.

We will propose a single-stage selection procedure for the case of normal populations with a common known variance, and for the case of Bernoulli (binomial) populations, these being the simplest cases which arise in the context of medical trials. The procedure will guarantee (1) that the probability of selecting the control treatment is at least a specified value P_0^* whenever none of the k test treatments is superior to the control treatment, and (2) that the probability of selecting the best of the k test treatments is at least a specified value P_1^* whenever it is sufficiently better than the control treatment and the other $k-1$ test treatments. The procedure is formulated in terms of three constants

(δ_0^*, δ_1^* and δ_2^*) in addition to the two probabilities (P_0^* and P_1^*). These constants, to be defined in the next section, represent certain differences between the population parameters, defining "indifference zones."

For a review of selection procedures proposed for treatment trials in which one of the treatments is a standard or control, see Gupta and Panchapakesan (1979, Chapter 20). However, the procedure to be proposed here is not identical with any of these. It is actually an extension of earlier work by Paulson (1952, 1962), the procedure proposed by Paulson (1952) being a special case of ours in which $\delta_0^* = 0$ and $\delta_1^* = \delta_2^*$. It is closely related to a selection procedure proposed by Bechhofer and Turnbull (1978), the main difference being that they assumed a specified standard value, not subject to experimentation, against which the k treatment categories are compared, whereas we assume this value must be estimated from the data by including a standard or control treatment in the experimental design. Other authors have proposed subset selection approaches to the problem considered here. Such procedures avoid the specification of the constants δ_1^* and δ_2^* but at the cost of possibly selecting more than one treatment. This might be appropriate if additional experiments can be done to identify the best treatment, or if the selection can be based on some alternative criterion for determining which treatment in the selected subset is the best.

As an example to illustrate how the selection procedure proposed could be applied to an actual medical trial, we will consider a large-scale clinical trial conducted by The Coronary Drug Project Research Group (1973a). This study involved five drug treatments and a placebo control, the object being to see whether any of the drugs would be effective in reducing mortality when administered over a five-year period to subjects who had previously suffered a heart attack. Sample sizes for this trial were determined on the basis of testing null hypotheses for each drug that its mortality rate was equal to that of the placebo control. We will pose the problem as a selection problem and calculate the sample sizes accordingly, and compare the results.

2. ASSUMPTIONS, GOAL AND PROBABILITY REQUIREMENT

We assume that we have k populations Π_i ($i = 1,\ldots,k$), corresponding to the test treatments, and a population Π_0 corresponding to the control treatment. Associated with population Π_i is an unknown parameter θ_i representing the mean efficacy of the ith treatment ($i = 0,1,\ldots,k$). We assume that the populations are either normal with mean θ_i and common known variance σ^2, or Bernoulli with parameter θ_i. For example, the populations might be decreases in blood pressure in hypertensive patients after various drug treatments, with θ_i representing the population mean decrease in blood pressure for the ith treatment. Or they might be successes and failures after various surgical procedures with θ_i representing the success rate for the ith procedure. We denote the ranked values of the θ's for the k test treatments by $\theta_{[1]} \leq \theta_{[2]} \leq \dots \leq \theta_{[k]}$. We assume that $\theta_{[k]}$ represents the best treatment, unless $\theta_{[k]} \leq \theta_0$ in which case the control treatment is best. (The situation where $\theta_{[1]}$ instead of $\theta_{[k]}$ is the best of the k test treatments can be handled in a similar manner.) The case of normal populations with σ^2 known is considered in Section 3. The Bernoulli (binomial) case is considered in Section 4.

In planning an experiment to compare the treatments, the first thing to be decided is what the goal of the experiment should be. Suppose the parameter space can be partitioned by two constants, δ_0^* and δ_1^*, satisfying $-\infty < \delta_0^* < \delta_1^*$ and $\delta_1^* > 0$, such that a treatment with $\theta_i \leq \theta_0 + \delta_0^*$ is considered to be "ineffective," and a treatment with $\theta_i \geq \theta_0 + \delta_1^*$ is considered to be "effective," relative to the control treatment. Treatments with $\theta_0 + \delta_0^* < \theta_i < \theta_0 + \delta_1^*$ may have some effect relative to the control, but not enough to be considered clinically important: we call these treatments "marginal."

A natural goal for the experiment might be to classify the k test treatments into two categories: "effective" and "ineffective" treatments. We could specify a probability requirement stated in terms of a minimum probability of correctly classifying the effective and the ineffective treatments. The interval $\theta_0 + \delta_0^*$ to $\theta_0 + \delta_1^*$ corresponding to the marginal treatments would represent an indifference zone and it would not matter much to which category such treatments were classified. Then the experiment could be designed and a selection rule specified that would meet the stated probability requirement. For one possible way of doing this, see Tong (1969).

However, suppose that more than one treatment is judged to be "effective." Then the question may be asked as to which one of these is best. For instance, medical practitioners other than the clinicians actually involved in the experiment might use the results of the experiment to choose a treatment to prescribe for their own patients. Thus, it may be important to plan an experiment with a more ambitious goal than merely partitioning the k treatments into two categories consisting of effective and ineffective treatments. In other words, we may wish to design the experiment to actually select the best of the effective treatments.

Accordingly, we propose to state the goal of the experiment as follows:

> <u>Goal</u>: To select the treatment associated with $\theta_{[k]}$ provided that $\theta_{[k]} > \theta_0$; if there is no treatment with $\theta_i > \theta_0$, then the control treatment is to be selected (i.e., none of the k test treatments is to be selected). (2.1)

To define the probability requirements, an additional constant $\delta_2^* > 0$ needs to be specified in addition to the constants δ_0^* and δ_1^* already mentioned. This constant represents the minimum difference between the best and second best treatments, i.e., between $\theta_{[k]}$ and $\theta_{[k-1]}$, that is considered clinically important to be able to detect.

The constants δ_0^*, δ_1^* and δ_2^*, together with two probability values P_0^* and P_1^*, comprise a set of five constants satisfying $0 < \delta_1^*$, $\delta_2^* < \infty$, $-\infty < \delta_0^* < \delta_1^*$, $2^{-k} < P_0^* < 1$, $(1-2^{-k})/k < P_1^* < 1$, the values of which must be specified prior to doing the experiment. These determine the following probability requirements for the selection procedure:

$$\Pr\{\text{select } \Pi_0\} \geq P_0^* \text{ whenever } \theta_{[k]} \leq \theta_0 + \delta_0^* \tag{2.2a}$$

and

$$\Pr\{\text{select } \Pi_{[k]}\} \geq P_1^* \text{ whenever } \theta_{[k]} \geq \theta_0 + \delta_1^*$$

$$\text{and}$$

$$\theta_{[k]} \geq \theta_{[k-1]} + \delta_2^* \tag{2.2b}$$

where $\Pi_{[k]}$ denotes the population associated with $\theta_{[k]}$. It may be noted that the goal stated in (2.1) and the probability requirements (2.2a) and (2.2b) are identical with those specified by Bechhofer and Turnbull (1978), except that they were concerned with problems in which θ_0 was a known standard value.

3. THE SELECTION PROCEDURE FOR THE CASE OF NORMAL POPULATIONS WITH COMMON KNOWN VARIANCE

The following single-stage selection procedure is proposed for (2.1) to guarantee (2.2a) and (2.2b) when observations in the treatment categories are normally distributed with common known variance σ^2:

> Take N_0 observations X_{0j} $(j = 1,2,\ldots,N_0)$ on the control treatment and N_i observations X_{ij} on the ith test treatment $(i = 1,\ldots,k;\ j = 1,2,\ldots,N_i)$. Compute $\bar{X}_0 = \sum_j X_{0j}/N_0$ and $\bar{X}_i = \sum_j X_{ij}/N_i$ $(i = 1,\ldots,k)$. If $\bar{X}_i < \bar{X}_0 + c\sigma\sqrt{1/N_i + 1/N_0}$ for all i, select the control treatment. Otherwise, any treatment with $\bar{X}_i \geq \bar{X}_0 + c\sigma\sqrt{1/N_i + 1/N_0}$ may be classified as effective, from which we select the treatment that corresponds to the largest value of $(\bar{X}_i - \bar{X}_0)/\sqrt{1/N_i + 1/N_0}$ as the best treatment. (3.1)

Note that, if the sample sizes N_i are equal $(i = 1,\ldots,k)$, the last sentence of the selection rule in (3.1) may be simplified to read "otherwise, select the treatment that corresponds to the largest value of $\bar{X}_i$." If the sample sizes are unequal, the treatment selected by the selection rule in (3.1) and the treatment corresponding to the largest value of $\bar{X}_i$ do not necessarily coincide. The selection rule given in (3.1), which selects the treatment with the largest standardized difference from the control, is intuitively reasonable, but no claim is made that it is necessarily optimal in unequal sample size situations.

The selection procedure in (3.1) requires that values be assigned for the design constants (N_0, N_i, c). These are to be chosen in such a way as to guarantee the requirements (2.2a) and (2.2b) above. We propose that the experiment be designed to have an equal number of observations on each of the test treatments, say $N_i = N_1$ $(i = 1,\ldots,k)$, and $N_0 = RN_1$ observations on the control. The value of $R = N_0/N_1$, representing the ratio of control to test treatment sample sizes, is arbitrary. It may be chosen to make the total sample size, $N_t = N_0 + kN_1$, as small as possible, subject to any restrictions imposed by practical considerations.

<u>Theorem 1</u>: For any specified $(N_0, N_i;\ i = 1,\ldots,k)$, taking $c = h + \delta_0^*/\sigma\sqrt{1/N_i + 1/N_0}$, where h is the solution to

$$\int_{-\infty}^{+\infty} \prod_{i=1}^{k} \Phi\left(z\sqrt{N_i/N_0} + h\sqrt{1 + N_i/N_0}\right) \phi(z)\,dz = P_0^*, \tag{3.2a}$$

and where $\Phi(\cdot)$ and $\phi(\cdot)$ denote the cumulative distribution function and the probability density function, respectively, of the standard normal distribution, guarantees (2.2a).

Theorem 2: For any specified R and for $N_i = N_1$ $(i = 1,\ldots,k)$ and $N_0 = RN_1$, and taking h to satisfy (3.2a) with $N_i/N_0 = 1/R$, choosing the total sample size N_t to satisfy

$$\int_{-\infty}^{+\infty} \Phi^{k-1}\left[z + \frac{\delta_2^*\sqrt{N_t}}{\sigma\sqrt{R+k}}\right] \Phi\left[\sqrt{R}\left(z - h\sqrt{1 + R^{-1}} + \frac{\delta_1^*\sqrt{N_t}}{\sigma\sqrt{R+k}}\left(1 - \frac{\delta_0^*}{\delta_1^*}\right)\right)\right] \times \phi(z)dz = P_1^*, \tag{3.2b}$$

guarantees (2.2a) and (2.2b).

The proofs of these two theorems are given in the appendix. It should be noted that h in (3.2a) is the equicoordinate 100 P_0^* percentage point of the multivariate normal distribution $N_k(\underset{\sim}{0},P)$, where $P = \{\rho_{ij}\}$, $\rho_{ij} = 1/\sqrt{(1 + N_0/N_i)(1 + N_0/N_j)}$, is the correlation matrix. Taking $N_i = N_1$ and $N_0 = RN_1$ as specified in Theorem 2 makes $\rho_{ij} = 1/(1 + R)$, $i \neq j$. Table 1 gives the required values of h for $k = 2(1)5$; $P_0^* = .90, .95, .99$; and $R = 0.5(.1)\,3.0$.

It should be noted also that if the sample sizes actually realized in the experiment are not exactly equal to the values N_0 and N_1 satisfying (3.2b), we may still choose h to satisfy (3.2a) based on the actual sample sizes and the requirement (2.2a) would be satisfied. Also, after doing the experiment, if σ^2 is replaced by an unbiased estimate s^2 obtained from the data, h may be chosen as the 100 P_0^* percentage point of the corresponding multivariate Student t-distribution with the appropriate degrees of freedom corresponding to s^2, and the requirement (2.2a) would be satisfied. However, requirement (2.2b) may no longer be guaranteed in these situations. (In fact, for σ^2 replaced by s^2, no single-stage procedure can guarantee (2.2b) for all values of σ^2.)

To apply (3.2b), one would first determine the value of h satisfying (3.2a) for any specified value of R, and then enter values for the parameters $\sqrt{N_t}\delta_1^*/\sigma$, δ_0^*/δ_1^* and δ_2^*/δ_1^* into the integral expression in (3.2b), adjusting the value of the first parameter until the specified value of P_1^* is obtained. The required total sample size N_t is then obtained from the value found for $\sqrt{N_t}\delta_1^*/\sigma$ by substituting the value of δ_1^*/σ and solving, rounding upwards to the nearest value for which $N_t/(R + k) = N_1$ and $RN_1 = N_0$ are integers.

A computer program is given in the appendix for evaluating the integral in (3.2b).

3.1 Least Favorable Configuration

The integral in (3.2b) above assumes that the population parameters have the configuration $\theta_{[k]} = \theta_0 + \delta_1^*$, $\theta_{[1]} = \cdots = \theta_{[k-1]} = \theta_{[k]} - \delta_2^*$. This is the "least favorable configuration" for the requirement (2.2b); choosing sample sizes so that (3.2b) is satisfied under this configuration assures that requirement (2.2b) is achieved for all other configurations of the θ's outside of the "indifference zone." In some situations, the experimenter may consider that this configuration is so unlikely to arise that it would be overly conservative to base the sample sizes for the trial upon it. Instead he may be willing to assume a priori that no more than m $(0 \le m \le k-1)$ of the k-1 test treatments other than the best one could achieve a value of θ equal to $\theta_{[k]} - \delta_2^*$, the remaining $k-m-1$ test treatments having a value of θ equal to the control value, θ_0. In that case, the probability requirement could be achieved by replacing (3.2b) by the following:

$$\int_{-\infty}^{+\infty} \Phi^m\left[z + \frac{\delta_2^*\sqrt{N_t}}{\sigma\sqrt{R+k}}\right] \Phi^{k-m-1}\left[z + \frac{\delta_1^*\sqrt{N_t}}{\sigma\sqrt{R+k}}\right] \Phi\left[\sqrt{R}\left(z - h\sqrt{1+R^{-1}} + \frac{\delta_1^*\sqrt{N_t}}{\sigma\sqrt{R+k}}\left(1 - \frac{\delta_0^*}{\delta_1^*}\right)\right)\right] \times \phi(z)dz = P_1^*. \qquad (3.3)$$

The integral in (3.3) is a generalization of the integral in (3.2b), reducing to that integral when $m = k-1$. It may be noted also that taking $m = 0$ is equivalent to $\delta_1^* = \delta_2^*$.

The computer program given in the appendix actually evaluates the integral in (3.3), so that the value $m = k-1$ must be entered in order to use it to evaluate the integral in (3.2b).

4. THE SELECTION PROCEDURE FOR THE BERNOULLI (BINOMIAL) CASE

For the Bernoulli case, the observations are "successes" and "failures" and the samples are binomial, with the parameters θ_0 and θ_i representing the population proportions of successes in the control and ith treatment groups $(i = 1,\dots,k)$, respectively. To apply the results of Section 3, we use the arcsine transformation of proportions which (when multiplied by the square root of the sample size) achieves approximate normality with a constant variance, $\sigma^2 = 0.25$. This results in the following selection procedure for this case:

> Take N_0 observations X_{0j} $(j = 1,2,\ldots,N_0)$ on the control treatment and N_i observations X_{ij} $(j = 1,2,\ldots,N_i)$ on the ith test treatment $(i = 1,\ldots,k)$. Compute $p_0 = \sum_j X_{0j}/N_0$ and $p_i = \sum_j X_{ij}/N_i$, which are the observed proportions of successes. Define $z_0 = \arcsin\sqrt{p_0}$ and $z_i = \arcsin\sqrt{p_i}$. If $z_i < z_0 + c\sqrt{.25(1/N_i + 1/N_0)}$ for all i, select the control treatment. Otherwise, select the treatment corresponding to the largest value of $(z_i - z_0)/\sqrt{1/N_i + 1/N_0}$. (4.1)

The choice of the design constants (N_0, N_i, c) is made according to (3.2a) and (3.2b), but with δ_0^*, δ_1^*, and δ_2^* redefined to represent parameter differences on the transformed scale. In other words, replace δ_0^*, δ_1^* and δ_2^* in (3.2a) and (3.2b) by

$$\delta_0^{*\prime} = \arcsin\sqrt{\theta_0 + \delta_0^*} - \arcsin\sqrt{\theta_0}\,,$$

$$\delta_1^{*\prime} = \arcsin\sqrt{\theta_0 + \delta_1^*} - \arcsin\sqrt{\theta_0}\,,$$

$$\delta_2^{*\prime} = \arcsin\sqrt{\theta_{[k]}} - \arcsin\sqrt{\theta_{[k]} - \delta_2^*}\,, \qquad (4.2)$$

respectively. The values of θ_0 and $\theta_{[k]}$ required in the above computation must be specified based on design considerations. Taking c $= h + \delta_0^{*\prime}/\sqrt{.25(1/N_i + 1/N_0)}$, where h is the solution to (3.2a), will achieve the desired P_0^* specified in (2.2a). The study should then be designed with N_1 observations on each test treatment and $N_0 = RN_1$ observations on the control, where R is arbitrary, and N_t (where $N_t = N_0 + kN_1$ is the total sample size) is chosen to satisfy (3.2b).

5. AN EXAMPLE: THE CORONARY DRUG PROJECT

The study conducted by The Coronary Drug Project Research Group (1973a) was a randomized controlled trial in which subjects who had previously had a myocardial infarction were administered one of five drugs or a placebo over a five-year period to see if any of the drugs would reduce mortality relative to the placebo group. The sample sizes for the study were determined by formulating null hypotheses for comparing each drug with the placebo. The primary end-point was chosen to be the five-year mortality: denote the population values by θ_0 for the placebo group and θ_i for the ith drug ($i = 1,\ldots,5$). The null hypothesis to be tested was H_0: $\theta_i = \theta_0$ against the alternative $\theta_i = \theta_0 - \Delta$, for $i = 1,\ldots,5$. The values specified for the Type I and Type II error rates were $\alpha = .01$ and $\beta = .05$. The value of Δ was chosen to represent a 25% reduction in mortality; since a 30% mortality was anticipated for the control group, the values $\theta_0 = .3$ and $\Delta = .075$ were used for calculating the sample sizes. Denote $\theta_1 = \theta_0 - \Delta = .225$. Sample sizes in the ratio $R = N_0/N_1 = 2.5$ for the placebo and each drug group were specified, which led to a total of $N_t = 5{,}865$ subjects required, to be allocated $N_0 = 1{,}955$ to the placebo group and $N_1 = 782$ to each drug. (These numbers were then increased by a factor 10/7 to allow for an estimated 30% dropout rate.)

Let us first interpret the hypothesis tests as defining a selection rule that selects the control group if none of the drugs achieve a significance level of $\alpha = .01$ when each observed mortality is compared with that in the control group. Since the one-sided .01 point of the standard normal distribution is 2.3263, we can determine the probability of selecting the control when all the drugs have the same population mortality rates as the control by substituting into the left-side of (3.2a) the values $h = 2.3263$ and $N_i/N_0 = 1/R = 0.4$ (since $R = 2.5$). The numerical value of the integral can be determined by numerical integration. The result obtained for this probability is $P_0^* = .955$.

To calculate P_1^*, representing the probability that a drug able to reduce mortality by an amount δ_1^* on the arcsine scale will be declared to be significantly better than the control, and in addition will have an observed mortality which is less than that of any of the other four drugs when its population mortality is an amount δ_2^* less than that of the others on the arcsine scale, we substitute the relevant parameter values into the integral on the left-side of (3.2b). We set $\delta_1^* = \arcsin\sqrt{.3} - \arcsin\sqrt{.225} = .0854$, representing the stipulated reduction in mortality from 30 to 22.5 percent, $\sigma^2 = 0.25$ which is the variance on the arcsine scale, $N_t = 5{,}865$ for the total sample size, and $R = 2.5$ which was the chosen sample size ratio N_0/N_1. The remaining variables to be specified are the magnitudes of δ_0^* and δ_2^* relative to δ_1^*. We take $\delta_0^* = 0$ and calculate P_1^* as a function of the ratio δ_2^*/δ_1^*. The following table shows the results, along with the population mortality θ corresponding to each value of δ_2^*/δ_1^* for $\theta_0 = .30$.

δ_2^*/δ_1^*	θ	P_1^*
1.0	.300	.956
0.8	.284	.950
0.6	.269	.909
0.5	.262	.853
0.4	.254	.759
0.3	.247	.630
0.2	.239	.478

The first row indicates that the sample sizes chosen for the trial will achieve a probability $P_1^* = .956$ of correctly identifying a treatment with a population mortality of .225 relative to .30 in the control when the other four treatments have the same mortality as the control. The succeeding rows show how this probability decreases as the assumed mortalities in the four remaining treatments take the values of θ shown. As their θ values approach the value $\theta = .225$ assumed for the best treatment, the probability of correctly identifying the best treatment decreases, as may be expected.

Now suppose the study is to be designed as a selection procedure. First, we have to set the values of the design constants. Suppose we take $\delta_0^* = 0$, since there is no reason to introduce any "handicap" into the treatment versus control comparisons. The value of δ_1^* should be set at $\delta_1^* = \arcsin\sqrt{.3} - \arcsin\sqrt{.225} = .0854$ to conform with the CDP specification that a reduction in mortality from .30 to .225 represented the desired therapeutic effect. For the value of δ_2^*, we designate a value to represent an effect that is less than the desired level, and sufficiently different from it to be considered important that a drug producing the desired effect would not have its observed effect surpassed by that of any less effective treatments at this stipulated value of θ. After deciding upon this value of θ, we would set $\delta_2^* = \arcsin\sqrt{\theta} - \arcsin\sqrt{.225}$. Finally, we need to specify the two probabilities; say we decide to use $P_0^* = .95$ and $P_1^* = .90$. We would then choose a particular value R for the control/drug group sample size ratio, read the value of h from Table 1 corresponding to $P_0^* = .95$ and to the chosen value of R, and then substitute this value along with the other specified constants and a trial value for $\sqrt{N_t}\ \delta_1^*/\sigma$ into the program given in the appendix for evaluating the integral in (3.2b). Then the value of $\sqrt{N_t}\ \delta_1^*/\sigma$ is adjusted until $P_1^* = .90$ is achieved. Since δ_1^* has been specified and $\sigma^2 = .25$, this determines N_t.

Table 2 shows a number of solutions for different values specified for the ratio δ_2^*/δ_1^* and for the sample size ratio R. The upper part of

Table 1

Values of h to Satisfy (3.2a)

	k = 2			k = 3			k = 4			k = 5		
R	.90	.95	.99	.90	.95	.99	.90	.95	.99	.90	.95	.99
.5	1.5389	1.8855	2.5382	1.6741	2.0135	2.6532	1.7641	2.0992	2.7312	1.8308	2.1630	2.7896
.6	1.5500	1.8947	2.5444	1.6913	2.0279	2.6632	1.7856	2.1173	2.7438	1.8555	2.1840	2.8044
.7	1.5588	1.9019	2.5490	1.7051	2.0393	2.6707	1.8027	2.1316	2.7534	1.8753	2.2005	2.8157
.8	1.5660	1.9077	2.5526	1.7163	2.0484	2.6765	1.8167	2.1431	2.7609	1.8914	2.2138	2.8244
.9	1.5719	1.9124	2.5555	1.7256	2.0559	2.6811	1.8284	2.1525	2.7668	1.9049	2.2247	2.8314
1.0	1.5770	1.9163	2.5578	1.7335	2.0621	2.6849	1.8383	2.1603	2.7716	1.9162	2.2338	2.8370
1.1	1.5813	1.9197	2.5597	1.7403	2.0673	2.6879	1.8467	2.1670	2.7755	1.9259	2.2415	2.8416
1.2	1.5850	1.9225	2.5613	1.7461	2.0718	2.6904	1.8539	2.1726	2.7787	1.9343	2.2481	2.8454
1.3	1.5883	1.9249	2.5626	1.7511	2.0757	2.6926	1.8603	2.1775	2.7815	1.9416	2.2537	2.8486
1.4	1.5911	1.9271	2.5637	1.7556	2.0790	2.6944	1.8658	2.1817	2.7838	1.9480	2.2586	2.8513
1.5	1.5936	1.9289	2.5647	1.7595	2.0820	2.6959	1.8707	2.1854	2.7857	1.9537	2.2629	2.8536
1.6	1.5959	1.9306	2.5655	1.7630	2.0846	2.6973	1.8751	2.1887	2.7875	1.9587	2.2667	2.8556
1.7	1.5979	1.9320	2.5662	1.7661	2.0869	2.6984	1.8790	2.1916	2.7889	1.9632	2.2701	2.8574
1.8	1.5997	1.9334	2.5669	1.7689	2.0889	2.6994	1.8825	2.1941	2.7902	1.9673	2.2731	2.8589
1.9	1.6013	1.9345	2.5674	1.7715	2.0908	2.7003	1.8857	2.1965	2.7914	1.9709	2.2757	2.8602
2.0	1.6028	1.9356	2.5679	1.7738	2.0924	2.7011	1.8886	2.1985	2.7924	1.9742	2.2782	2.8614
2.1	1.6042	1.9365	2.5684	1.7759	2.0939	2.7018	1.8912	2.2004	2.7933	1.9773	2.2803	2.8625
2.2	1.6054	1.9374	2.5688	1.7778	2.0953	2.7025	1.8936	2.2021	2.7941	1.9800	2.2823	2.8634
2.3	1.6065	1.9382	2.5691	1.7796	2.0965	2.7030	1.8958	2.2037	2.7948	1.9826	2.2841	2.8642
2.4	1.6076	1.9389	2.5694	1.7812	2.0977	2.7035	1.8978	2.2051	2.7954	1.9849	2.2858	2.8650
2.5	1.6086	1.9396	2.5697	1.7827	2.0987	2.7040	1.8997	2.2064	2.7960	1.9870	2.2873	2.8657
2.6	1.6095	1.9402	2.5700	1.7841	2.0997	2.7044	1.9014	2.2077	2.7966	1.9890	2.2887	2.8663
2.7	1.6103	1.9408	2.5703	1.7854	2.1006	2.7048	1.9030	2.2088	2.7970	1.9909	2.2900	2.8669
2.8	1.6111	1.9413	2.5705	1.7866	2.1014	2.7052	1.9045	2.2098	2.7975	1.9926	2.2912	2.8674
2.9	1.6118	1.9418	2.5707	1.7877	2.1022	2.7055	1.9059	2.2108	2.7979	1.9942	2.2923	2.8678
3.0	1.6125	1.9423	2.5709	1.7888	2.1029	2.7058	1.9072	2.2116	2.7983	1.9957	2.2933	2.8683

Table 2

Total Sample Sizes, $N_t = N_0 + 5N_1$, for Coronary Drug Project to Achieve $P_0^* = .95$, $P_1^* = .90$ When $\theta_0 = .30$, $\theta_1 = .225$

δ_2^*/δ_1^*	Sample size ratio $R = N_0/N_1$ 1.0	1.5	2.0	2.5
	Case: m = 4 contenders at δ_2^*			
∞	5,082	4,667	4,564	4,590
1.0	5,088	4,680	4,571	4,605
0.8	5,142	4,771	4,690	4,740
0.7	5,268	4,940	4,907	4,995
0.6	5,604	5,382	5,432	5,610
0.5	6,456	6,487	6,755	7,125
0.4	8,832	9,438	10,143	10,860
0.0†	(1,389.2)	(1,505.0)	(1,620.8)	(1,736.6)
	Case: m = 1 contender at δ_2^*			
1.0	5,088	4,680	4,571	4,605
0.8	5,100	4,706	4,606	4,650
0.7	5,136	4,745	4,662	4,710
0.6	5,232	4,875	4,809	4,875
0.5	5,484	5,161	5,138	5,235
0.4	6,102	5,889	5,971	6,165
0.0†	(675.2)	(731.5)	(787.7)	(844.0)

†Values shown in parentheses for $\delta_2^*/\delta_1^* \to 0$ are the limiting values of N_t divided by $(\delta_2^*/\delta_1^*)^2$.

the table pertains to the population parameters being in the "least favorable" configuration (to which the integral in (3.2b) applies), namely, the population mortalities are .30 for the placebo, .225 for the best drug, and the remaining $m = 4$ contending drugs have a value corresponding to a difference of δ_2^* from the value .225 on the arcsine scale. The lower part of the table is for the situation where only $m = 1$ of the four contending drugs is a real contender at the value given by δ_2^*, the other three having the same mortality rate as the placebo control.

For $\delta_2^*/\delta_1^* = 1.0$, a total sample size of 4,571 subjects, allocated between control and treatment groups in a 2:1 ratio (i.e., 1306 subjects in the control group and 653 in each of the five drug groups) are needed. However, note that $\delta_2^* = \delta_1^*$ implies that four of the drug groups have the same effect as the placebo, i.e., only one of the five has the desired activity. On the other hand, if we take $\delta_2^*/\delta_1^* = 0.5$, we are saying that it is possible for four of the drugs to have roughly one-half of the desired activity, in which case we would want assurance that a drug which has the desired activity will not turn out to have a higher observed mortality than any of them. To achieve the desired values of P_0^* and P_1^* in this situation, we would take $R = 1$ since this sample size ratio corresponds to the smallest total sample size, and require $N_t = 6{,}456$ subjects, allocated 1,076 to the control group and the same number to each of the treatment groups.

The results discussed in the preceding paragraph are for the "least favorable" case where there are $m = 4$ contenders. If the experimenter is willing to assume a priori that it is extremely unlikely for there to be more than $m = 1$ contenders, the lower part of the table would apply and, as may be seen, the necessary sample sizes are reduced.

The values of N_t tabulated in Table 2 approach a lower limit for each specified value of the sample size ratio R as $\delta_2^*/\delta_1^* \to \infty$ which, it may be seen, is only slightly smaller than the values shown for $\delta_2^*/\delta_1^* = 1$. In the limit, the optimum value of R (viz., the value of R for which N_t is a minimum) is $\sqrt{k}$, or $\sqrt{5} = 2.24$ in this case; see Dunnett (1955) and Bechhofer (1969). As the value of δ_2^*/δ_1^* decreases to values less than 1.0, the minimum value of N_t occurs at progressively smaller values of R, indicating that as $\delta_2^* \to 0$ (viz., as the response of the marginal treatments approaches that of the best treatment), the proportion of observations that should be allocated to the control should be steadily decreased. The limiting values of N_t may be obtained by equating $\sqrt{N_t}\delta_2^*/\sigma$ to the appropriate value from Bechhofer's (1954) Table I for $P^* = .90$ (i.e., 2.5997 for $k = 5$, $t = 1$ corresponding to $m = 4$ contenders, and 1.8124 for $k = 2$, $t = 1$ corresponding to $m = 1$ contender) multiplied by $\sqrt{R + k}$. This indicates that, in the limit as $\delta_2^* \to 0$, it is the difference between the best treatment and the contending treatments that determines the sample sizes needed, whereas at the other extreme as $\delta_2^* \to \infty$ it is the difference between the best treatment and the control treatment that determines the sample sizes.

It may be of interest to note how the CDP study actually turned out. Three of the drugs were dropped from the study early, due to the accumulating results indicating that a decrease in mortality relative to the control group was unlikely and that a slight increase might even be possible: see The Coronary Drug Project Research Group (1970, 1972, 1973b). The remaining two drug groups had almost identical mortalities to that of the control group at the end of the five-year study period: see The Coronary Drug Project Research Group (1975). Thus, in retrospect, $\delta_2^* = \delta_1^*$ (or, equivalently, $m = 0$) would have defined the correct configuration of the population parameters on which to base the sample sizes but, of course, this would not have been known at the time of the planning of the study.

The reader should see Canner (1981) for a detailed description of the reasons for the early termination of three of the five treatments in the CDP study and especially for considerable insight into the decision-making processes that are involved in monitoring a major medical trial.

SUMMARY

Sample sizes in a medical trial are usually determined by formulating a null hypothesis and specifying the Type I and Type II error risks. In some applications, however, a formulation as a selection problem may be more realistic. We consider a situation where $k+1$ treatments, consisting of a control or standard treatment and k test treatments, are to be compared. Sample sizes are determined to satisfy the specifications of a selection problem for the cases: (1) normal populations with common, known variance, and (2) Bernoulli populations. The results are illustrated by applying them to a major medical trial, and the required sample sizes compared with those that were actually recommended.

APPENDIX A - PROOFS OF THE THEOREMS

Proof of Theorem 1: Consider (2.2a) using (3.1). We find that

$$\Pr\{\text{select } \Pi_0\} = \Pr\{\bar{X}_i \le \bar{X}_0 + c\sigma\sqrt{1/N_i + 1/N_0};\ i = 1,\ldots,k\}$$

$$= \int_{-\infty}^{+\infty} \prod_{i=1}^{k} \Phi\left[z\sqrt{N_i/N_0} + c\sqrt{1 + N_i/N_0} - \sqrt{N_i}(\theta_i - \theta_0)/\sigma\right] \phi(z)dz$$

$$= \int_{-\infty}^{+\infty} \prod_{i=1}^{k} \Phi\left[z\sqrt{N_i/N_0} + c\sqrt{1 + N_i/N_0} - \sqrt{N_i}\ \delta_0^*/\sigma\right] \phi(z)dz$$

whenever $\theta_{[k]} \leq \theta_0 - \delta_0^*$. Defining c as stipulated in Theorem 1 makes this last integral equal to P_0^* and hence (2.2a) is guaranteed.

Proof of Theorem 2: Next, consider (2.2b) using (3.1) with $N_i = N_1$ $(i = 1,\ldots,k)$ and $N_0 = RN_1$, and with c chosen in accordance with Theorem 1. Denote by $\bar{X}_{[i]}$ the sample mean corresponding to the population mean $\theta_{[i]}$. We find that

$$\Pr\{\text{select } \Pi_{[k]}\} = \Pr\{\bar{X}_{[k]} > \bar{X}_0 + h\sigma\sqrt{1/N_1 + 1/N_0} - \delta_0^*;\ \bar{X}_{[k]} > \bar{X}_{[i]};\ i \neq k\}$$

$$= \int_{-\infty}^{+\infty} \Phi\left[z\sqrt{N_0/N_1} - h\sqrt{1 + N_0/N_1} + \sqrt{N_0}(\theta_{[k]} - \theta_0 - \delta_0^*)/\sigma\right]$$

$$\times \prod_{i=1}^{k-1} \Phi\left[z + \sqrt{N_1}(\theta_{[k]} - \theta_{[i]})/\sigma\right] \phi(z)dz$$

$$\geq \int_{-\infty}^{+\infty} \Phi\left[z\sqrt{N_0/N_1} - h\sqrt{1 + N_0/N_1} + \sqrt{N_0}(\delta_1^* + \delta_0^*)/\sigma\right] \Phi^{k-1}\left(z + \sqrt{N_1}\ \delta_2^*/\sigma\right) \phi(z)dz$$

whenever $\theta_{[k]} \geq \theta_0 + \delta_1^*$ and $\theta_{[k]} \geq \theta_{[k-1]} + \delta_2^*$. Putting $N_0/N_1 = R, N_1 = N_t/(R + k)$ and $N_0 = RN_t/(R + k)$, it may be seen that (2.2b) is guaranteed by choosing N_t to satisfy (3.2b).

APPENDIX B - COMPUTER PROGRAM TO EVALUATE THE INTEGRALS IN (3.2b) and (3.3)

A Fortran listing is given below for a subroutine that can be used to evaluate the integrals in (3.2b) and (3.3). The program uses a 64-point Hermite series to perform the numerical quadrature; see Stroud and Secrest (1966).

The subroutine HERM generates the zeros and the weights for the Hermite series; it must be called once in the calling program before the first time that the subroutine PROB, which calculates the value of the integral, is called. The function FNRM is to determine the left-hand tail area of the univariate normal distribution. The listing shows the IMSL subroutine MDNOR to obtain this, but any other equivalent subroutine available to the user could be substituted.

To use the subroutine PROB, the user must write his own calling program to generate the values of the parameters, and then use a CALL PROB statement to have the value of the integral determined.

The following are the definitions used for the variables required in the program:

Input:

X, W	-	zeros and weights of the Hermite series
NHERM	-	number of terms in the Hermite series
K	-	number of test treatments, k
M	-	number of "contenders", $m = 0$ to $k - 1$
H	-	value of h from Table 1
RATN	-	sample size ratio, $N_0/N_1 = R$
DØ	-	value of ratio δ_0^*/δ_1^* (usually zero)
D2	-	value of ratio δ_2^*/δ_1^*
DNT	-	value of argument, $\sqrt{N_t}\ \delta_1^*/\sigma$

Output:

PRB	-	value of integral in (3.3) or, if $m = k - 1$, the integral in (3.2b)
IER	-	error indicator (91: m or k out of range; 92: R non-positive; 93: D2 or DNT non-positive).

```
      SUBROUTINE PROB(X,W,NHERM,K,M,H,RATN,D0,D2,DNT,PRB,IER)
C
C     COMPUTE INTEGRAL IN (3.3)
C
      DIMENSION X(64),W(64)
      DATA ZERO, ONE, TWO  / 0.0, 1.0, 2.0 /
      SQRT2    =  SQRT(TWO)
      KP       =  K - 1
      IER      =  0
      IF(M.LT.0.OR.KP.LT.0.OR.M.GT.KP)  IER  =  91
      IF(RATN.LE.ZERO)       IER  =  92
      IF(D2.LE.ZERO.OR.DNT.LE.ZERO)   IER  =  93
      PRB      =  ZERO
      IF(IER.NE.0)  RETURN
      SQRTM    =  SQRT(RATN)
      G        =  DNT/SQRT(RATN + FLOAT(K))
      HTERM    =  H*SQRT(ONE + ONE/RATN)
```

```
      ONEPL    =  ONE - D0
      DO 3 I = 1,NHERM
      IF(W(I).LE.ZERO)  GO TO 3
      SQ2X     =  SQRT2*X(I)
      ARG      =  SQRTM*(SQ2X + G*ONEPL - HTERM)
      PP       =  FNRM(ARG)
      IF(KP.LT.1)  GO TO 2
      ARG      =  SQ2X + G*D2
      P        =  FNRM(ARG)
      ARG      =  SQ2X + G
      Q        =  FNRM(ARG)
      DO 1 J = 1,KP
      PP       =  PP * P
    1 CONTINUE
    2 PRB      =  PRB + W(I)*PP
    3 CONTINUE
      PRB      =  PRB/(SQRT(ATAN(ONE))*TWO)*FLOAT(KT)
      RETURN
      END


      FUNCTION FNRM(ARG)
C
C     LEFT-HAND TAIL AREA OF NORMAL DISTRIBUTION
C
      CALL MDNOR(ARG,P)
      FNRM     =   P
      RETURN
      END


      SUBROUTINE  HERM (X,W,NHERM)
C
C     DEFINE ABSCISSAE AND WEIGHTS FOR HERMITE POLYNOMIALS
C
      DIMENSION X(64),W(64)
      NHERM = 64
      X(16)  =  -4.701815647
      X(17)  =  -4.399917168
      X(18)  =  -4.101634475
      X(19)  =  -3.806571514
      X(20)  =  -3.514375936
      X(21)  =  -3.224731292
      X(22)  =  -2.937350823
      X(23)  =  -2.651972435
      X(24)  =  -2.368354589
      X(25)  =  -2.086272880
      X(26)  =  -1.805517171
      X(27)  =  -1.525889140
      X(28)  =  -1.247200157
      X(29)  =  -.9692694231
      X(30)  =  -.6919223058
      X(31)  =  -.4149888241
      X(32)  =  -.1383022450
      DO 1 I=1, 15
    1 W(I)   =   .0
      W(16)  =   .0000000001
      W(17)  =   .0000000012
```

```
      W(18)  =  .0000000147
      W(19)  =  .0000001496
      W(20)  =  .0000012583
      W(21)  =  .0000087885
      W(22)  =  .0000512593
      W(23)  =  .0002509837
      W(24)  =  .0010363291
      W(25)  =  .0036225870
      W(26)  =  .0107560405
      W(27)  =  .0272031290
      W(28)  =  .0587399820
      W(29)  =  .1084983493
      W(30)  =  .1716858423
      W(31)  =  .2329947861
      W(32)  =  .2713774249
      DO 2 J=1,32
      L = 65 - J
      IF(W(J).GT.0.0) X(L) = -X(J)
    2 W(L) = W(J)
      RETURN
      END
```

ACKNOWLEDGMENTS

The author is indebted to the Natural Sciences and Engineering Research Council of Canada for research support, to Halina Wallace for typing the manuscript and, above all, to Bob Bechhofer for his help and encouragment over the years.

REFERENCES

Bechhofer, Robert E. (1954), "A Single-Sample Multiple Decision Procedure for Ranking Means of Normal Populations with Known Variances," Annals of Mathematical Statistics, 25, 16-39.

Bechhofer, Robert E. (1969), "Optimal Allocation of Observations When Comparing Several Treatments with a Control," Multivariate Analysis-II (P. R. Krishnaiah, Editor), Academic Press, New York, 463-473.

Bechhofer, Robert E. and Turnbull, Bruce W. (1978), "Two (k+1)-Decision Selection Procedures for Comparing k Normal Means with a Specified Standard," Journal of the American Statistical Association, 73, 385-392.

Canner, Paul (1981), "Practical Aspects of Decision Making in Clinical Trials," Controlled Clinical Trials, 1, 363-376.

The Coronary Drug Project Research Group (1970), "The Coronary Drug Project: Initial Findings Leading to Modifications of its Research Protocol," Journal of the American Medical Association, 214, 1303-1313.

The Coronary Drug Project Research Group (1972), "The Coronary Drug Project: Findings Leading to Further Modifications of its Protocol with respect to Dextrothyroxide," Journal of the American Medical Association, 220, 996-1008.

The Coronary Drug Project Research Group (1973a), "The Coronary Drug Project: Design, Methods and Baseline Results," Circulation, 47 (suppl. 1), I1-I50.

The Coronary Drug Project Research Group (1973b), "The Coronary Drug Project: Findings Leading to Discontinuation of the 2.5-mg/day Estrogen Group," Journal of the American Medical Association, 226, 652-657.

The Coronary Drug Project Research Group (1975), "Clofibrate and Niacin in Coronary Heart Disease," Journal of the American Medical Association, 231, 360-381.

Dunnett, Charles W. (1955), "A Multiple Comparison Procedure for Comparing Several Treatments with a Control," Journal of the American Statistical Association, 50, 1096-1121.

Gupta, Shanti S. and Panchapakesan, S. (1979), Multiple Decision Procedures: Theory and Methodology of Selecting and Ranking Populations, John Wiley & Sons, Inc., New York.

Paulson, Edward (1952), "On the Comparison of Several Experimental Categories with a Control," Annals of Mathematical Statistics, 23, 239-246.

Paulson, Edward (1962), "A Sequential Procedure for Comparing Several Experimental Categories with a Standard or Control," Annals of Mathematical Statistics, 33, 438-443.

Stroud, A. H. and Secrest, D. (1966), Gaussian Quadrature Formulas, Prentice-Hall, Englewood Cliffs, New Jersey.

Tong, Yung Liang (1969), "On Partitioning a Set of Normal Populations by their Locations with Respect to a Control," Annals of Mathematical Statistics, 40, 1300-1324.

SELECTION PROCEDURES FOR OPTIMAL SUBSETS OF REGRESSION VARIABLES*

Shanti S. Gupta

Department of Statistics
Purdue University
West Lafayette, Indiana

D.Y. Huang and C.L. Chang†

Department of Mathematics
National Taiwan Normal University
Taipei, Republic of China

1. INTRODUCTION

In recent years, a number of authors have studied the problem of selecting the "best" or a "good" subset of regression variables from the viewpoint of selection and ranking theory. Arvesen and McCabe (1975) considered a procedure for selecting the best model from among all reduced models involving r (fixed) out of p independent variables. Huang and Panchapakesan (1982) discussed the problem of eliminating all inferior models using the criterion of expected residual sum of squares to define inferior models. Hsu and Huang (1982) investigated a sequential procedure for selecting good regression models. In this paper, we deal with selection of an optimal subset of variables in a linear regression model. Based on the criterion of expected residual mean squares, we reject "inferior regression models". The derivation of the rule is different from those of earlier papers in that here we use the simultaneous tests of a family of hypotheses. A real-data example is provided to illustrate the application of the proposed procedure.

Consider the usual linear model

---

* This research was supported by the Office of Naval Research Contract N00014-75-C-0455 at Purdue University. Reproduction in whole or in part is permitted for any purpose of the United States Government.

† Current affiliation: Examination Yuan, Taipei, Republic of China.

$$\underline{Y} = X\underline{\beta} + \varepsilon \tag{1.1}$$

where $X = [\underline{1}, \underline{X}_1, \ldots, \underline{X}_{p-1}]$ is an n×p matrix of known constants, $\underline{\beta} = (\beta_0, \beta_1, \ldots, \beta_{p-1})$ is a 1×p vector of unknown parameters and $\underline{\varepsilon} \sim N(\underline{0}, \sigma_0^2 I_n)$. Here I_n denotes the identity matrix of order n×n and $\underline{a} = (a, a, \ldots, a)$. The model (1.1) having p-1 independent variables is considered the true model; all expectations and probabilities are calculated under it. Any reduced model whose "X matrix" has r columns is obtained by retaining any r-1 of the p-1 independent variables, where $2 \leq r \leq p-2$. There are $k_r = \binom{p-1}{r-1}$ such models. These k_r reduced models of "size" r are indexed arbitrarily by i going from 1 to k_r. We will refer to a typical model in this context as Model ri. A reduced model of size r can be written as

$$E[\underline{Y}] = X_{ri}\underline{\beta}_{ri}, \quad i = 1, 2, \ldots, k_r \tag{1.2}$$

where X_{ri} and $\underline{\beta}_{ri}$ are obtained from X and $\underline{\beta}$ corresponding to the variables that are retained in the model.

It should be first noted that our comparisons of models are made under the true model assumptions. Any reduced model with associated error variance σ_{ri}^2 defined by (2.3) is called inferior if $\sigma_{ri}^2 \geq \Delta\sigma_0^2$ where $\Delta (> 1)$ is a specified constant. Our goal is to eliminate all inferior models from the set of $2^{p-1}-1$ regression models including the true model. For this purpose, we consider a family of hypotheses testing problems, namely, H_{ri} against K_{ri}, $i = 1, \ldots, k_r$; $r = 2, \ldots, p$. Rejecting H_{ri} will mean declaring Model ri to be inferior. We derive our rule in Section 3 subject to controlling errors as explained therein. We shall give an example to see that the proposed procedures are easy to apply to obtain our desired models. We also guarantee the most conservative power for any model we selected.

Our rule is designed to select all models which are good in the sense of having adequate precision in predicting compared to the true model. When the final selection consists of several models, the experimenter may be guided by practical considerations in choosing one of these. For example, a model with smallest number of variables may be a consideration. In practice, it may be easier to obtain information on some variables than the others; thus a model with easy-to-handle variables will be preferable. Alternatively, it may be preferable to build a cost factor in the problem.

2. PRELIMINARIES

For any r, $2 \leq r \leq p$, we know that

$$\begin{aligned} SS_{ri} &= \underline{Y}'\{I-X_{ri}(X'_{ri}X_{ri})^{-1}X'_{ri}\}\underline{Y} \\ &= \underline{Y}'Q_{ri}\underline{Y}, \text{ say,} \end{aligned} \tag{2.1}$$

and

$$\frac{SS_{ri}}{\sigma_0^2} \sim \chi^2\{\nu_r,\lambda_{ri}\} \text{ under (1.1),} \tag{2.2}$$

where the degrees of freedom $\nu_r = n-r$, and the noncentrality parameter $\lambda_{ri} = (X\underline{\beta})'Q_{ri}(X\underline{\beta})/2\sigma_0^2$, $1 \leq i \leq k_r$. We note that λ_{ri} is not, in general, zero. Define

$$\sigma_{ri}^2 \equiv E[SS_{ri}/\nu_r] = \sigma_0^2 + \frac{2}{\nu_r}\sigma_0^2\lambda_{ri}. \tag{2.3}$$

It is clear from (2.3) that λ_{ri} should not be large for a good model.

For convenience, we use $\tilde{\underline{\beta}}_{ri}$ to denote the vector obtained from $\underline{\beta} = (\beta_0,\beta_1,\ldots,\beta_{p-1})$ replacing with zeros the β_j associated with variables that are dropped from the full model. For example, when p = 6, r = 4 and the variables that are retained are X_1, X_3 and X_4, we have $\underline{\beta} = (\beta_0,\beta_1,\beta_2,\beta_3,\beta_4,\beta_5)$, $\underline{\beta}_{ri} = (\beta_0,\beta_1,\beta_3,\beta_4)$, and $\tilde{\underline{\beta}}_{ri} = (\beta_0,\beta_1,0,\beta_3,\beta_4,0)$.

Finally, we let

$$\Omega_{0,ri} = \{\tilde{\underline{\beta}}_{ri}|\lambda_{ri} = 0\} \tag{2.4}$$

and

$$\Omega_{1,ri} = \{\tilde{\underline{\beta}}_{ri}|\lambda_{ri} \geq \lambda_r\}, \tag{2.5}$$

where $i = 1,2,\ldots,k_r$; $r = 2,\ldots,p$, where $\lambda_r = \nu_r(\Delta-1)/2$.

3. DERIVATION OF THE RULE

As explained in Section 1, our rule is based on a family of hypotheses testing problems. Let $\Omega_1 = \bigcup_{r=2}^{p}\bigcup_{i=1}^{k_r}\Omega_{1,ri}$, and $\Omega_0 = \bigcap_{r=2}^{p}\bigcap_{i=1}^{k_r}\Omega_{0,ri}$. Consider the following family of hypotheses testing problems.

$$H_{ri}: \tilde{\underline{\beta}}_{ri} \in \Omega_{0,ri} \text{ vs } K_{ri}: \tilde{\underline{\beta}}_{ri} \in \Omega_{1,ri}; \tag{3.1}$$

$i = 1,\ldots,k_r$, $r = 2,\ldots,p$. Let ϕ_{ri} be the test function for H_{ri} vs K_{ri}. Then the simultaneous test of all the hypotheses in (3.1) is defined by the vector $\underline{\phi}$ whose components are ϕ_{ri}, $i = 1,\ldots,k_r$; $r = 2,\ldots,p$. The power function of the test is a vector of the power functions $P_{ri}(\underline{\tilde{\beta}}_{ri})$ of the individual tests, where

$$P_{ri}(\underline{\tilde{\beta}}_{ri}) = E_{\underline{\tilde{\beta}}_{ri}} \phi_{ri}(\underline{Y}),$$

$i = 1,\ldots,k_r$; $r = 2,\ldots,p$.

Let $S(\alpha)$ be the set of all tests ϕ_{ri}, $i = 1,\ldots,k_r$; $r = 2,\ldots,p$ such that

$$E_{\underline{\tilde{\beta}}_{ri}} \phi_{ri}(\underline{Y}) \leq \alpha, \ \underline{\tilde{\beta}}_{ri} \in \Omega_0, \tag{3.2}$$

where α is the specified value, $i = 1,\ldots,k_r$; $r = 2,\ldots,p$.

Case 1: σ_0^2 known.

Since

$$\nu_r S_{ri} = \frac{SS_{ri}}{\sigma_0^2},$$

is distributed with noncentral chi-square $\chi^2(\nu_r,\lambda_{ri})$, we denote the probability density of S_{ri} as $g_{\lambda_{ri}}(s_{ri})$, $i = 1,\ldots,k_r$; $r = 2,\ldots,p$. For any r and i, we define

$$\phi_{ri}^0(\underline{y}) = \begin{cases} 1, & \text{if } g_{\lambda_{ri}}(s_{ri}) \geq c g_0(s_{ri}), \\ 0, & \text{if } g_{\lambda_{ri}}(s_{ri}) < c g_0(s_{ri}), \end{cases}$$

such that $E_{\underline{\tilde{\beta}}_{ri}} \phi_{ri}^0(\underline{Y}) = \alpha$, $\underline{\tilde{\beta}}_{ri} \in \Omega_0$, where s_{ri} is the observed value of S_{ri} and $g_0(s_{ri})$ is the central chi-square probability density. It can be shown (Gupta and Huang, 1977) that ϕ_{ri}^0, $i = 1,\ldots,k_r$; $r = 2,\ldots,p$, maximize

$$\min_{\substack{i=1,\ldots,k_r \\ r=2,\ldots,p}} \ \inf_{\underline{\tilde{\beta}}_{ri} \in \Omega_{1,ri}} E_{\underline{\tilde{\beta}}_{ri}} \phi_{ri}(\underline{Y})$$

among all tests ϕ_{ri} in $S(\alpha)$, $i = 1,\ldots,k_r$; $r = 2,\ldots,p$.

Since

$$g_{\lambda_{ri}}(s_{ri}) = \nu_r e^{-\lambda_{ri}} \sum_{\ell=0}^{\infty} \frac{\lambda_{ri}^{\ell} (\nu_r s_{ri})^{\frac{1}{2}\nu_r + \ell - 1} e^{-\frac{1}{2}(\nu_r s_{ri})}}{\ell! 2^{\frac{1}{2}\nu_r + \ell} \Gamma(\frac{1}{2}\nu_r + \ell)},$$

for $\lambda_{ri} > 0$, and

$$g_0(s_{ri}) = \frac{\nu_r(\nu_r s_{ri})^{\frac{1}{2}\nu_r - 1}}{\Gamma(\frac{\nu_r}{2})2^{\frac{1}{2}\nu_r}} e^{-\frac{1}{2}\nu_r s_{ri}},$$

hence

$$\frac{g_{\lambda_{ri}}(s_{ri})}{g_0(s_{ri})} = \sum_{\ell=0}^{\infty} \frac{e^{-\lambda_{ri}}\lambda_{ri}^{\ell}}{\ell!} \left(\frac{\nu_r s_{ri}}{2}\right)^{\ell} \frac{\Gamma(\frac{1}{2}\nu_r)}{\Gamma(\frac{1}{2}\nu_r + \ell)}$$

is a strictly increasing function of s_{ri}.

The rule ϕ_{ri}^0 is equivalent to the following

$$\phi_{ri}^0(\underline{y}) = \begin{cases} 1, & \text{if } s_{ri} \geq c_{ri} \\ 0, & \text{if } s_{ri} < c_{ri}, \end{cases} \tag{3.3}$$

where c_{ri} is determined by

$$P\{S_{ri} \geq c_{ri} | \lambda_{ri} = 0\} = \alpha.$$

<u>Case 2</u>: σ_0^2 unknown.

Since

$$V_{ri} = \frac{(SS_{ri} - SS_{p1})/(p-r)}{SS_{p1}/(n-p)}$$

is distributed with noncentral F distribution $F(p-r, n-p; \lambda_{ri})$ with noncentrality parameter λ_{ri}, we denote the probability density of V_{ri} as $f_{\lambda_{ri}}(v_{ri})$, and

$$\frac{f_{\lambda_{ri}}(v_{ri})}{f_0(v_{ri})} = \sum_{j=0}^{\infty} \frac{\lambda_{ri}^j e^{-\lambda_{ri}}\Gamma(\frac{2j+n-r}{2})\Gamma(\frac{n-r}{2})}{j!\Gamma(\frac{2j+p-r}{2})\Gamma(\frac{n-r}{2})} \left(\frac{p-r}{n-p}\right)^j \left(\frac{v_{ri}}{1 + \frac{p-r}{n-p} v_{ri}}\right)^j$$

is strictly increasing in v_{ri}, for $i = 1,\ldots,k_r$; $r = 2,\ldots,p$.

We define

$$\psi_{ri}^0(\underline{y}) = \begin{cases} 1 & \text{if } f_{\lambda_{ri}}(v_{ri}) \geq df_0(v_{ri}), \\ 0 & \text{if } f_{\lambda_{ri}}(v_{ri}) < df_0(v_{ri}), \end{cases}$$

such that $E_{\underline{\tilde{\beta}}_{ri}}\psi_{ri}^0(\underline{Y}) = \alpha$, $\underline{\tilde{\beta}}_{ri} \in \Omega_0$, where v_{ri} is the observed value of V_{ri} and $f_0(v_{ri})$ is the central F probability density. It can also be shown

(Gupta and Huang, 1977) that ψ_{ri}^{0}, $i = 1,\ldots,k_r$; $r = 2,\ldots,p$, maximizes

$$\min_{\substack{i=1,\ldots,k_r \\ r=2,\ldots,p}} \inf_{\tilde{\underline{\beta}}_{ri}\in\Omega_{1,ri}} E_{\tilde{\underline{\beta}}_{ri}} \psi_{ri}(\underline{Y})$$

among all tests ψ_{ri} in $S(\alpha)$, $i = 1,\ldots,k_r$; $r = 2,\ldots,p$.

The rule ψ_{ri}^{0} is equivalent to the following

$$\psi_{ri}^{0}(\underline{y}) = \begin{cases} 1 & \text{if } V_{ri} \geq d_{ri}, \\ 0 & \text{if } V_{ri} < d_{ri}, \end{cases} \tag{3.4}$$

where d_{ri} is determined by

$$P\{V_{ri} \geq d_{ri} | \lambda_{ri} = 0\} = \alpha.$$

Note that the power functions of (3.3) and (3.4) are increasing in λ_{ri}.

Example:

We use the Hald data (Draper and Smith, 1981, Appendix B, page 629) to discuss the procedure as follows:

No.	X_1	X_2	X_3	X_4	Y
1	7	26	6	60	78.5
2	1	29	15	52	74.3
3	11	56	8	20	104.3
4	11	31	8	47	87.6
5	7	52	6	33	95.9
6	11	55	9	22	109.2
7	3	71	17	6	102.7
8	1	31	22	44	72.5
9	2	54	18	22	93.1
10	21	47	4	26	115.9
11	1	40	23	34	83.8
12	11	66	9	12	113.3
13	10	68	8	12	109.4

Assume the following regression model proposed by Draper and Smith (1981)

$$Y_i = \beta_0 + \beta_1 X_{1i} + \beta_2 X_{2i} + \beta_3 X_{3i} + \beta_4 X_{4i} + \varepsilon_i,$$

where $\varepsilon_i \sim N(0,\sigma_0^2)$, $i = 1,2,\ldots,13$ is the true model. Let

$$p_1 = \min_{2\le r\le 5} \min_{1\le i\le k_r} \inf_{\lambda_{ri}\ge\lambda_r} E[\phi^0_{ri}(\underline{Y})] = \min_{2\le r\le p} \min_{1\le i\le k_r} P\{S_{ri} \ge c_{ri} | \lambda_{ri} = \lambda_r\},$$

and

$$p_2 = \min_{2\le r\le 5} \min_{1\le i\le k_r} \inf_{\lambda_{ri}\ge\lambda_r} E[\psi^0_{ri}(\underline{Y})] = \min_{2\le r\le p} \min_{1\le i\le k_r} P\{V_{ri} \ge d_{ri} | \lambda_{ri} = \lambda_r\}.$$

Case 1: For illustrative purposes we use the residual mean square $s^2 = \frac{1}{12}\sum_{i=1}^{13} (Y_i-\bar{Y})^2 = 5.98$ as the known value of σ_0^2. Testing

$$H_{ri}: \underline{\tilde{\beta}} \in \Omega_{0,ri} \text{ vs } K_{ri}: \underline{\tilde{\beta}} \in \Omega_{1,ri}$$

$i = 1,\dots,k_r$; $r = 2,3,4,5$.

The procedure ϕ^0_{ri} is

$$\phi^0_{ri}(\underline{y}) = \begin{cases} 1 & \text{if } S_{ri} \ge c_{ri}, \\ 0 & \text{if } S_{ri} < c_{ri}. \end{cases}$$

We list the process to reject inferior models as follows: For $\alpha = 0.05$,

Order	Model	$\nu_r = 13-r$	$\nu_r s_{ri}$	$\nu_r c_{ri}$	Result
1	X_1	11	211.65	19.675	reject
2	X_2	11	151.56	19.675	reject
3	X_3	11	324.31	19.675	reject
4	X_4	11	147.81	19.675	reject
5	X_1, X_2	10	9.68	18.307	not reject

Continuing in this fashion, the final decision is: Retain any of the models $\{X_1,X_2\}$, $\{X_1,X_4\}$, $\{X_1,X_2,X_3\}$, $\{X_1,X_2,X_4\}$, $\{X_1,X_3,X_4\}$ and $\{X_2,X_3,X_4\}$ as the desired reduced models.

The following table shows the behavior of p_1 when Δ is changed.

$\alpha = 0.05$				
Δ	1.6	2.0	2.4	2.8
p_1	0.30	0.51	0.69	0.82

It shows that p_1 increases rapidly as Δ increases.

Case 2: σ_0^2 unknown.

Testing

$$H_{ri}: \underset{\sim}{\beta} \in \Omega_{0,ri} \text{ vs } K_{ri}: \underset{\sim}{\beta} \in \Omega_{1,ri}$$

$i = 1,\ldots,k_r$; $r = 2,\ldots,5$.

The procedure ψ_{ri}^0 is

$$\psi_{ri}^0(\underline{Y}) = \begin{cases} 1 & \text{if } V_{ri} \geq d_{ri}, \\ 0 & \text{if } V_{ri} < d_{ri}. \end{cases}$$

We list the process to reject inferior models as follows: For $\alpha = 0.05$,

Order	Model	V_{ri}	$(p-r,n-p)$	d_{ri}	Result
1	X_1	67.88	(3,8)	4.07	reject
2	X_2	47.85	(3,8)	4.07	reject
3	X_3	107.23	(3,8)	4.07	reject
4	X_4	46.60	(3,8)	4.07	reject
5	X_1,X_2	0.84	(2,8)	4.46	not reject

Continuing in this fashion, the final decision is: Retain any of the models $\{X_1,X_2\}$, $\{X_1,X_4\}$, $\{X_1,X_2,X_3\}$, $\{X_1,X_2,X_4\}$, $\{X_1,X_3,X_4\}$ and $\{X_2,X_3,X_4\}$ as the desired reduced models.

The table of the relation between Δ and p_2 is as follows:

	$\alpha = 0.05$			
Δ	1.6	2.0	2.4	2.8
p_2	0.37	0.57	0.74	0.84

SUMMARY

This paper studies selection of an optimal subset of variables in a linear regression model. Based on the criterion of expected residual mean squares, we reject inferior regression models. The derivation of the rule is different from those of the earlier papers in that here we use simultaneous tests of a family of hypotheses. A real-data example is provided to illustrate the application of the proposed procedure.

REFERENCES

Arvesen, J. N. and McCabe, G. P. (1975). Subset selection problems for variances with applications to regression analysis. J. Am. Statist. Assoc., 70, 166-170.

Draper, N. and Smith, H. (1981). Applied Regression Analysis (2nd ed.). New York: Wiley.

Gupta, S. S. and Huang, D. Y. (1977). Some multiple decision problems in analysis of variance. Comm. Statist. A - Theor. Meth., 6(11), 1035-1054.

Hsu, T. A. and Huang, D. Y. (1982). Some sequential selection procedures for good regression models. Comm. Statist. A. - Theor. Meth., 11(4), 411-421.

Huang, D. Y. and Panchapakesan, S. (1982). On eliminating inferior regression models. Comm. Statist. A. - Theor. Meth., 11(7), 751-759.

A TWO-STAGE ELIMINATION TYPE PROCEDURE FOR SELECTING THE LARGEST OF SEVERAL NORMAL MEANS WITH A COMMON UNKNOWN VARIANCE

Shanti S. Gupta*

Department of Statistics
Purdue University
West Lafayette, Indiana

Woo-Chul Kim†

Department of Computer Science and Statistics
Seoul National University
Seoul, Korea

1. INTRODUCTION

Following the work of Bechhofer (1954), considerable amount of research has been carried out for the problem of selecting the normal population with the largest mean from k ($k \geqq 2$) normal populations using the so-called indifference-zone approach.

For the case of common known variance, Alam (1970) proposed a two-stage elimination type procedure, in which he used Gupta's (1956, 1965) subset selection procedure in the first stage to screen out non-contending populations and Bechhofer's (1954) indifference-zone approach to all populations retained in the second stage. However, most of his work was limited to the special case of k=2 populations. Recently, Tamhane and Bechhofer (1977, 1979) studied in detail a two-stage elimination type procedure using a minimax design criterion, and they found it more efficient than the single-stage procedure of Bechhofer (1954). For relevant results regarding various two-stage procedures, see Gupta and Miescke (1982).

---

* Research supported by ONR contract N00014-75-C-0455 at Purdue University, West Lafayette, IN., U.S.A. Reproduction in whole or in part is permitted for any purpose of the United States Government.

† Research supported in part by the Ministry of Education, Korean Government, through the Research Institute of Basic Sciences, Seoul National University, Seoul, Korea.

For the case of common unknown variance, it is known that there does not exist a single-stage procedure which guarantees a "minimum probability requirement," if one adopts the indifference-zone approach. Bechhofer, Dunnett and Sobel (1954) (henceforth referred to as BDS) proposed a two-stage non-elimination type procedure, in which the observations in the first stage are only used to obtain an estimate of the common unknown variance. Tamhane (1976) proposed a three-stage elimination type procedure, and carried out a Monte Carlo study to compare his procedure with the non-elimination type procedure of BDS (1954). However, as pointed out by Hochberg and Marcus (1981), Tamhane's argument is not quite correct. Tamhane (1975) also considered a two-stage elimination type procedure for the case of common unknown variance. However, in his Monte Carlo sampling results, the performance of his two-stage elimination type procedure was found to be inferior to that of the non-elimination type procedure of BDS (1954).

In this paper, we propose and study a two-stage elimination type procedure for the case of common unknown variance. In Section 2, a precise formulation of the problem and a two-stage elimination type procedure are given. Section 3 discusses some possible design criteria, and the tables to implement the procedure are given. Section 4 gives the results of the Monte Carlo study in which we compare the two-stage elimination type procedure with the non-elimination type procedure of BDS (1954). The results show that the two-stage elimination type procedures perform much better than the non-elimination type procedure of BDS (1954).

It is worth noting that Paulson (1964) studied a sequential elimination type procedure for the case of common unknown variance, and Hochberg and Marcus (1981) considered three-stage elimination type procedures for the case of unequal and unknown variances.

For recent reviews of the literature, see Miescke (1982). The reader is referred to the book by Gupta and Panchapakesan (1979) for a complete overview.

2. A TWO-STAGE PROCEDURE

Let $\pi_1,\ldots,\pi_k$ $(k \geq 2)$ be k normal populations with unknown means $\mu_1,\ldots,\mu_k$ and a common unknown variance σ^2. The ordered values of $\mu_1,\ldots,\mu_k$ are denoted by $\mu_{[1]} \leq \cdots \leq \mu_{[k]}$. No prior knowledge is assumed concerning the correct pairing between π_i and $\mu_{[j]}$ $(1 \leq i,j \leq k)$.

The goal of the experimenter is to select the "best" population, i.e., the one associated with the largest mean $\mu_{[k]}$. Following the indifference-zone approach of Bechhofer (1954), the experimenter, prior to the experiment, specifies two constants $\delta^* > 0$ and P^* $(1/k < P^* < 1)$ which are incorporated into a probability requirement

$$P_{\underline{\theta}}\{CS\} \geqq P^* \quad \text{for all } \underline{\theta} \in \Omega(\delta^*) \tag{2.1}$$

where CS denotes the event of selecting the best population and

$$\Omega(\delta^*) = \{\underline{\theta} = (\mu_1,\ldots,\mu_k;\sigma) \mid \mu_{[k]} - \mu_{[k-1]} \geqq \delta^*\}.$$

The two-stage non-elimination type procedure of BDS (1954) and Paulson's (1964) sequential elimination type procedure satisfy the probability requirement (2.1). However, Tamhane's (1976) three-stage procedure needs to be re-examined for the probability requirement (2.1) because of the error as pointed out previously.

The following two-stage elimination type procedure P can be viewed as a generalization of the non-elimination type procedure of BDS (1954).

Stage 1: Take n_1 $(n_1 \geqq 2)$ independent observations X_{ij} $(j = 1,\ldots,n_1)$ from each π_i $(i = 1,\ldots,k)$ and compute the sample means

$$\overline{X}_i^{(1)} = \sum_{j=1}^{n_1} X_{ij}/n_1 \quad (i = 1,\ldots,k)$$

and the pooled sample variance

$$S^2 = \sum_{i=1}^{k} \sum_{j=1}^{n_1} (X_{ij} - \overline{X}_i^{(1)})^2/k(n_1-1).$$

Determine a subset I of $\{1,\ldots,k\}$ where

$$I = \{i \mid \overline{X}_i^{(1)} \geqq \max_{1 \leqq j \leqq k} \overline{X}_j^{(1)} - (dS/\sqrt{n_1} - \delta^*)^+\},$$

the symbol a^+ denoting the positive part of a, i.e., $a^+ = \max(a,0)$; here d is a positive constant chosen to satisfy (2.1).

(a) If I has only one element, then stop sampling and assert that the population associated with $\max_{1 \leqq j \leqq k} \overline{X}_j^{(1)}$ is the best.

(b) If I has more than one element, then proceed to the second stage.

Stage 2: Take $N-n_1$ additional observations X_{ij} from each π_i $(i \in I)$ where

$$N = \max\{n_1, [(hS/\delta^*)^2]\},$$

the symbol [y] denoting the smallest integer equal to or greater than y;

here h is a positive constant chosen to satisfy (2.1). Then compute the overall sample means

$$\overline{X}_i = \sum_{j=1}^{N} X_{ij}/N \quad (i \in I)$$

and assert that the population associated with $\max_{i \in I} \overline{X}_i$ is the best.

Remark 2.1. This procedure is essentially of the same type as the one in Tamhane (1975) even though we have a different type of screening procedure. However, it is important to note that we are using an improved lower bound on $P_{\underline{\theta}}\{CS\}$ to determine the constants d and h so as to satisfy the probability requirement (2.1).

Remark 2.2. Note that in the limiting case $d = \infty$, this procedure reduces to that of BDS (1954).

In the sequel, $\Phi(\cdot)$ denotes the cdf of the standard normal distribution and $G_\nu(\cdot)$ denotes the cdf of $(\chi^2(\nu)/\nu)^{1/2}$ with $\chi^2(\nu)$ denoting the chi-square random variable with $\nu = k(n_1-1)$ degrees of freedom.

Theorem 2.1. For the procedure P, we have the following inequalities:

$$\inf_{\Omega(\delta^*)} P_{\underline{\theta}}\{CS\} \geq \int_0^\infty \int_{-\infty}^\infty \{\Phi^{k-1}(x+dw)d\Phi(x)\}\{\int_{-\infty}^\infty \Phi^{k-1}(y+hw)d\Phi(y)\}dG_\nu(w)$$

$$\geq \{\int_0^\infty \int_{-\infty}^\infty \Phi^{k-1}(x+dw)d\Phi(x)dG_\nu(w)\}\{\int_0^\infty \int_{-\infty}^\infty \Phi^{k-1}(x+hw)d\Phi(x)dG_\nu(w)\}.$$

Proof. Let $\delta_{ki} = \mu_{[k]} - \mu_{[i]}$, and let $\overline{X}_{(i)}^{(1)}$ and $\overline{X}_{(i)}$ denote respectively the first stage sample mean and the overall sample mean of the population with mean $\mu_{[i]}$, $i = 1,\ldots,k$. Then, from Tamhane and Bechhofer (1979), we have the following inequalities for all $\underline{\theta} = (\mu_1,\ldots,\mu_k;\sigma) \in \Omega(\delta^*)$;

$$P_{\underline{\theta}}\{CS\} \geq P_{\underline{\theta}}\{\overline{X}_{(i)}^{(1)} \geq \overline{X}_{(k)}^{(1)} - (dS/\sqrt{n_1} - \delta^*)^+,\ \overline{X}_{(k)} \geq \overline{X}_{(i)},\ i = 1,\ldots,k-1\}$$

$$= P_{\underline{\theta}}\{U_i \leq \frac{\sqrt{n_1}}{\sigma}\delta_{ki} + (d\frac{S}{\sigma} - \frac{\sqrt{n_1}}{\sigma}\delta^*)^+,\ V_i \leq \frac{\sqrt{N}}{\sigma}\delta^*,\ i = 1,\ldots,k-1\}$$

where

$$U_i = (\overline{X}_{(i)}^{(1)} - \overline{X}_{(k)}^{(1)} + \delta_{ki})\sqrt{n_1}/\delta \text{ and } V_i = (\overline{X}_{(i)} - \overline{X}_{(k)} + \delta_{ki})\sqrt{N}/\sigma,$$

$i = 1,\ldots,k-1$. Since $\sqrt{N} \geq hS/\delta^*$ and $\frac{\sqrt{n_1}}{\sigma}\delta_{ki} + (d\frac{S}{\sigma} - \frac{\sqrt{n_1}}{\sigma}\delta^*)^+ \geq d\frac{S}{\sigma}$ for $i = 1,\ldots,k-1$, it follows that for all $\theta \in \Omega(\delta^*)$,

$$P_{\underline{\theta}}\{CS\} \geq P_{\underline{\theta}}\{U_i \leq d\frac{S}{\sigma},\ V_i \leq h\frac{S}{\sigma},\ i = 1,\ldots,k-1\}.$$

Now, conditioning on $S/\sigma = w$ and applying Slepian's inequality as in Tamhane and Bechhofer (1979), we obtain the first inequality in the theorem. We note that the lower bound on $P_{\underline{\theta}}\{CS\}$ by the first inequality can be written as $E\{F_d(W)F_h(W)\}$ where $F_d(w) = \int_{-\infty}^{\infty} \Phi^{k-1}(x+dw)d\Phi(x)$, $F_h(w) = \int_{-\infty}^{\infty} \Phi^{k-1}(x+hw)d\Phi(x)$ and W has cdf $G_\nu(w)$. Since F_d and F_h are non-decreasing in w, it follows from Tchebysheff's inequality (Hardy, Littlewood and Pólya (1934), Theorem 43) that

$$E\{F_d(W)F_h(W)\} \geqq E\{F_d(W)\}E\{F_h(W)\}$$

which gives the second inequality in the theorem.

Now, the constants d and h in the procedure P can be chosen by setting either of the two lower bounds in Theorem 2.1 to be P* so as to satisfy the probability requirement (2.1).

Even though the second lower bound is less sharp than the first, it is much easier to evaluate. Furthermore, the functions in the second lower bound are the cdf's at equicoordinate point of a multivariate central t-distribution, which have been studied and tabulated by Dunnett (1955), Gupta and Sobel (1957) and Krishnaiah and Amritage (1966). Thus, we used the second lower bound to compare our two-stage elimination type procedure P with the two-stage non-elimination type procedure of BDS (1954).

We note that the lower bound on $P_{\theta}\{CS\}$ we are using is sharper than the one in Tamhane (1975), as can be seen in Section 11 of Tamhane and Bechhofer (1977).

3. A DESIGN CRITERION

The constants n_1, d and h in the procedure P can be considered as design constants. As in the case of known σ^2, one might try to minimize the maximum of the expected total sample size wrt $\mu_1,\ldots,\mu_k$ which depends on unknown σ^2 and increases indefinitely as σ^2 increases. Thus, as an alternative, we might ask for the procedure P to have a high probability of including the best in the first stage.

More precisely, we ask for the procedure P to satisfy the following requirement in addition to (2.1): For all $\underline{\theta} \in \Omega(\delta^*)$,

$$P_{\underline{\theta}}\{\text{including the best in the first stage}\} \geqq P_1^* \tag{3.1}$$

where P_1^* $(P^* < P_1^* < 1)$ is a preassigned constant.

It can be easily shown that, for the procedure P,

$$\inf_{\Omega(\delta^*)} P_{\underline{\theta}}\{\text{including the best in the first stage}\} = \int_0^\infty \int_{-\infty}^\infty \Phi^{k-1}(x+dw)\,d\Phi(x)\,dG_\nu(w).$$

Therefore, if d > 0 and h > 0 are chosen to satisfy

$$\begin{aligned} &\int_0^\infty \int_{-\infty}^\infty \Phi^{k-1}(x+dw)\,d\Phi(x)\,dG_\nu(w) = P_1^* \\ &\int_0^\infty \int_{-\infty}^\infty \Phi^{k-1}(x+hw)\,d\Phi(x)\,dG_\nu(w) = P^*/P_1^*, \end{aligned} \tag{3.2}$$

then the procedure P satisfies (2.1) and (3.1).

Note that the additional requirement (3.1) for the procedure P makes the choice of d and h unique once n_1 is given, and it can be regarded as a design criterion for d and h.

From now on, we will denote by P_1 (P_2) the procedure P satisfying (3.1) with $P_1^* = 0.99$ (0.975). When $P^* = 0.90$ Table 3.1 and Table 3.2 give the constants d and h for P_1 and P_2, respectively, for selected values of n_1 and k. In making these tables, the integrals on the right side of (3.2) were computed by using the result of Hartley (1943-46) and the Gauss-Hermite quadrature formula.

Another simple way to make the choice of d and h unique is to assume that d = h. In such a case, d is chosen to satisfy

$$\int_0^\infty \int_{-\infty}^\infty \Phi^{k-1}(x+dw)\,d\Phi(x)\,dG_\nu(w) = \sqrt{P^*}. \tag{3.3}$$

Note that this is equivalent to taking $P_1^* = \sqrt{P^*}$ in (3.2). In fact, this was the choice made in Tamhane's (1975) Monte Carlo study even though Tamhane used a lower bound different from the one used in (3.3). We will call by P_3 the procedure P satisfying (3.3). The constants d (=h) in procedure P_3 are given in Table 3.3 for selected values of n_1, k and P^*.

The constants h in the procedure P (BDS) of BDS (1954) can be found in Dunnett (1955), Gupta and Sobel (1957), Krishnaiah and Armitage (1966) and Gupta, Panchapakesan and Sohn (1983).

Finally, we remark that we, by no means, claim that (3.2) and (3.3) are the only possible design criteria. For example, the expected number of populations, say $E_{\underline{\theta}}(M)$, to be sampled at the second stage can be used in devising a design criterion. Using the minimax criterion one may wish to minimize the supremum of $E_{\underline{\theta}}(M)$ over Ω. Following the arguments in

Table 3.1

Constants d and h for the Procedure P_1

(a) constants d $P_1^* = 0.99$

k \ n_1	3	5	7	9	10	15	20	25
2	5.064	4.140	3.807	3.660	3.614	3.490	3.435	3.404
3	5.026	4.277	4.030	3.917	3.881	3.782	3.737	3.712
4	4.953	4.330	4.135	4.045	4.016	3.935	3.897	3.876
5	4.896	4.368	4.207	4.131	4.106	4.038	4.006	3.988
10	4.791	4.502	4.413	4.370	4.356	4.316	4.297	4.286
15	4.795	4.596	4.533	4.502	4.492	4.463	4.450	4.442

(b) constants h $P^* = 0.90$

k \ n_1	3	5	7	9	10	15	20	25
2	2.291	2.068	2.005	1.974	1.964	1.936	1.923	1.916
3	2.677	2.473	2.413	2.384	2.375	2.348	2.335	2.328
4	2.847	2.672	2.620	2.594	2.586	2.562	2.551	2.544
5	2.953	2.802	2.756	2.733	2.726	2.704	2.694	2.689
10	3.225	3.134	3.105	3.090	3.085	3.071	3.065	3.061
15	3.368	3.301	3.279	3.269	3.265	3.255	3.250	3.247

Table 3.2

Constants d and h for the Procedure $\mathcal{P}_2$

(a) constants d $P_1^* = 0.975$

k \ n_1	3	5	7	9	10	15	20	25
2	4.152	3.284	3.087	3.000	2.972	2.897	2.863	2.844
3	4.177	3.553	3.396	3.324	3.301	3.237	3.208	3.191
4	4.151	3.675	3.547	3.488	3.468	3.415	3.390	3.375
5	4.139	3.755	3.648	3.597	3.580	3.534	3.512	3.500
10	4.178	3.978	3.916	3.886	3.876	3.848	3.835	3.828
15	4.243	4.104	4.060	4.038	4.031	4.010	4.001	3.995

(b) constants h $P^* = 0.90$

k \ n_1	3	5	7	9	10	15	20	25
2	2.503	2.229	2.153	2.117	2.106	2.073	2.058	2.049
3	2.862	2.619	2.550	2.516	2.505	2.474	2.460	2.452
4	3.014	2.810	2.750	2.720	2.711	2.684	2.671	2.664
5	3.108	2.934	2.881	2.855	2.847	2.823	2.812	2.805
10	3.356	3.253	3.220	3.204	3.198	3.183	3.176	3.172
15	3.491	3.415	3.391	3.379	3.375	3.363	3.358	3.355

Table 3.3

Constants d (=h) for the Procedure $\mathcal{P}_3$

$P^* = 0.90$

k \ n_1	3	5	7	9	10	15	20	25
2	3.070	2.610	2.500	2.449	2.432	2.386	2.365	2.353
3	3.329	2.963	2.867	2.821	2.806	2.765	2.746	2.735
4	3.422	3.132	3.050	3.011	2.998	2.963	2.946	2.936
5	2.482	3.241	3.171	3.137	3.126	3.094	3.080	3.071
10	3.663	3.528	3.485	3.464	3.457	3.438	3.429	3.423
15	3.774	3.677	3.646	3.631	3.626	3.611	3.604	3.600

Section 6 of Tamhane and Bechhofer (1977) it can be shown that this supremum is given by

$$\sup_{\Omega} E_{\underline{\theta}}\{M\} = k \int_0^{\infty} \int_{-\infty}^{\infty} \{\Phi^{k-1}(x+dw) - \Phi^{k-1}(x-dw)\} d\Phi(x) dG_{\nu}(w). \tag{3.4}$$

4. RESULTS OF MONTE CARLO COMPARISON

The elimination type procedure P in Section 2 can, in fact, be regarded as a class of procedures, since there are many ways of choosing the constants d and h satisfying (2.1). In particular, the procedures P_1, P_2, P_3 in Section 3 can be regarded as various versions of the elimination type procedure P. Hence, a comparison of the procedures P_1, P_2, P_3 with the non-elimination type procedure P (BDS) can shed light on the effect of elimination in the first stage.

This section gives the results of Monte Carlo comparisons of the elimination type procedures P_1, P_2, P_3 with the non-elimination type procedure P (BDS) in terms of the expected total sample sizes.

We will use respectively T and M as generic notations for the total sample size and the number of populations to be sampled in the second stage in using the elimination type procedure. Then, the total sample size of the elimination type procedure is given by

$$T = \begin{cases} kn_1 + M(N-n_1) & \text{if } M > 1 \\ kn_1 & \text{if } M = 1 \end{cases} \tag{4.1}$$

where M is an integer-valued random variable which can take values 1 to k, inclusive. Now the total sample size of the non-elimination type procedure P (BDS) is given by T = kN. Expressions for the expected total sample sizes can be found in BDS (1954) and Tamhane (1975).

For the elimination type procedures in this simulation study, the estimates of the expected total sample sizes $E_{\underline{\theta}}\{T\}$ were computed at the so-called equal means configuration (EMC), $\mu_{[1]} = \cdots = \mu_{[k]}$, and at least favorable configuration (LFC) $\mu_{[1]} = \cdots = \mu_{[k-1]} = \mu_{[k]} - \delta^*$. Note that $E_{\underline{\theta}}\{T\}$ of the procedure P (BDS) does not depend on $\mu_1, \ldots, \mu_k$.

The Monte Carlo comparisons were carried out for k = 5,10,15, $P^* = 0.90$, 0.95 and $\delta^* = 0.5$, 1.0. The choice of n_1, the first sample size for each population, is optional and we considered the procedures with $n_1 = 5,10,15,20,25$ to get some idea about the relationship between $E_{\underline{\theta}}\{T\}$ and n_1. For each (k,P^*,δ^*,n_1)-combination, we considered the cases of $\sigma = 1,2,3$ and 1000 simulations were carried out. In each simulation, k

Table 4.1

Monte Carlo Estimates of the Expected Total Sample Size[a]

(a) $k = 5$, $\delta^* = 1.0$, $P^* = 0.90$

n_1	σ	$E\{T\}$	$E_{LFC}\{T\}$			$E_{EMC}\{T\}$		
		$\mathcal{P}$ (BDS)	$\mathcal{P}_1$	$\mathcal{P}_2$	$\mathcal{P}_3$	$\mathcal{P}_1$	$\mathcal{P}_2$	$\mathcal{P}_3$
5	1	54.58 (0.52)	35.40 (0.38)	33.94 (0.37)	33.25 (0.41)	39.51 (0.39)	39.69 (0.43)	40.84 (0.54)
	2	210.70 (2.09)	148.27 (1.78)	152.00 (2.08)	164.89 (2.63)	152.49 (1.71)	158.93 (1.98)	177.77 (2.55)
	3	470.74 (4.64)	340.84 (3.75)	358.02 (4.38)	406.26 (5.76)	344.06 (3.68)	364.42 (4.29)	418.22 (5.67)
10	1	54.27 (0.21)	50.07 (0.02)	50.11 (0.03)	50.09 (0.03)	50.24 (0.04)	50.41 (0.05)	50.60 (0.07)
	2	196.48 (1.31)	127.16 (1.30)	121.93 (1.50)	123.44 (1.88)	139.67 (1.18)	140.22 (1.41)	147.63 (1.83)
	3	441.16 (2.98)	309.84 (2.95)	316.83 (3.51)	346.34 (4.53)	319.76 (2.70)	330.52 (3.21)	366.72 (4.24)
15	1†	75.02 (0.01)	75.00 (0.00)	75.00 (0.00)	75.00 (0.00)	75.00 (0.00)	75.00 (0.00)	75.00 (0.00)
	2	192.64 (1.03)	119.17 (1.03)	114.86 (1.12)	113.16 (1.34)	134.06 (0.95)	132.92 (1.12)	136.49 (1.43)
	3	430.63 (2.23)	289.97 (2.50)	284.76 (3.06)	295.71 (4.04)	307.90 (2.20)	312.37 (2.72)	335.92 (3.64)
20	1†	100.00 (0.00)	100.00 (0.00)	100.00 (0.00)	100.00 (0.00)	100.00 (0.00)	100.00 (0.00)	100.00 (0.00)
	2	191.39 (0.88)	120.44 (0.68)	117.62 (0.70)	116.23 (0.81)	134.88 (0.72)	134.36 (0.82)	136.53 (1.06)
	3	426.21 (1.91)	274.68 (2.38)	263.81 (2.79)	266.61 (3.64)	300.35 (2.06)	301.03 (2.51)	316.96 (3.43)
25	1†	125.00 (0.00)	125.00 (0.00)	125.00 (0.00)	125.00 (0.00)	125.00 (0.00)	125.00 (0.00)	125.00 (0.00)
	2	192.48 (0.79)	132.87 (0.38)	131.81 (0.39)	131.51 (0.47)	141.72 (0.50)	143.15 (0.57)	145.58 (0.78)
	3	425.19 (1.69)	264.75 (2.28)	253.33 (2.67)	252.59 (3.33)	295.78 (1.93)	292.60 (2.37)	303.53 (3.19)

[a] The numbers in parentheses are the estimated standard errors.

Table 4.1 (Continued)

Monte Carlo Estimates of the Total Expected Sample Size

(b) $k = 15$, $\delta^* = 1.0$, $P^* = 0.90$

n_1	σ	$E\{T\}$	$E_{LFC}\{T\}$			$E_{EMC}\{T\}$		
		$\mathcal{P}$(BDS)	$\mathcal{P}_1$	$\mathcal{P}_2$	$\mathcal{P}_3$	$\mathcal{P}_1$	$\mathcal{P}_2$	$\mathcal{P}_3$
5	1	210.78 (1.13)	126.81 (1.16)	117.35 (1.13)	112.85 (1.16)	146.52 (1.14)	138.25 (1.20)	134.84 (1.31)
	2	808.71 (4.66)	597.24 (4.93)	589.65 (5.85)	607.75 (7.90)	611.84 (4.62)	612.01 (5.55)	640.31 (6.89)
	3	1821.95 (10.60)	1411.98 (9.96)	1438.80 (11.76)	1550.68 (14.88)	1422.77 (9.75)	1455.05 (11.43)	1576.57 (14.39)
10	1	203.87 (0.74)	152.77 (0.19)	152.13 (0.16)	151.92 (0.17)	158.21 (0.32)	157.59 (0.29)	157.26 (0.31)
	2	795.60 (3.10)	536.59 (4.20)	504.62 (4.81)	493.40 (5.60)	578.50 (3.60)	559.16 (4.26)	561.12 (5.21)
	3	1769.78 (6.61)	1323.44 (7.88)	1314.81 (9.75)	1367.52 (12.47)	1347.80 (7.27)	1350.95 (9.17)	1413.69 (11.90)
15	1†	226.05 (0.14)	225.00 (0.00)	225.00 (0.00)	225.00 (0.00)	225.00 (0.00)	225.00 (0.00)	225.03 (0.01)
	2	787.41 (2.33)	485.11 (3.86)	447.10 (4.04)	430.15 (4.35)	543.54 (3.18)	514.51 (3.73)	503.75 (4.36)
	3	1772.42 (5.50)	1276.49 (7.65)	1233.22 (9.59)	1234.36 (11.90)	1324.92 (6.69)	1303.19 (8.52)	1328.29 (10.97)
20	1†	300.00 (0.00)	300.00 (0.00)	300.00 (0.00)	300.00 (0.00)	300.00 (0.00)	300.00 (0.00)	300.00 (0.00)
	2	785.07 (2.12)	458.84 (3.10)	426.93 (2.97)	412.68 (3.02)	530.50 (2.72)	500.33 (2.99)	485.76 (3.29)
	3	1757.37 (4.49)	1199.67 (8.30)	1133.44 (9.71)	1118.41 (11.36)	1281.32 (6.66)	1239.73 (8.42)	1245.92 (10.59)
25	1†	375.00 (0.00)	375.00 (0.00)	375.00 (0.00)	375.00 (0.00)	375.00 (0.00)	375.00 (0.00)	375.00 (0.00)
	2	784.77 (1.85)	465.40 (2.25)	444.78 (2.07)	433.39 (2.07)	532.02 (2.16)	508.85 (2.24)	496.63 (2.47)
	3	1753.50 (4.03)	1136.27 (8.13)	1061.76 (9.27)	1027.48 (10.73)	1241.22 (6.43)	1185.11 (7.91)	1172.15 (9.83)

independent standard normal random variables were generated by the random number generator of H. Rubin and C. Hinkle at Purdue University. In this paper, we have reported the results only for $k = 5, 15$, $P^* = 0.90$ and $\delta^* = 1.0$ and the results for other cases indicate similar conclusions and are available upon request.

The estimates of the expected total sample sizes of the procedures P_1, P_2, P_3 and P (BDS) are given in Table 4.1 along with their respective estimated standard errors. Throughout all the cases considered, the elimination type procedures P_1, P_2, P_3 have substantially smaller E{T} values than corresponding E{T} values of the non-elimination type procedure P (BDS), except when the procedures stop at the first stage (these cases are denoted by a symbol † in Table 4.1 and Table 4.2). It can also be observed from Table 4.1 that the reduction in E{T} by using the elimination type procedure increases with k for fixed n_1 and σ.

Remark 4.1: The superiority of the two-stage elimination type procedure to the two-stage non-elimination type procedure is what one might expect intuitively; however, this was not revealed in Tamhane's (1975) Monte Carlo study. This can be partly explained, aside from using a better bound on $P_{\underline{\theta}}(CS)$, by the fact that we are considering the values of k and δ^* larger than those in Tamhane (1975).

From Table 4.1, we also observe that, as n_1 increases, E{T} values decrease at the beginning and then increase. Such a relationship between n_1 and E{T} for the procedure P (BDS) has been given in BDS (1954). A glance at the results in Table 4.1 shows what n_1 should be for the range of σ-values considered.

To summarize the comparison, we computed the estimated relative savings RSAVE expressed in percentages in using an elimination type procedure P over the non-elimination type procedure P (BDS), i.e.,

$$\text{RSAVE}\,\{P \mid P\,(\text{BDS})\} = \frac{E\{T \mid P\,(\text{BDS})\} - E\{T \mid P\}}{E\{T \mid P\,(\text{BDS})\}} \times 100\ (\%).$$

These values are based on Table 4.1, and are given in Table 4.2.

It is observed from Table 4.2 that there is a more pronounced improvement of the elimination type procedure over the procedure P (BDS) at the LFC than at the EMC. In fact, it can be shown that, for fixed σ^2, $E_{\underline{\theta}}\{T\}$ of the elimination type procedure is maximized at the EMC when $\underline{\theta} = (\mu_1, \ldots, \mu_k; \sigma)$ varies over the whole parameter space. Thus, even greater savings can be obtained by using the elimination type procedure than those indicated by the values in Table 4.2. Finally, it should be

Table 4.2

Estimated Relative Savings Expressed in Percentages in Using Procedures P_1, P_2, P_3 over the Procedure P (BDS)

		(a) k=5, δ*=1.0, P*=0.90						(b) k=15, δ*=1.0, P*=0.90					
		LFC			EMC			LFC			EMC		
n_1	σ	P_1	P_2	P_3	P_1	P_2	P_3	P_1	P_2	P_3	P_1	P_2	P_3
5	1	35.1	37.8	39.1	27.6	27.3	25.2	39.8	44.3	46.5	30.5	34.4	36.0
	2	29.6	27.9	21.7	27.6	24.6	15.6	26.1	27.1	24.9	24.3	24.3	20.8
	3	27.6	23.9	13.7	26.9	22.6	11.2	22.5	21.0	14.9	21.9	20.1	13.5
10	1	7.7	7.7	7.7	7.4	7.1	6.7	25.1	25.3	25.9	22.4	22.7	22.9
	2	35.3	37.9	37.2	28.9	28.6	24.9	32.6	36.6	38.0	27.3	29.7	29.5
	3	29.8	28.2	21.5	27.5	25.1	16.9	25.2	25.7	22.7	23.8	23.7	20.1
15	1†	0.03	0.03	0.03	0.02	0.02	0.02	0.5	0.5	0.5	0.5	0.5	0.5
	2	38.1	40.4	41.3	30.4	31.0	29.1	38.4	43.2	45.3	31.0	34.7	36.0
	3	32.7	33.9	31.3	28.5	27.5	22.0	28.0	30.4	30.4	25.2	26.5	25.1
20	1†	0.0	0.0	0.0	0.0	0.0	0.0	0.0	0.0	0.0	0.0	0.0	0.0
	2	37.1	38.5	39.3	29.5	29.8	28.7	41.6	45.6	47.4	32.4	36.3	38.1
	3	35.6	38.1	37.5	29.5	29.4	25.6	31.7	35.5	36.4	27.1	29.5	29.1
25	1†	0.0	0.0	0.0	0.0	0.0	0.0	0.0	0.0	0.0	0.0	0.0	0.0
	2	31.0	31.5	31.7	26.4	25.6	24.4	41.0	43.3	44.8	32.2	35.2	36.7
	3	37.7	40.4	40.6	30.4	31.2	28.6	35.2	39.4	41.4	29.2	32.4	33.2

Table 4.3

Monte Carlo Estimates of the Expected Number of Populations to Be Sampled in the Second Stage[a]

(a) $k = 5$, $\delta^* = 1.0$, $P^* = 0.90$

n_1	σ	$E_{LFC}\{M\}$ $\mathcal{P}_1$	$E_{LFC}\{M\}$ $\mathcal{P}_2$	$E_{LFC}\{M\}$ $\mathcal{P}_3$	$E_{EMC}\{M\}$ $\mathcal{P}_1$	$E_{EMC}\{M\}$ $\mathcal{P}_2$	$E_{EMC}\{M\}$ $\mathcal{P}_3$
5	1	2.40 (0.06)	1.70 (0.05)	1.06 (0.04)	3.80 (0.04)	3.00 (0.05)	2.10 (0.05)
	2	4.50 (0.03)	4.11 (0.04)	3.56 (0.05)	4.69 (0.02)	4.37 (0.03)	3.91 (0.04)
	3	4.76 (0.02)	4.51 (0.03)	4.14 (0.04)	4.82 (0.02)	4.61 (0.03)	4.28 (0.03)
10	1	0.39 (0.03)	0.16 (0.02)	0.03 (0.01)	1.99 (0.05)	0.92 (0.04)	0.20 (0.02)
	2	3.72 (0.04)	3.03 (0.05)	2.38 (0.05)	4.38 (0.03)	3.84 (0.04)	3.19 (0.05)
	3	4.47 (0.03)	4.12 (0.04)	3.69 (0.05)	4.66 (0.02)	4.36 (0.03)	3.96 (0.04)
15	1†	0.03 (0.01)	0.00 (0.00)	0.00 (0.00)	0.53 (0.03)	0.03 (0.01)	0.00 (0.00)
	2	2.86 (0.05)	2.17 (0.05)	1.51 (0.05)	3.91 (0.04)	3.21 (0.05)	2.46 (0.05)
	3	4.16 (0.03)	3.62 (0.04)	3.02 (0.05)	4.53 (0.02)	4.12 (0.03)	3.60 (0.04)
20	1†	0.00 (0.00)	0.00 (0.00)	0.00 (0.00)	0.00 (0.00)	0.00 (0.00)	0.00 (0.00)
	2	1.98 (0.05)	1.34 (0.05)	0.82 (0.04)	3.49 (0.04)	2.66 (0.05)	1.85 (0.05)
	3	3.80 (0.04)	3.15 (0.05)	2.50 (0.05)	4.37 (0.03)	3.88 (0.04)	3.27 (0.04)
25	1†	0.00 (0.00)	0.00 (0.00)	0.00 (0.00)	0.00 (0.00)	0.00 (0.00)	0.00 (0.00)
	2	1.36 (0.05)	0.79 (0.04)	0.41 (0.03)	3.20 (0.05)	2.26 (0.05)	1.36 (0.04)
	3	3.40 (0.05)	2.72 (0.05)	2.07 (0.05)	4.18 (0.03)	3.59 (0.04)	2.92 (0.05)

[a] The numbers in parentheses are the estimated standard errors.

Table 4.3 (Continued)

Monte Carlo Estimates of the Expected Number of Populations to Be Sampled in the Second Stage

(b) $k = 15$, $\delta^* = 1.0$, $P^* = 0.90$

n_1	σ	$E_{LFC}\{M\}$			$E_{EMC}\{M\}$		
		P_1	P_2	P_3	P_1	P_2	P_3
5	1	7.55 (0.13)	5.47 (0.12)	3.87 (0.10)	10.63 (0.10)	8.29 (0.11)	6.19 (0.10)
	2	13.38 (0.07)	12.14 (0.09)	10.63 (0.10)	13.79 (0.06)	12.70 (0.08)	11.30 (0.10)
	3	14.29 (0.04)	13.52 (0.06)	12.48 (0.08)	14.41 (0.04)	13.69 (0.06)	12.72 (0.08)
10	1	1.67 (0.08)	0.88 (0.05)	0.42 (0.03)	5.52 (0.10)	3.32 (0.08)	1.65 (0.06)
	2	11.55 (0.10)	9.71 (0.11)	7.83 (0.11)	12.83 (0.07)	11.22 (0.09)	9.40 (0.10)
	3	13.62 (0.06)	12.54 (0.08)	11.18 (0.10)	13.91 (0.05)	12.93 (0.07)	11.61 (0.09)
15	1†	0.17 (0.02)	0.04 (0.01)	0.00 (0.00)	2.12 (0.06)	0.59 (0.04)	0.03 (0.01)
	2	9.23 (0.12)	7.14 (0.12)	5.38 (0.11)	11.36 (0.09)	9.33 (0.10)	7.32 (0.10)
	3	12.90 (0.07)	11.43 (0.09)	9.70 (0.10)	13.52 (0.06)	12.25 (0.08)	10.63 (0.09)
20	1†	0.01 (0.01)	0.00 (0.00)	0.00 (0.00)	0.19 (0.02)	0.00 (0.00)	0.00 (0.00)
	2	6.89 (0.12)	4.89 (0.11)	3.42 (0.09)	10.07 (0.10)	7.75 (0.10)	5.65 (0.09)
	3	11.88 (0.10)	10.13 (0.11)	8.36 (0.11)	12.97 (0.07)	11.43 (0.09)	9.66 (0.10)
25	1†	0.00 (0.00)	0.00 (0.00)	0.00 (0.00)	0.00 (0.00)	0.00 (0.00)	0.00 (0.00)
	2	4.95 (0.11)	3.29 (0.09)	2.06 (0.07)	8.77 (0.10)	6.43 (0.09)	4.38 (0.08)
	3	10.78 (0.10)	8.90 (0.11)	7.03 (0.11)	12.29 (0.08)	10.52 (0.09)	8.61 (0.10)

pointed out that the superiority of one procedure over other among the procedures P_1, P_2, P_3 are not quite clear from Table 4.2. Thus, it is recommended that the choice of d and h satisfying (2.1) may be made to depend on the experimenter's other requirements such as those described in Section 3. In this respect, we have computed the expected numbers of populations to be sampled in the second stage, $E\{M\}$, as by-products of our simulation study, which are given in Table 4.3.

SUMMARY

The problem of selecting the normal population with the largest mean from several populations with a common unknown variance is considered. A new simple two-stage elimination type procedure is proposed which guarantees the same probability requirement using the indifference-zone approach as does the two-stage non-elimination type procedure of Bechhofer, Dunnett and Sobel (1954). The observations in the first stage are used not only to estimate the common unknown variance but also to screen out non-contending populations. The proposed procedure has some similarities with the procedure of Tamhane (1975), but uses a new design criterion and a sharper lower bound on the probability of a correct selection. A table of constants necessary to implement the procedure is provided. The results of a Monte Carlo study of comparison of the elimination type procedure with the non-elimination type procedure are also given. The results strongly indicate that the proposed elimination type procedure performs much better than the non-elimination type procedure of Bechhofer, Dunnett and Sobel (1954) in terms of the expected total sample size.

ACKNOWLEDGMENTS

The authors wish to thank Mr. Joong Kweon Sohn for much help in computing the tables in Section 3.

REFERENCES

Alam, K. (1970). A two-sample procedure for selecting the populations with largest mean from k normal populations. Ann. Inst. Statist. Math. 22, 127-136.

Bechhofer, R. E. (1954). A single-sample multiple decision procedure for ranking means of normal populations with known variances. Ann. Math. Statist. 25, 16-39.

Bechhofer, R. E., Dunnett, C. W., and Sobel, M. (1954). A two-sample multiple decision procedure for ranking means of normal populations with a common unknown variance. Biometrika 41, 170-176.

Dunnett, C. W. (1955). A multiple comparison procedure for comparing several treatments with a control. J. Amer. Statist. Assoc. 50, 1096-1121.

Gupta, S. S. (1956). On a decision rule for a problem in ranking means. Ph.D. Thesis (Mimeo. Ser. No. 150). Inst. of Statist., Univ. of North Carolina, Chapel Hill.

Gupta, S. S. (1965). On some multiple decision (selection and ranking) rules. Technometrics 7, 225-245.

Gupta, S. S., and Miescke, K. J. (1982). On the least favorable configurations in certain two-stage selection procedures. Statistics and Probability: Essays in Honor of C. R. Rao (Eds. Kallianpur, G., et al.), Amsterdam-New New York-Oxford, North-Holland, pp. 295-305.

Gupta, S. S., and Panchapakesan, S. (1979). Multiple Decision Procedures: Theory and Methodology of Selecting and Ranking Populations. New York, John Wiley.

Gupta, S. S., Panchapakesan, S. and Sohn, J. K. (1983). On the distribution of the studentized maximum of equally correlated normal random variables. Technical Report #83-31, Dept. of Statist., Purdue University, W. Lafayette, IN.

Gupta, S. S., and Sobel, M. (1957). On a statistic which arises in selection and ranking problems. Ann. Math. Statist. 28, 957-967.

Hardy, G. H., Littlewood, J. E., and Pólya, G. (1934). Inequalities, Cambridge University Press, Cambridge.

Hartley, H. O. (1943-46). Studentization or the elimination of standard deviation of the parent population from the random sample distribution of statistics. Biometrika 33, 173-180.

Hochberg, Y., and Marcus, R. (1981). Three stage elimination type procedures for selecting the best normal population when variances are unknown. Commun. Statist. A10, 597-612.

Krishnaiah, P. R., and Armitage, J. V. (1966). Tables for multivariate t-distribution. Sankhyā, Ser. B 28, 31-56.

Miescke, K. J. (1982). Recent results on multi-stage selection procedures. Technical Report #82-25, Dept. of Statist., Purdue University, W. Lafayette, IN.

Paulson, E. (1964). A sequential procedure for selecting the population with the largest mean from k normal populations. Ann. Math. Statist. 35, 174-180.

Tamhane, A. C. (1975). On minimax multistage elimination type rules for selecting the largest normal mean. Technical Report No. 259, Dept. of Operations Research, Cornell University, Ithaca, NY.

Tamhane, A. C. (1976). A three-stage elimination type procedure for selecting the largest normal mean (common unknown variance). Sankhyā, Ser. B38, 339-349.

Tamhane, A. C., and Bechhofer, R. E. (1977). A two-stage minimax procedure with screening for selecting the largest normal mean. Commun. Statist.-Theor. Meth. A6, 1003-1033.

Tamhane, A. C., and Bechhofer, R. E. (1979). A two-stage minimax procedure with screening for selecting the largest normal mean (II): An improved PCS lower bound and associated tables. Commun. Statist.-Theor. Meth. A8, 337-358.

NONPARAMETRIC ISOTONIC SELECTION RULES UNDER A PRIOR ORDERING

Wen-Tao Huang

Institute of Mathematics
Academia Sinica
Taipei, Republic of China

1. INTRODUCTION

Consider k+1 populations $\pi_0, \pi_1, \ldots \pi_k$ such that the jth observation from π_i is denoted by Y_{ij} which is a continuous random variable and can be expressed as $Y_{ij} = \theta_i + \beta_i x_{ij} + \varepsilon_{ij}$ $(i = 1,2,\ldots,k,\ j = 1,2,\ldots,n)$ where θ_i and β_i are both unknown parameters, x_{ij} is a known value and ε_{ij} is an error term. For example, in calibrating a spring, Y_{ij} denotes the position of the scale pointer with mass x_{ij} on the pan, θ_i is the initial position of the pointer, and ε_{ij} denotes the measurement error. Or, Y_{ij} may describe the weight of an animal fed a certain kind of ration for a given time x_{ij} (>0) with initial weight of the animal being θ_i.

In many cases, ε_{ij} may not be a normal random variable; here we assume that ε_{ij} is distributed according to a common unknown cdf $G(\cdot)$. Hence, the random variable Y_{ij} is distributed according to $G(y - \theta_i - \beta_i x_{ij})$ with unknown parameters θ_i and β_i, a controllable value x_{ij} (assumed not all equal), and unspecified absolutely continuous G.

In many occasions, for example in medical treatments, the unknown parameters θ_i are simply ordered, i.e. $\theta_1 \leq \theta_2 \leq \cdots \leq \theta_k$. This type of prior information concerning the complete ordering may come from established scientific knowledge or some past experience. For instance, the higher dose level results in a higher blood pressure, etc. In this situation we consider θ_0 as a control, and are interested in selecting those populations whose intercepts θ_i are all larger than θ_0; the $\{\beta_i\}$ are nuisance parameters. For this problem, we propose an isotonic procedure based on ranks which is given in Section 3 and in Section 4 we propose an alternative procedure with a preliminary test. Some properties of the proposed procedures are studied and some efficiencies are calculated. When $\beta_i = 0$ $(i = 0,1,\ldots,k)$ and G is known, this problem has been studied in Gupta and Yang (1981); when G is binomially distributed, an isotonic procedure is proposed in Gupta and Huang (1983). And when G is exponentially distributed, some isotonic procedures are proposed and studied in Gupta and Leu (1983). In this paper, we consider the case of unknown G in a class which is described in Section 2.

2. NOTATIONS AND DEFINITIONS

Let $\mathcal{G}$ denote the class of all absolutely continuous distribution functions symmetric about zero having continuous density with finite Fisher information. For real values x_{ij}, define

$$\bar{x}_{in} = \frac{1}{n} \sum_{j=1}^{n} x_{ij} \tag{2.1}$$

$$s_{in}^2 = \frac{1}{n} \sum_{j=1}^{n} (x_{ij} - \bar{x}_{in})^2$$

$$\text{for } i = 0,1,\ldots,k \text{ and } n \geq 2 \tag{2.2}$$

Let A denote the class of sequences $\{x_{ij}\}$ such that both the limits (2.1) and (2.2) exist and are finite.

Let π_i $(i = 0,1,\ldots,k)$ denote a population such that the jth observation Y_{ij} from π_i follows a distribution $G(y - \theta_i - \beta_i x_{ij})$ where $G \in \mathcal{G}$ and $\{x_{ij}\} \in A$. Let $\phi(u)$ $(0 < u < 1)$ be a nondecreasing, skew-symmetric and square integrable (score) function. Define the function

$$\alpha(u) = \phi\left(\frac{1+u}{2}\right), \quad 0 < u < 1 \tag{2.3}$$

and for each n = 1,2,... define

$$a_n(i) = \phi\left(\frac{i}{n+1}\right), \quad i = 1,\ldots,n \tag{2.4}$$

$$b_n(i) = \alpha\left(\frac{i}{n+1}\right), \quad i = 1,\ldots,n. \tag{2.5}$$

Also define for $i = 0,1,\ldots,k$ and $n = 1,\ldots,$

$$T_{in}(a,b) = \frac{1}{n}\sum_{j=1}^{n} \operatorname{sgn}(Y_{ij} - a - bx_{ij})b_n(R_{ij}^{+}(a,b)) \tag{2.6}$$

and

$$U_{in}(a,b) = \frac{1}{n}\sum_{j=1}^{n} (x_{ij} - \bar{x}_{in})a_n(R_{ij}(a,b)) \tag{2.7}$$

where $R_{ij}(a,b)$ ($R_{ij}^{+}(a,b)$) denotes the rank of $(Y_{ij} - a - bx_{ij})$ ($|Y_{ij} - a - bx_{ij}|$) among $Y_{i1} - a - bx_{i1}, Y_{i2} - a - bx_{i2}, \ldots, Y_{in} - a - bx_{in}$, ($|Y_{i1} - a - bx_{i1}|, \ldots, |Y_{in} - a - bx_{in}|$) and $\bar{x}_{in}$ is defined by (2.1). Then the estimators of β_i and θ_i proposed in Adichie (1967) are given by $\hat{\beta}_i$ and $\hat{\theta}_i$ defined as follows. For $i = 0,1,2,\ldots,k$, let

$$\beta_{i1} = \sup\{b\colon U_{in}(a,b) > 0 \text{ for all } a\},$$

$$\beta_{i2} = \inf\{b\colon U_{in}(a,b) < 0 \text{ for all } a\},$$

$$\hat{\beta}_i = \frac{1}{2}(\beta_{i1} + \beta_{i2}), \tag{2.8}$$

$$\theta_{i1} = \sup\{a\colon T_{in}(a,\hat{\beta}_i) > 0\},$$

$$\theta_{i2} = \inf\{a\colon T_{in}(a,\hat{\beta}_i) < 0\},$$

and

$$\hat{\theta}_i = \frac{1}{2}(\theta_{i1} + \theta_{i2}). \tag{2.9}$$

It is known (Adichie, 1967a) that the estimator $\hat{\theta}_i$ is translation-invariant, and consistent for θ_i. Since we have assumed that $\theta_1 \le \theta_2 \le \cdots \le \theta_k$, it is desirable that the estimators for θ_i also possess the property of ascending order. The isotonic regression of $g(i) = \hat{\theta}_i$ defined by (2.9) with weights $w(i) = 1$ is given by (see Barlow et al., 1972, p. 19)

$$\tilde{\theta}_i = \max_{s\le i}\min_{t\ge s}\left\{\sum_{r=s}^{t} \hat{\theta}_r/(t-s+1)\right\}$$

$$= \max_{s\le i} \hat{\hat{\theta}}_s \tag{2.10}$$

where

$$\hat{\hat{\theta}}_s = \min\{\hat{\theta}_s, \frac{1}{2}(\hat{\theta}_s + \hat{\theta}_{s+1}), \ldots, \frac{1}{k-s+1}\sum_{j=s}^{k}\hat{\theta}_j\}. \tag{2.11}$$

The isotonic estimator $\tilde{\theta}_i$ defined by (2.10) is clearly translation-invariant and consistent by the translation-invariance and consistency of $\{\hat{\theta}_i\}$.

3. AN ISOTONIC SELECTION PROCEDURE δ_I

3.1. The Procedure δ_I

Recall that population π_j $(1 \leq j \leq k)$ is considered good if and only if $\theta_j \geq \theta_0$. By a correct selection (CS) we mean a selection of any subset which contains all good populations.

Let

$$\Omega = \{(\theta_0,\theta_1,\ldots,\theta_k;\ \beta_0,\beta_1,\ldots,\beta_k)|\theta_1 \leq \theta_2 \cdots \leq \theta_k,$$

$$-\infty < \theta_i,\ \beta_i < \infty,\ i = 0, 1, \ldots, k\} \tag{3.1}$$

$$A_i = \{i, i+1, \ldots, k\},\ 1 \leq i \leq k \tag{3.2}$$

$$A_0 \equiv \emptyset, \text{ empty set.} \tag{3.3}$$

If action A_i is taken, it means the subset $\{\pi_i, \pi_{i+1}, \ldots, \pi_k\}$ is selected.

For a preassigned positive integer n, draw n observations from each of the (k+1) populations. Let $\underset{\sim}{d} = (d_1, d_2, \ldots, d_k)$ be a vector of non-negative real values. Let X denote the sample space. We define a non-randomized δ_I as follows. For $x \in X$ and $\underset{\sim}{d}$,

$$\delta_I(\underset{\sim}{x};\underset{\sim}{d}) = A_{I(\underset{\sim}{x})} \qquad \text{where} \tag{3.4}$$

$$I(\underset{\sim}{x}) = \min\{i|\tilde{\theta}_i(x) \geq \hat{\theta}_0(\underset{\sim}{x}) - \frac{d_i}{\sqrt{n}}\} \tag{3.5}$$

where $\tilde{\theta}_i(\underset{\sim}{x})$ is defined by (2.10) and $\hat{\theta}_0(\underset{\sim}{x})$ is defined by (2.9). Our main criterion for finding δ_I is the following P*-condition.

$$P_{\underset{\sim}{\omega}}[CS|\delta_I] \geq P^* \quad \text{for all} \quad \underset{\sim}{\omega} \in \Omega \tag{3.6}$$

where P* is preassigned $(\frac{1}{k+1} < P^* < 1)$ and $[CS|\delta_I]$ is the event of correct selection using δ_I.

Let

$$\Omega_i = \{\underset{\sim}{\omega} \in \Omega|\theta_{k-i} < \theta_0 \leq \theta_{k-i+1}\} \qquad i = 1,2,\ldots,k-1 \tag{3.7}$$

$$\Omega_k = \{\underset{\sim}{\omega} \in \Omega | \theta_0 \leq \theta_i\} \tag{3.8}$$

$$\Omega_0 = \{\omega \in \Omega | \theta_k < \theta_0\}. \tag{3.9}$$

Then, $\Omega_0, \Omega_1, \ldots, \Omega_k$ are disjoint and exhaustive. We note that for $\underset{\sim}{\omega} \in \Omega_i$, $i = 1,2,\ldots,k$,

$$P_{\underset{\sim}{\omega}}(CS|\delta_I) = P_{\underset{\sim}{\omega}}\{\bigcup_{j=1}^{k-i+1} (\tilde{\theta}_j(x) \geq \hat{\theta}_0(x) - \frac{d_j}{\sqrt{n}})\}$$

$$= P_{\underset{\sim}{\omega}}\{\bigcup_{j=1}^{k-i+1} \bigcup_{r=1}^{j} (\hat{\hat{\theta}}_r(\underset{\sim}{x}) \geq \hat{\theta}_0(x) - \frac{d_j}{\sqrt{n}})\}$$

$$\geq P_{\underset{\sim}{\omega}}\{\hat{\hat{\theta}}_{k-i+1}(\underset{\sim}{x}) \geq \hat{\theta}_0(\underset{\sim}{x}) - \frac{d_{k-i+1}}{\sqrt{n}}\} \tag{3.10}$$

where

$$\hat{\hat{\theta}}_{k-i+1}(\underset{\sim}{x}) = \min\{\hat{\theta}_{k-i+1}(\underset{\sim}{x}), \frac{1}{2}(\hat{\theta}_{k-i+1}(\underset{\sim}{x})+\hat{\theta}_{k-i+2}(\underset{\sim}{x})),\ldots, \frac{1}{i}\sum_{j=k-i+1}^{k} \hat{\theta}_j(x)\}.$$

For our convenience, we write $\hat{\theta}_i$ instead of $\hat{\theta}_i(\underset{\sim}{x})$. Since $\hat{\theta}_j$ is translation invariant, it follows that, conditioning on $\hat{\theta}_0 = \alpha$, the right hand side of (3.10) is increasing in θ_{k-i+1} keeping all other θ_r's $(r > k-i+1)$ fixed. By repeating this argument, we conclude that the right hand side of (3.10) attains its minimum at $\theta_{k-i+1} = \theta_{k-i+2} = \cdots = \theta_k = \theta_0$. On the other hand, we note that by taking the special point

$$\underset{\sim}{\omega}_0 = (\theta_0, \overbrace{-\infty, -\infty, \ldots, -\infty}^{k-i}, \theta_0, \theta_0, \ldots, \theta_0; \beta_0, \beta_1, \ldots, \beta_k) \in \bar{\Omega}_i, \tag{3.11}$$

the closure of Ω_i, we have

$$P_{\underset{\sim}{\omega}_0}\{\bigcup_{j=1}^{k-i+1} \bigcup_{r=1}^{j} (\hat{\hat{\theta}}_r \geq \hat{\theta}_0 - \frac{d_j}{\sqrt{n}})\} = P_{\underset{\sim}{\omega}_0}(\hat{\hat{\theta}}_{k-i+1} \geq \hat{\theta}_0 - \frac{d_{k-i+1}}{\sqrt{n}}). \tag{3.12}$$

It follows from (3.10), (3.11) and (3.12) that $\inf_{\Omega_i} P_{\underset{\sim}{\omega}}(CS|\delta_I)$ occurs at $\underset{\sim}{\omega} = \underset{\sim}{\omega}_0$. Noting that, on Ω_0, any action is correct, we have therefore

$$\inf_{\Omega} P(CS|\delta_I) = \min_{1\leq i\leq k} \{\inf_{\Omega_i} P(CS|\delta_I)\}. \tag{3.13}$$

In order to satisfy the P*-condition of (3.6), it suffices to have, for each $i = 1,2,\ldots,k$,

$$\inf_{\Omega_i} P(CS|\delta_I) \geq P^*. \tag{3.14}$$

Now, we have

$$\inf_{\Omega_i} P(CS|\delta_I) = P_{\underset{\sim}{\omega}_0}(\hat{\hat{\theta}}_{k-i+1} \geq \hat{\theta}_0 - \frac{d_{k-i+1}}{\sqrt{n}})$$

$$= P_{\underset{\sim}{\omega}_0}\{\hat{\theta}_r \geq \hat{\theta}_0 - \frac{d_r}{\sqrt{n}}, \hat{\theta}_r + \hat{\theta}_{r+1} \geq 2(\hat{\theta}_0 - \frac{d_r}{\sqrt{n}}), \ldots,$$

$$\sum_{j=r}^{k} \hat{\theta}_j \geq i(\hat{\theta}_0 - \frac{d_r}{\sqrt{n}})\} \tag{3.15}$$

where $r = k-i+1$ and $\hat{\theta}_0, \hat{\theta}_r, \hat{\theta}_{r+1}, \ldots, \hat{\theta}_k$ are independent but not identically distributed, in general. However, when $\beta_0 = \beta_1 = \ldots = \beta_k = \beta$, known, $\hat{\theta}_0, \hat{\theta}_{k-i+1}, \ldots, \hat{\theta}_k$ are iid. Since G is not specified, for given d, (3.15) cannot, in general, be exactly calculated. However, we consider an asymptotic estimate of (3.15) for large n.

It follows from Adichie (1967a) that when $\beta_i = 0$ $(i = 1,2,\ldots,k)$,

$$\sqrt{n}(\hat{\theta}_i - \theta_i) \overset{L}{\to} N(0, v^2(\psi)) \tag{3.16}$$

where $N(\mu,\sigma^2)$ denotes the normal distribution with mean μ and variance σ^2 and

$$v^2(\psi) = (\int_0^1 \psi^2(u)du)/(\int_0^1 \psi(u)\xi(u)du)^2 \tag{3.17}$$

$$\xi(u) = -[g'(G^{-1}(u))/g(G^{-1}(u)] \tag{3.18}$$

$$\psi(u) = -[f'(F^{-1}(u))/f(F^{-1}(u)] \quad 0 < u < 1 \tag{3.19}$$

where F^{-1} is the inverse of some $F \in G$ and f is the density of F. We have then the following

Theorem 1. If $\beta_i = 0$ for $i = 0,1,2,\ldots,k$, then, when n is large,

$$\inf_{\Omega_i} P(CS|\delta_I) \approx P\{Z_1 \geq (-d_r/\sqrt{2}\, v(\psi)), Z_1 + Z_2 \geq (-2d_r/\sqrt{2}\, v(\psi)),$$

$$\ldots, \sum_{j=1}^{i} Z_j \geq (-id_r/\sqrt{2}\, v(\psi))\}$$

for $i = 1,\ldots,k$, $r = k-i+1$, $v(\psi)$ is defined by (3.17) and $Z' = (Z_1, Z_2, \ldots, Z_i)$ is equally-correlated multivariate normal random vector with mean vector zero, variances 1 and common correlation 1/2.

Proof: It follows from (3.15) that

$$\inf_{\Omega_i} P(CS|\delta_I(d)) = P_{\underset{\sim}{\omega}_0}\Big((\hat{\theta}_1-\hat{\theta}_0 \geq \frac{-d_r}{\sqrt{n}},\ (\hat{\theta}_1-\hat{\theta}_0) + (\hat{\theta}_2-\hat{\theta}_0) \geq \frac{-2d_r}{\sqrt{n}},$$

$$\ldots, \sum_{j=1}^{i} (\hat{\theta}_j-\hat{\theta}_0) \geq \frac{-id_r}{\sqrt{n}}\Big)$$

$$= P_{\underset{\sim}{\omega}_0}\{Z_1 \geq (-d_r/v(\psi)),\ Z_1 + Z_2 \geq (-2d_r/v(\psi)),$$

$$\ldots, \sum_{j=1}^{i} Z_j \geq (-id_r/v(\psi))\}$$

where $Z_j = \sqrt{n}\,((\hat{\theta}_j-\theta_0) - (\hat{\theta}_0-\theta_0))/v(\psi)$ and Z_j is asymptotically normal with mean 0, variance 2 and $\rho(Z_i,Z_j) = 1/2$ by (3.16) and (3.17). Note that the result still holds true for $\beta_i = \beta$ $(i = 0,1,2,\ldots,k)$, known or not, when all $\bar{x}_{in}$ are equal $(i = 0,1,2,\ldots,k)$. It is well known that when G is logistic, $\psi(u) = u$ (Wilcoxon) and when G is normal, $\psi(u) = \Phi^{-1}(\frac{1}{2}u + \frac{1}{2})$, (Van der Waerden). For these special choices of $\psi(u)$, we have the following

$$1/v^2(\psi) = \begin{cases} 12\left(\int_{-\infty}^{\infty} g^2(y)dy\right)^2 & \text{if } \psi(u) = u \\ \int_{-\infty}^{\infty} \dfrac{g^2(y)dy}{\Phi'(\Phi^{-1}(G(y)))} & \text{if } \psi(u) = \Phi^{-1}(\frac{u}{2} + \frac{1}{2}). \end{cases}$$

It is known (Hodges and Lehmann, 1956) that $12(\int_{-\infty}^{\infty} g^2(y)dy)^2 \geq 0.864$ for all g and also (Puri and Sen, 1971) that $\int_{-\infty}^{\infty}\{g^2(y)/\Phi'(\Phi^{-1}(G(y))\}dy \geq 1$ for all G. Hence, we have the following conservative results which are useful to estimate $\underset{\sim}{d}$ in $\delta(\underset{\sim}{d})$ satisfying the P*-condition of (3.6).

Corollary 1. If $\beta_i = 0$ $(i = 0,1,2,\ldots,k)$ and n is larger, then for $i = 1,2,\ldots,k$,

$$\inf_{\Omega_i} P(CS|\delta_I(\underset{\sim}{d})) \overset{\bullet}{\geq} \begin{cases} \Pr\{Z_1 \geq -\sqrt{0.864}\,d_r,\ Z_1+Z_2 \geq -2\sqrt{0.864}\,d_r, \ldots, \\ \sum_{j=1}^{i} Z_j \geq -i\sqrt{0.864}\,d_r\}, \text{ if } \psi(u) = u \\ \Pr\{Z_1 \geq -\,d_r,\ Z_1+Z_2 \geq -2d_r, \ldots, \\ \sum_{j=1}^{i} Z_j \geq -id_r\}, \text{ if } \psi(u) = \Phi^{-1}((u+1)/2) \end{cases}$$

where $r = k-i+1$, Z_i are defined in Theorem 1, and $\dot{\geq}$ means that $\geq$ holds asymptotically.

If n is large, set the right hand side of the asymptotic inequality of Corollary 1 equalling to P*, to obtain a conservative d_r.

When the β_j are not all known equal to zero then the right hand side of Corollary 1 cannot be used to determine the $\{d_r\}$ values but rather the following corollary gives an approximate larger expression for the infimum of the probability of correct selection over Ω_i.

Corollary 2. If n is large, then for $i = 1,2,\ldots,k$,

(i) when $\psi(u) = u$

$$\inf_{\Omega_i} P(CS|\delta_I(\underset{\sim}{d}))$$
$$\dot{\geq} P\{U_1 \geq -\sqrt{0.864}\, d_r,\ U_1+U_2 \geq -2\sqrt{0.864}\, d_r, \ldots \sum_{j=1}^{i} U_j \geq -\, i\sqrt{0.864}\, d_r\}$$

and

ii) when $\psi(u) = \Phi^{-1}((u+1)/2)$

$$\inf_{\Omega_i} P(CS|\delta_I(d_1))$$
$$\dot{\geq} P\{U_1 \geq -d_r,\ U_1+U_2 \geq -2d_r,\ \ldots,\ \sum_{j=1}^{i} U_j \geq -id_r\}$$

where $\underset{\sim}{U}' = (U_1, U_2, \ldots, U_k)$ is multivariate normal with mean 0,

$$\mathrm{Var}(U_j) = (\alpha_j^2 + \alpha_0^2),\ \rho(U_r, U_s) = \alpha_0^2/\{(\alpha_r^2+\alpha_o^2)(\alpha_s^2+\alpha_0^2)\}^{1/2}$$

where

$$\alpha_j^2 = [1 + (x_j/s_j^2)^2] \tag{3.20}$$

where x_j and s_j^2 are limits of $\overline{x}_{jn}$ and s_{jn}^2 which are defined by (2.1) and (2.2).

Proof. When β_i are not all zero, $\sqrt{n}(\hat{\theta}_i-\theta_i)$ is asymptotically normal with mean zero and variance $(1 + \alpha_i^2)v^2(\psi)$, where α_i^2 and $v(\psi)$ are defined by (3.20) and (3.17). Following analogous arguments as in Corollary 1, and taking $U_i = \sqrt{n}[(\hat{\theta}_i-\theta_0) + (\hat{\theta}_0-\theta_0)]/v(\psi)$, we obtain the result.

In order to estimate the probabilities given in Theorem 1 and Corollaries 1 and 2, we apply a recursive method. Firstly, suppose θ_0 is known. By following an argument analogous to that in Theorem 1, we have, by (3.15),

$$\inf_{\Omega_i} P(CS|\delta_I(d)) = P_{\underset{\sim}{\omega}_0}\{\hat{\theta}_r \geq \theta_0 - \frac{d_r}{\sqrt{n}}, \hat{\theta}_r + \hat{\theta}_{r+1} \geq 2(\theta_0 - \frac{d_r}{\sqrt{n}}), \ldots, \sum_{j=1}^{i} \hat{\theta}_{r+j-1} \geq i(\theta_0 - \frac{d_r}{\sqrt{n}})\}$$

$$= P_{\underset{\sim}{\omega}_0}\{\zeta_1 \geq 0, \zeta_1 + \zeta_2 \geq 0, \ldots, \zeta_1 + \zeta_2 + \ldots + \zeta_i \geq 0\}$$

$$= P_{\underset{\sim}{\omega}_0}\{\eta_i \geq \theta_0 - d_i'\} \tag{3.21}$$

where

$$\zeta_j = \hat{\theta}_{k-i+j} - \theta_0 + \frac{d_{k-i+j}}{\sqrt{n}} \quad \text{for } j = 1,2,\ldots,i, \text{ and} \tag{3.22}$$

$$\eta_i = \hat{\hat{\theta}}_{k-i+1} \text{ where } \hat{\hat{\theta}}_s \text{ is defined in (2.12), and} \tag{3.23}$$

$$d_i' = d_{k-i+1}/\sqrt{n} \tag{3.24}$$

Define

$$S_j = \zeta_1 + \zeta_2 + \ldots + \zeta_j \qquad j = 1,2,\ldots,k \tag{3.25}$$

and let A_c denote the subset of A such that for any $\{x_{ij}\}$ in A_c,

$$\lim_{n\to\infty} \{n\bar{x}_{in}^2 / \sum_{j=1}^{n} (x_{ij} - \bar{x}_{in})^2\} = c^2 \tag{3.26}$$

where $\bar{x}_{in}$ is defined in (2.1). Then we have the following.

<u>Theorem 2</u>. If $\{x_{ij}\} \in A_c$, then

$$P_{\underset{\sim}{\omega}_0}(\eta_i \geq \theta_0 - d_i') \approx \frac{1}{i}\sum_{j=1}^{i-1} P_{\underset{\sim}{\omega}_0}(\eta_j \geq \theta_0 - d_i')\cdot\Phi((i-j)^{1/2}d_i'/(1+c^2)^{1/2}v(\psi))$$

where η_i and d_i' are defined by (3.23) and (3.24), $\Phi(x)$ denotes the standard normal cdf and $P(\eta_0 \geq a) = 1$ for all a.

<u>Proof</u>. It follows from (3.21) and (3.22) and (3.25) that

$$\inf_{\Omega_i} P(CS|\delta_I(\underset{\sim}{d})) = P\{S_1 \geq 0, S_2 \geq 0, \ldots, S_i \geq 0\}$$

where $\zeta_r = \hat{\theta}_{k-i+r} - \theta_0 + d_i'$. We note that $\zeta_1, \zeta_2, \ldots, \zeta_i$ are asymptotically iid since $\sqrt{n}\,(\hat{\theta}_j - \theta_j) \xrightarrow{L} N(0, v^2(\psi)(1+c^2))$ (see Adichie, 1967a) where $v(\psi)$ and c^2 are defined by (3.17) and (3.26). Let

$$p_n = P(S_1 \geq 0, S_2 \geq 0, \ldots, S_n \geq 0). \tag{3.27}$$

Then, it follows from the Sparre-Andersen Theorem (see, for example, Feller, 1969, p. 395), the generating function of $\{p_n\}$ is given by

$$p(s) \equiv \sum_{n=1}^{\infty} p_n s^n = \exp\{\sum_{n=1}^{\infty} \frac{s^n}{n} P(S_n \geq 0)\}. \tag{3.28}$$

Now, the i-th derivative of p(s) is given by

$$p^{(i)}(s) = \sum_{j=0}^{i} \binom{n}{j} p^{(j)}(s)\lambda^{(i-j)}(s) \quad \text{where} \tag{3.29}$$

$$\lambda(s) = \sum_{n=1}^{\infty} \frac{s^n}{n} P(S_n \geq 0). \tag{3.30}$$

Therefore, we have

$$p_n = \frac{1}{i!} \lim_{s \to 0^+} p^{(i)}(s) = \frac{1}{i!} \lim_{s \to 0^+} \sum_{j=0}^{i-1} \binom{n}{j} p^{(j)}(s)\lambda^{(i-j)}(s)$$

$$= \frac{1}{i!} \sum_{j=0}^{i-1} \frac{p^{(j)}(0)}{j!} P(S_{ij} \geq 0)$$

$$= \frac{1}{i!} \sum_{j=0}^{i-1} P(\eta_j \geq \theta_0 - d_i')P(S_{i-j} \geq 0). \tag{3.31}$$

Noting that

$$\sqrt{n}\, \zeta_j/\sqrt{1+c^2}\, v(\psi) = \{\sqrt{n}(\hat{\theta}_{k-i+j} - \theta_0)/(1+c^2)^{1/2} v(\psi)\} + \{d_i'/(1+c^2)^{1/2} v(\psi)\}$$

and since the former term is asymptotically standard normal, it follows that

$$P(S_{i-j} \geq 0) = P(\sqrt{n}\, S_{i-j}/\sqrt{(i-j)(1+c)^2}\, v(\psi) \geq 0)$$

$$\approx \Phi(\sqrt{(i-j)}\, d_i'/\sqrt{1+c^2}\, v(\psi))$$

as n large. It follows from (3.31) that the proof is complete.

By conditioning on $u = \sqrt{n}(\hat{\theta}_0 - \hat{\theta}_0)/(1+c^2)^{1/2}v(\psi)$, we have immediately

Corollary 3. If $\{x_{ij}\} \in A_c$ and n is large, then

$$\inf_{\Omega_i} P(CS|\delta_I(\underset{\sim}{d})) \approx \int_{-\infty}^{\infty} P_{\underset{\sim}{\omega}_0}\{\eta_i \geq u - (d_i'/(1+c^2)^{1/2}v(\psi))\}d\Phi(u)$$

$$= \frac{1}{i}\sum_{j=0}^{i-1}\Phi((i-j)^{1/2}d_i'/(1+c^2)^{1/2}v(\psi)) \int_{-\infty}^{\infty} P_{\underset{\sim}{\omega}_0}\{\eta_j \geq u - \frac{d_i'}{\sqrt{1+c^2}\ v(\psi)}\}\ d\Phi(u)$$

where $\underset{\sim}{\omega}_0 = (\theta_0, \overbrace{-\infty, -\infty, \ldots, -\infty}^{k-i}, \theta_0, \theta_0, \ldots, \theta_0; \beta_0, \beta_1, \ldots, \beta_k) \in \bar{\Omega}_i$ and η_j and d_i' are defined by (3.23) and (3.24).

We also have the following conservative approximations.

Corollary 4. If θ_0 is known, $\{x_{ij}\} \in A_c$ and n is large, then,

(i) when $\psi(u) = u$

$$\inf_{\Omega_i} P(CS|\delta_I(\underset{\sim}{d})) \stackrel{\bullet}{\geq} \frac{1}{i}\sum_{j=0}^{i-1}\Phi(\sqrt{(i-j)}\ d_i'/\sqrt{0.864(1+c^2)})\ P_{\underset{\sim}{\omega}_0}(\eta_j \geq \theta_0 - d_i')$$

(ii) and when $\psi(u) = \Phi^{-1}((u+1)/2)$

$$\inf_{\Omega_i} P(CS|\delta_I(\underset{\sim}{d})) \stackrel{\bullet}{\geq} \frac{1}{i}\sum_{j=0}^{i-1}\Phi(\sqrt{(i-j)}\ d_i'/\sqrt{1+c^2})\ P_{\underset{\sim}{\omega}_0}(\eta_j \geq \theta_0 - d_i')$$

where $P(\eta_0 \geq a) = 1$ for all a and $\underset{\sim}{\omega}_0$ is defined by (3.11).

3.2. Some Properties and the Asymptotic Relative Efficiency of δ_I.

Definition 1. A selection procedure δ is isotonic if $\theta_i < \theta_j$ and δ selects π_i (associated with θ_i), then δ selects π_j (associated with θ_j). δ is translation-invariant if for any real c, $\delta(\underset{\sim}{x}+\underset{\sim}{c}) = \delta(\underset{\sim}{x})$ where $\underset{\sim}{c} = (c,c,\ldots,c)$. δ is consistent if the probability of correct decision tends to 1 as n, the sample size, tends to infinity. It follows from definitions of $\tilde{\theta}_i$ and procedure δ_I that:

Theorem 3. δ_I is isotonic, translation-invariant and consistent. Let S_B denote the number of populations selected by δ_I with $\theta_i < \theta_0$. Then, when all $\beta_i = 0$, $i = 0,1,2,\ldots,k$,

$$\sup_{\Omega_{k-i}} E_{\underset{\sim}{\omega}}(S_B|\delta_I) = \sup_{\Omega_{k-i}} \sum_{t=1}^{i} P_{\underset{\sim}{\omega}}(\bigcup_{r=1}^{t}\{\hat{\hat{\theta}}_r \geq \hat{\theta}_0 - \frac{d_r}{\sqrt{n}}\})$$

$$= \sum_{t=1}^{i} P_{\underset{\sim}{\omega}_0}(\bigcup_{r=1}^{t}\{\hat{\hat{\theta}}_r \geq \hat{\theta}_0 - \frac{d_r}{\sqrt{n}}\}) \tag{3.32}$$

where $\underset{\sim}{\omega}_0 = (\theta_0, \theta_0, \ldots, \theta_0) \in \overline{\Omega}_{k-i}$. Since d_r is decreasing in r, we have an immediate upper bound on $E(S_B|\delta_I)$ over Ω_{k-i} which is given by

Corollary 5. When $\beta_0 = \beta_1 = \ldots = \beta_k = 0$ then

$$\sup_{\Omega_{k-i}} E(S_B|\delta_I) \leq i\, P_{\underset{\sim}{\omega}_0}\left(\bigcup_{r=1}^{i}(\hat{\hat{\theta}}_r \geq \hat{\theta}_0 - \frac{d_1}{\sqrt{n}})\right)$$

Let $B_r = \{\hat{\hat{\theta}}_r \geq \theta_0 - \frac{d_1}{\sqrt{n}}\}$. Since it can be shown that $P_{\omega_0}(B_r) \leq P_{\omega_0}(B_s)$ for $r \leq s$, hence, we have a conservative upper bound given by

Corollary 6. When $\beta_0 = \beta_1 = \ldots = \beta_k = 0$ then

$$\sup_{\Omega_{k-i}} E(S_B|\delta_I) \leq i^2 P_{\underset{\sim}{\omega}_0}\{\hat{\theta}_i \geq \hat{\theta}_0 - \frac{d_1}{\sqrt{n}},\ (\hat{\theta}_i+\hat{\theta}_{i+1}) \geq 2(\hat{\theta}_0 - \frac{d_1}{\sqrt{n}}),$$

$$(\hat{\theta}_i+\hat{\theta}_{i+1}+\ldots+\hat{\theta}_k) \geq (k-i+1)(\hat{\theta}_0 - \frac{d_1}{\sqrt{n}})\}$$

where the right hand side is independent of θ_0 when the θ_i are location parameters.

Suppose G is a normal distribution with location parameter θ_i and common variance 1. Let the selection procedure associated with δ_I taking $\hat{\theta}_i(\underset{\sim}{x}) = \frac{1}{n}\sum_{j=1}^{n} X_{ij}$, the usual sample mean, be denoted by δ_N. We define the efficiency of δ_I to δ_N, denoted by $\mathrm{Eff}(\delta_I\colon \delta_N|i, P^*)$ by

$$\lim_{n\to\infty}\{\sup_{\Omega_{k-i}} E(S_B|\delta_I)/\sup_{\Omega_{k-i}} E(S_B|\delta_N)\} \tag{3.33}$$

for some fixed i ($1 \leq i \leq k$) satisfying the criterion (3.6) for both δ_I and δ_N. Then, we have

Corollary 7. $\mathrm{Eff}(\delta_I\colon\delta_N|1, P^*) = 1$.

4. A RANDOMIZED PROCEDURE δ_R

Since the estimator $\hat{\theta}_i$ involves an estimate, $\hat{\beta}_i$, of β_i, the infimum of the probability of correct selection given by (3.15) involves the β_j's. Suppose that $H_0\colon \underset{\sim}{\beta} = \beta \underset{\sim}{1}_k$ is true. When the data support the hypotheses H_0, it is desirable to consider some statistics depending on H_0 for our selection procedure and when H_0 is not tenable, we consider other statistics which are distribution-free from $\underset{\sim}{\beta}$. When treating $\underset{\sim}{\beta}$ as a nuisance parameter, the elimination of $\underset{\sim}{\beta}$ generally entails some loss of information. This loss may be recovered (to some extent) when $\underset{\sim}{\beta}$ is not specified but partial information is known about $\underset{\sim}{\beta}$, as for example, a test of H_0. Hence, it is useful to consider a preliminary decision on the nuisance parameters $\underset{\sim}{\beta}$ before a final selection or ranking is made. When a selection procedure is based on a test, it is a randomized selection rule. We propose such a rule and study its behavior.

Let

$$A_\phi^2 = \int_0^1 \phi^2(u)du - (\int_0^1 \phi(u)du)^2, \tag{4.1}$$

$$U_n(b_1, b_2, \ldots, b_k) = \sum_{i=1}^{k} U_{in}(a, b_i), \tag{4.2}$$

$$\tilde{\beta}_n = \frac{1}{2} (\sup\{b: U_n(b\, \underset{\sim}{1}_k) > 0\} + \inf\{b: U_n(b\, \underset{\sim}{1}_k) < 0\}), \text{ and} \tag{4.3}$$

$$T_n = A_\phi^{-2} \sum_{i=1}^{k} \{U_i^2(\tilde{\beta}_n)/\sum_{j=1}^{n} (x_{ij} - \bar{x}_{in})^2\}. \tag{4.4}$$

A preliminary test of the parallelism of the regressions, i.e. H_0: $\beta_1 = \ldots = \beta_k$, is based on the statistic T, introduced in Sen (1969); the test is given by

$$t(\underset{\sim}{x}) = \begin{cases} 0 & \text{if } T_n < T_{n,\alpha} \\ 1 & \text{if } T_n \geq T_{n,\alpha} \end{cases} \tag{4.5}$$

where $T_{n,\alpha}$ is a constant chosen to make the test size α. It is known (Saleh and Sen, 1982; Sen, 1969) that under H_0, $T_n \xrightarrow{L} \chi^2_{k-1}$ a Chi-Squared random variable with k-1 degrees of freedom as $n \to \infty$ and thus $T_{n,\alpha} \to \chi^2_{k-1,\alpha}$, the 100α % point of χ^2_{k-1}.

Let

$$\theta_i^* = \frac{1}{2} (\sup\{a: T_{in}(a,\tilde{\beta}_n) > 0\} + \inf\{a: T_{in}(a,\tilde{\beta}_n) < 0\}) \tag{4.6}$$

$$\tilde{\theta}_i^* = \max_{s \leq i} \hat{\hat{\theta}}_s^* \qquad i = 1,2,\ldots,k \tag{4.7}$$

where

$$\hat{\hat{\theta}}_s^* = \min\{\theta_s^*, \frac{1}{2} (\theta_s^* + \theta_{s+1}^*), \ldots, \frac{1}{k-s+1} \sum_{j=s}^{k} \theta_j^*\}. \tag{4.8}$$

Thus, $\tilde{\theta}_i^*$ is a translation-invariant consistent isotonic estimator. We define the randomized rule $\delta_R(\alpha)$ as follows:

$$\delta_R(\underset{\sim}{x};c,\alpha) = \begin{cases} A_{J(\underset{\sim}{x})} & \text{if } T_n < T_{n,\alpha} \\ A_{I(\underset{\sim}{x})} & \text{if } T_n \geq T_{n,\alpha} \end{cases} \tag{4.9}$$

where

$$J(\underset{\sim}{x}) = \min\{i \mid \tilde{\theta}_i^*(\underset{\sim}{x}) \geq \hat{\theta}_0(\underset{\sim}{x}) - \frac{c_i}{\sqrt{n}}\} \tag{4.10}$$

and $I(\underset{\sim}{x})$ is defined by (3.5).

Note that $T_n(\underset{\sim}{x}) \geq 0$. Hence, when we take the special value $T_{n,\alpha}(\underset{\sim}{x}) = 0$, δ_R becomes δ_I.

Let

$$\Omega_{00} = \{(\underset{\sim}{\theta}_1', \underset{\sim}{\beta}): \beta_1 = \ldots = \beta_k\} \tag{4.11}$$

and let Ω_{00}^c denote the complement of Ω_{00} with respect to Ω. Then, we note that

$$\lim_{n\to\infty} \inf_{\Omega_i} P(CS|\delta_R) = \min\{\lim_{n\to\infty} \inf_{\Omega_i\cap\Omega_{00}} P(CS|\delta_R), \lim_{n\to\infty} \inf_{\Omega_i\cap\Omega_{00}^c} P(CS|\delta_R)\}. \tag{4.12}$$

Following analogous arguments which yield (3.15) shows that by taking $T_{n,\alpha} = \infty$,

$$\lim_{n\to\infty} \inf_{\Omega_i\cap\Omega_{00}} P(CS|\delta_R) = P_{\underset{\sim}{\omega}_0'}\{\theta_r^* \geq \hat{\theta}_0 - c_r, \theta_r^* + \theta_{r+1}^* \geq 2(\hat{\theta}_0 - c_r), \ldots, \sum_{j=r}^{k} \theta_j^* \geq i(\hat{\theta}_0 - c_r)\} \tag{4.13}$$

where

$$\omega_0' = (\theta_0, \overbrace{-\infty, -\infty, \ldots, -\infty}^{k-i}, \theta_0, \theta_0, \ldots, \theta_0; \beta_0, \beta_1 = \beta_2 = \ldots = \beta_k). \tag{4.14}$$

Also, noting that the test T_n is consistent, we have

$$\lim_{n\to\infty} \inf_{\Omega_i\cap\Omega_{00}^c} P(CS|\delta_R) = P_{\underset{\sim}{\omega}_0}\{\hat{\theta}_r \geq \hat{\theta}_0 - d_r, \ldots, \sum_{i}^{k} \hat{\theta}_j \geq i(\hat{\theta}_0 - d_r)\} \tag{4.15}$$

where $\hat{\theta}_i$ is defined in (3.15) and $\underset{\sim}{\omega}_0$ is defined by (3.11). From Sen (1969) it is known that

$$\sqrt{n}\,(\underset{\sim}{\theta}_n^* - \underset{\sim}{\theta}) \xrightarrow{L} N(\underset{\sim}{0}, \tau^2 \ddagger_1)$$

where

$$\tau^2 = A_\phi^2 / (\int_0^1 \xi(u)\phi(u)du) \tag{4.16}$$

$$\Sigma_1 = \begin{bmatrix} \frac{1}{k} + \frac{x_1^2}{v} & \frac{x_1x_2}{v} & \cdots & \frac{x_1x_k}{v} \\ \frac{x_1x_2}{v} & \frac{1}{k} + \frac{x_2^2}{v} & \cdots & \frac{x_2x_k}{v} \\ & & \cdots & \\ \frac{x_kx_1}{v} & & \cdots & \frac{1}{k} + \frac{x_k^2}{v} \end{bmatrix} \tag{4.17}$$

where

$$v = \frac{1}{k}\sum_{1}^{k} s_i. \tag{4.18}$$

Let

$$Z_j^* = \sqrt{n}\,[(\theta_j^* - \theta_0) - (\hat{\theta} - \theta_0)] \qquad j = 1,2,\ldots,k-i+1 \tag{4.19}$$

$$\underset{\sim}{Z}_r^* = (Z_1^*, Z_2^*, \ldots, Z_r^*)' \text{ where } r = k-i+1. \tag{4.20}$$

Then, we have, by (4.13) that if

$$A(\underset{\sim}{x},v,c_r) = P(Z_1^* \geq -c_r, Z_1^* + Z_2^* \geq -2c_r, \ldots, \sum_{1}^{i} Z_j^* \geq -ic_r) \tag{4.21}$$

where $\underset{\sim}{Z}_r^*$ is multivariate normal $N(\underset{\sim}{0}, \Sigma_2)$, $\underset{\sim}{x}' = (x_1, x_2, \ldots, x_k)$,

$$\Sigma_2 = \tau^2 \Sigma_1 + v(\psi)(1+c^2)\underset{\sim}{1}_{i\times i}, \text{ and}$$

$$\underset{\sim}{1}_{i\times i} = \begin{bmatrix} 1 & 1 & \cdots & 1 \\ 1 & 1 & \cdots & 1 \\ 1 & 1 & \cdots & 1 \end{bmatrix}. \tag{4.22}$$

And if further

$$U_j^* = \sqrt{n}[\hat{\theta}_i-\theta_0) - (\hat{\theta}_0-\theta_0)],\ 1 \leq j \leq r, \tag{4.23}$$

$$\underset{\sim}{U}_r^* = (U_1, U_2, \ldots, U_r)', \quad \text{and} \quad r = k-i+1 \tag{4.24}$$

$$B(\underset{\sim}{x},\underset{\sim}{v},d_r) = P(U_1^* \geq -d_r, U_1^*+U_2^* \geq -2d_r, \ldots, \sum_{1}^{i} U_j^* \geq -id_r) \tag{4.25}$$

where

$\underset{\sim}{U}_r^*$ is multivariate normal $N(\underset{\sim}{0}, \Sigma_3)$, $\underset{\sim}{v}' = (v_1,v_2,\ldots,v_k)$, and

$$\ddagger_3 = \tau^2 \mathrm{diag}\left(\frac{1}{k} + \frac{v_1^2}{kv_1}, \frac{1}{k} + \frac{v_2^2}{kv_2}, \ldots, \frac{1}{k} + \frac{v_i^2}{kv_i}\right) + v(\psi)(1+c^2)\underset{\sim}{1}_{i\times i}. \quad (4.26)$$

Since under H_0 $T_n|H_0 \xrightarrow{L} \chi^2_{k-1}$ as $n \to \infty$ and $T_{n,\alpha} \to \chi^2_{k-1,\alpha}$, it follows from (4.13), (4.21) and (4.25) that

$$\lim_{n\to\infty} \inf_{\Omega_i \cap \Omega_{00}} P(CS|\delta_R) = (1-\alpha)A(\underset{\sim}{x},v,e_r) + \alpha B(\underset{\sim}{x},\underset{\sim}{v},d_r).$$

We therefore conclude the following.

Theorem 3.

$$\lim_{n\to\infty} \inf_{\Omega_i} P(CS|\delta_R) = \min\{(1-\alpha)A(\underset{\sim}{x},v,e_r) + \alpha B(\underset{\sim}{x},v,d_r),\ B(\underset{\sim}{x},\underset{\sim}{v},d_r)\}$$

where α is associated with $T_{n,\alpha}$ such that $T_{n,\alpha} \to \chi^2_{k-1}$, $(0 < \alpha < 1)$.

Remark. $\lim_{n\to\infty} \inf_{\Omega_i} P(CS|\delta_R) \geq \min\{A(\underset{\sim}{x},v,e_r),\ B(\underset{\sim}{x},v,e_r)\}$. Since

$(1-\alpha)A(\underset{\sim}{x},v,e_r) + \alpha B(\underset{\sim}{x},v,e_r) \geq B(\underset{\sim}{x},v,e_r)$ if $A(\underset{\sim}{x},v,e_r) \geq B(\underset{\sim}{x},v,e_r)$ and

$> A(\underset{\sim}{x},v,e_r)$ if $A(\underset{\sim}{x},v,e_r) < B(\underset{\sim}{x},v,e_r)$.

SUMMARY

Consider (k + 1) populations $\pi_0, \pi_1, \ldots, \pi_k$ such that the jth observation from π_i is denoted by $Y_{ij} = \theta_i + \beta_i x_{ij} + \varepsilon_{ij}$ ($i\leqslant i\leqslant k$, $1\leqslant j\leqslant n$) where θ_i and β_i are unknown parameters, x_{ij} is a known value and the ε_{ij} are iid with an unknown distribution function G. It is known that $\theta_1 \leqslant \ldots \leqslant \theta_k$. The problem is to select all the populations π_i ($1\leqslant i\leqslant k$) whose intercepts are larger than θ_0; the $\{\beta_i\}$ are nuisance parameters. An isotonic procedure based on ranks and an alternative based on a preliminary test are proposed. Properties of the proposed procedures are studied and some efficiencies are calculated.

REFERENCES

Adichie, J. N. (1967a). Estimates of regression parameters based on rank tests. Ann. Math. Statist. 38, 894-904.

Adichie, J. N. (1967b). Asymptotic efficiency of a class of non-parametric tests for regression parameters. Ann. Math. Statist. 38, 884-893.

Barlow, R. E., Bartholomew, D. J., Bremner, J. M. and Brunk, H. D. (1972). Statistical Inference under Order Restriction. John Wiley & Sons, New York.

Feller, W. (1969). An Introduction to Probability Theory and Its Applications, Vol. II, John Wiley & Sons, New York.

Gupta, S. S. and Huang, W. T. (1983). On isotonic selection rules for binomial populations better than a standard. (To appear in First Saudi Symposium on Statistics May 2-5, 1983).

Gupta, S. S. and Leu, L. Y. (1983). Isotonic procedures for selecting populations better than a standard: Two-parameter exponential distributions. Tech. Report #83-46, Dept. of Statist., Purdue University.

Gupta, S. S. and Yang, H. M. (1981). Isotonic procedures for selecting populations better than a control under ordering prior. Proceedings of Golden Jubilee Conference of the Indian Statistical Institute, Calcutta, India, December 1981.

Hodges, J. L. Jr. and Lehmann, E. L. (1956). The efficiency of some non-parametric competitors of the t-test. Ann. Math. Statist. 27, 324-335.

Puri, M. L. and Sen, P. K. (1971). Non-parametric Methods in Multivariate Analysis. Wiley, New York.

Saleh, A. K. M. E. and Sen, P. K. (1978). Non-parametric estimation of location parameter after a preliminary test on regression. Ann. Statist. 6, 154-168.

Saleh, A. K. M. E. and Sen, P. K. (1982). Non-parametric tests for location after a preliminary test on regression. Commun. Statist.-Theor. Meth. 11(6), 639-652.

Sen, P. K. (1969). On a class of rank order tests for the parallelism of several regression lines. Ann. Math. Statist. 40, 1668-1683.

OPTIMAL PROCEDURES FOR SELECTING THE BEST s OUT OF k BERNOULLI POPULATIONS

C. Jennison

Department of Mathematical Sciences
University of Durham
Durham, England

R.V. Kulkarni

SAS Institute
Cary, North Carolina

1. INTRODUCTION

Let Π_i $(1 \leq i \leq k)$ denote $k \geq 2$ Bernoulli populations with corresponding single-trial "success" probabilities p_i. Denote the ordered values of the p_i by $p_{[1]} \leq \cdots \leq p_{[k]}$; the pairing of the Π_i with the $p_{[j]}$ $(1 \leq i,j \leq k)$ is assumed to be completely unknown. The goal of the experimenter is to select the populations associated with the s $(1 \leq s \leq k-1)$ largest p-values $(p_{[k]},\ldots,p_{[k-s+1]})$ without regard to order. When these populations are selected we say that a correct selection (CS) has been made. These populations are called the s "best" populations.

Most of the papers in the context of selection problems restrict attention to the goal of selecting the "best" population (s = 1) or the "worst" population (s = k-1). Bechhofer (1954) described more general goals for selecting normal populations. Bofinger (1978) defined a truncated sampling scheme for selecting the populations with the t $(1 \leq t \leq k-1)$ smallest parameter values (without regard to order), for populations with non-negative observations. When applied to the Bernoulli selection problem, her goal is to select the populations associated with $p_{[1]},p_{[2]},\ldots,p_{[t]}$, which is equivalent to our goal with s = k-t.

Bechhofer and Kulkarni (1982) consider the problem of selecting the s best Bernoulli populations without regard to order (referred to as Goal I) and with regard to order (referred to as Goal II). For each goal they describe a class of closed sequential procedures which take no more than n observations per population and which employ a particular stopping rule and terminal decision rule; the procedures in each class differ only with respect to their sampling rules. Within each class, all of the procedures achieve the same probability of correct selection, P{CS}, uniformly in the parameter values. We shall restrict attention to the Goal I Bernoulli selection problem and denote the associated class of procedures by $C_{s,k}$ (the dependence on n, the maximum number of observations per population, is suppressed in the notation).

Let $\underset{\sim}{p} = (p_1,\ldots,p_k)$ and $\bar{\underset{\sim}{p}} = (p_{[1]},\ldots,p_{[k]})$. Define $N_{(i)}$ to be the number of observations taken from the population associated with $p_{[i]}$ $(1 \leq i \leq k)$ at the termination of sampling and define $N = \sum_{i=1}^{k} N_{(i)}$ to be the associated total number of observations. We denote by $E_P\{N|\bar{\underset{\sim}{p}}\}$ the expected total number of observations using a procedure P when $\bar{\underset{\sim}{p}}$ is known but the pairing of the Π_i with the $p_{[j]}$ is completely unknown. We are concerned with finding procedures in $C_{s,k}$ which minimize $E\{N|\bar{\underset{\sim}{p}}\}$ and/or minimize $E\{\sum_{i=1}^{k} \lambda_i N_{(i)}|\bar{\underset{\sim}{p}}\}$ for a vector $\underset{\sim}{\lambda} = (\lambda_1,\lambda_2,\ldots,\lambda_k)$ with $\lambda_1 \geq \lambda_2 \geq \ldots \geq \lambda_k \geq 0$.

Bechhofer and Kulkarni (1982) considered the above problem for s = 1 (in which case both Goals I and II coincide). They stated several theorems regarding the optimality of one particular procedure in $C_{1,k}$ (C in their notation), for k = 2. Kulkarni and Jennison (1983) strengthened some of these results and obtained generalizations to the case $k \geq 3$, for s = 1.

In Section 2 we define the class $C_{s,k}$ and state the key P{CS} theorem, namely that the P{CS} is the same for all procedures in $C_{s,k}$. In Section 3 we define a procedure $P^*_{s,k}$. We prove that $P^*_{s,k}$ minimizes $E\{N|\bar{\underset{\sim}{p}}\}$ among all procedures in $C_{s,k}$ in a particular region of the parameter space where all of the $p_{[i]}$ $(1 \leq i \leq k)$ are sufficiently "large". A conjugate procedure $\bar{P}^*_{s,k}$ (also defined in Section 3) minimizes $E\{N|\bar{\underset{\sim}{p}}\}$ when the $p_{[i]}$'s are sufficiently "small".

In Section 4 we consider the problem of minimizing $E\{\sum_{i=1}^{k} \lambda_i N_{(i)}|\bar{\underset{\sim}{p}}\}$ for a vector $\underset{\sim}{\lambda}$ with $\lambda_1 \geq \lambda_2 \geq \ldots \geq \lambda_k \geq 0$ and prove that $P^*_{s,k}$ is also optimal for this problem when all of the $p_{[i]}$ $(1 \leq i \leq k)$ are sufficiently "large".

In Section 5 we present approximations for $E\{N_{(i)}|\bar{\underset{\sim}{p}}\}$ under procedure $P^*_{s,k}$ and draw comparisons with other procedures.

2. A CLASS OF CLOSED SEQUENTIAL SELECTION PROCEDURES

A selection procedure is characterized by a sampling rule, a stopping rule and a terminal decision rule. When a total of m observations have been taken the experiment is said to be at stage m; the state of the experiment is then the complete history of the first m observations. This is denoted by

$$y_m = \{(i_1,v_1),(i_2,v_2),\ldots,(i_m,v_m)\},$$

where i_r is the index of the population from which the rth observation is taken, and v_r is the outcome of the rth observation (either a success or a failure); the notation y, without a subscript will be used to denote a general state of an experiment. The following variables are functions of y_m, although the dependence on m is suppressed in the notation:

n_i = number of observations taken from Π_i in the first m stages,
s_i = number of "successes" yielded by Π_i in the first m stages,
f_i = number of "failures" yielded by Π_i in the first m stages.

We restrict attention to procedures which take no more than n observations from any one of the k populations where $n \geq 1$ is a prespecified integer; thus $0 \leq m \leq kn$ and $n_i \leq n$ $(1 \leq i \leq k)$. Let the set of all possible histories at stage m be

$$\Omega_m = \{y_m: n_i \leq n,\ 1 \leq i \leq k;\ \sum_{i=1}^{k} n_i = m\}$$

and define the set of all possible states

$$\Omega = \cup_{m=0}^{kn} \Omega_m = \{y_m: 0 \leq m \leq kn;\ n_i \leq n,\ 1 \leq i \leq k;\ \sum_{i=1}^{k} n_i = m\}.$$

A sampling rule specifies (possibly at random) the index of the population from which the next observation is taken.

The stopping rule $S^*_{s,k}$ is: "Stop at the first stage m at which there exist disjoint sets $A_1, A_2 \subset \{1,2,\ldots,k\}$ with A_1 of order s and A_2 of order k-s such that

$$s_{i_1} + f_{i_2} \geq n \quad \text{for all } i_1 \in A_1,\ i_2 \in A_2." \tag{2.1}$$

The terminal decision rule $T^*_{s,k}$ is: "If there are r partitions $\{A_1,A_2\}$ satisfying (2.1), choose one of these at random and select populations with labels in A_1 as best."

We shall consider the class $C_{s,k}$ of sequential procedures using the stopping rule $S^*_{s,k}$, terminal decision rule $T^*_{s,k}$ and an arbitrary sampling rule R that takes at most n observations from any one population; we write $P = (R, S^*_{s,k}, T^*_{s,k})$ to denote the procedure in $C_{s,k}$ using the sampling rule R.

Theorem 2.1. If P_1 and P_2 are in $C_{s,k}$ then $P\{CS|\bar{\underset{\sim}{p}}\}$ is the same for both procedures, uniformly in $\bar{\underset{\sim}{p}}$.

Theorem 2.1 is a special case of a theorem proved by Jennison (1983a, Remark 2.3) for Bernoulli selection procedures. It can also be shown that the P{CS} of the single-stage procedure (a generalization of the procedure of Sobel and Huyett (1957), described in B-K) which takes exactly n observations per population is equal to the P{CS} for any $P \in C_{s,k}$, uniformly in $\bar{\underset{\sim}{p}}$. Questions regarding the choice of n have also been addressed in Sobel and Huyett (1957).

Our objective will be to select from the class $C_{s,k}$, a procedure $P = (R, S^*_{s,k}, T^*_{s,k})$ which is optimal in some sense. To stop sampling any earlier than indicated by $S^*_{s,k}$ reduces the P{CS} and hence, a procedure which is optimal within $C_{s,k}$ for some objective function is also optimal within the class of all procedures taking at most n observations from any one population, that achieve the same P{CS} as the single-stage procedure, uniformly in $\bar{\underset{\sim}{p}}$.

3. MINIMIZING THE EXPECTED TOTAL NUMBER OF OBSERVATIONS

We seek a procedure which is optimal within the class $C_{s,k}$ in that it minimizes $E\{N|\bar{\underset{\sim}{p}}\}$. First consider the case in which the vector of ordered success probabilities, $\bar{\underset{\sim}{p}}$ is fixed; assume that the vector $\underset{\sim}{p}$ has probability 1/k! of being any particular permutation of $\bar{\underset{\sim}{p}}$. We shall define the procedure $P^*_{s,k} = (R^*_{s,k}, S^*_{s,k}, T^*_{s,k})$ and show that it minimizes $E\{N|\bar{\underset{\sim}{p}}\}$ whenever all of the $p_{[i]}$ $(1 \leq i \leq k)$ are sufficiently "large".

Definition of Sampling Rule $R^*_{s,k}$

Construct an ordering $\Pi_{\{1\}}, \ldots, \Pi_{\{k\}}$ such that for $s+1 \leq j \leq k$ and $i < j$

$$f_{\{j\}} > f_{\{i\}} \quad \text{or} \quad (f_{\{j\}} = f_{\{i\}} \quad \text{and} \quad s_{\{j\}} \leq s_{\{i\}}),$$

and for $1 \leq i < j \leq s$

$$s_{\{i\}} > s_{\{j\}} \quad \text{or} \quad (s_{\{i\}} = s_{\{j\}} \quad \text{and} \quad f_{\{i\}} \leq f_{\{j\}}). \tag{3.1}$$

Let

$$W = \{\Pi_i: s_i = s_{\{s\}} \text{ and } f_i = f_{\{s\}}\}, \tag{3.2}$$

then $R^*_{s,k}$ samples from a population selected at random from W.

The procedure $P^*_{s,k}$ uses the sampling rule $R^*_{s,k}$, the stopping rule $S^*_{s,k}$ and the terminal decision rule $T^*_{s,k}$. We define $C^*_{s,k}$ to be the class of procedures in $C_{s,k}$ with non-randomized sampling rules, which always sample from a population in W. It is easily seen that all procedures in $C^*_{s,k}$ have the same $E\{N|\bar{p}\}$ since the distribution of $\underset{\sim}{p}$ is assumed to be permutationally invariant.

Remark 3.1. The sampling rule $R^*_{s,k}$ can be described simply in the following manner: Count off the k-s populations with the most failures, breaking ties according to least number of successes (further ties may be broken arbitrarily); find that one of the remaining s populations with the least number of successes, breaking ties according to most failures; call this population $\Pi_{\{s\}}$ and select at random from populations tied with $\Pi_{\{s\}}$.

Definition of conjugate procedure $\bar{P}^*_{s,k}$

If the success probabilities $p_{[i]}$ are small, then the failure probabilities $q_{[i]} = 1 - p_{[i]}$ $(1 \le i \le k)$ are high; identifying the s populations with the highest $p_{[i]}$'s is equivalent to identifying the k-s populations with the highest $q_{[i]}$'s. This motivates the definition $\bar{P}^*_{s,k} = (\bar{R}^*_{s,k}, S^*_{s,k}, T^*_{s,k})$ where $\bar{R}^*_{s,k}$ samples as $R^*_{k-s,k}$ would do if the labels "success" and "failure" were interchanged. $\bar{C}^*_{s,k}$ is defined accordingly.

It can be seen that the procedure P^*, defined in Bechhofer and Kulkarni (1982) and Kulkarni and Jennison (1983) is the same as $P^*_{1,k}$. Also, the procedure $\bar{P}^*$, defined in Kulkarni and Jennison (1983) is the same as $\bar{P}^*_{1,k}$. Consequently, there is equality between the classes of procedures with deterministic sampling rules: $C^* = C^*_{1,k}$ and $\bar{C}^* = \bar{C}^*_{1,k}$.

We now state the optimality results for the case s = 1, proved in Kulkarni and Jennison (1983)--which we shall henceforth refer to as K-J, and we state a new theorem for general s.

Result 3.1 (K-J, Theorem 6.1). Amongst all procedures in $C_{1,k}$, $P^*_{1,k}$ and procedures in $C^*_{1,k}$ minimize $E\{N|\underset{\sim}{p}\}$ if $p_{[1]} + (\sum_{i=2}^{k} p_{[i]})/(k-1) \ge 1$.

Result 3.2 (K-J, Theorem 6.2). Amongst all procedures in $C_{1,k}$, $\bar{P}^*_{1,k}$ and procedures in $\bar{C}^*_{1,k}$ minimize $E\{N|\underset{\sim}{p}\}$ if

$$p_{[k]}\{1 + \sum_{i=1}^{k-2} \Pi_{j=1}^{i} (1-p_{[j]})\} \le \Pi_{j=1}^{k-1} (1-p_{[j]}).$$

Sufficient conditions for this are:

$$\sum_{i=1}^{k-1} p_{[i]} + (k-1)p_{[k]} \leq 1 \quad \text{or} \quad p_{[k]} \leq 1 - (1/2)^{1/(k-1)}.$$

(The result is true in the case k = 2 if the sum $\sum_{i=1}^{0}$ is taken to be a null sum.)

Theorem 3.1. Amongst all procedures in $C_{s,k}$, $P^*_{s,k}$ and procedures in $C^*_{s,k}$ minimize $E\{N|\underset{\sim}{p}\}$ if

$$(1-p_{[1]})\{1 + \sum_{i=3}^{s+1} \Pi_{j=i}^{s+1} p_{[j]}\} \leq \Pi_{j=2}^{s+1} p_{[j]}. \tag{3.3}$$

Sufficient conditions for this are

$$sp_{[1]} + \sum_{i=2}^{s+1} p_{[i]} \geq 2s - 1 \tag{3.4}$$

or

$$p_{[1]} \geq (1/2)^{1/s}. \tag{3.5}$$

Remark 3.2. Replacing s by k-s and $p_{[i]}$ by $1-p_{[k+1-i]}$, it follows that $\overline{P}^*_{s,k}$ and procedures in $\overline{C}^*_{s,k}$ are optimal if

$$p_{[k]}\{1 + \sum_{i=s}^{k-2} \Pi_{j=s}^{i} (1-p_{[j]})\} \leq \Pi_{j=s}^{k-1} (1-p_{[j]}), \tag{3.6}$$

sufficient conditions for which are:

$$\sum_{i=s}^{k-1} p_{[i]} + (k-s)p_{[k]} \leq 1 \quad \text{or} \quad p_{[k]} \leq 1 - (1/2)^{1/(k-s)}.$$

Remark 3.3. Substituting s=1 in (3.3) and taking $\sum_{i=3}^{2}$ to be a null sum gives the sufficient condition for $P^*_{1,k}$ to be optimal: $p_{[1]} + p_{[2]} \geq 1$. This is a weaker result than K-J, Theorem 6.1. Substituting s=1 in (3.6) however, gives the same sufficient condition for $\overline{P}^*_{1,k}$ to be optimal as K-J, Theorem 6.2.

The proof of Theorem 3.1 and the proofs of other theorems in this paper are similar to those in K-J and for conciseness we shall omit details that can be found in K-J. The proof of Theorem 3.1 uses Lemma 3.1 below. We shall say that $\Pi_i \succ \Pi_j$ if $s_i + f_j \geq n$. Thus if $\Pi_i \succ \Pi_j$, population Π_j cannot finish with more successes than population Π_i, although the two populations could tie eventually if $s_i + f_j = n$. $S^*_{s,k}$ stops as soon as there is a partition $\{A_1,A_2\}$ with A_1 of order s, A_2 of order k-s and $\Pi_{i_1} \succ \Pi_{i_2}$ for all $i_1 \in A_1$ and $i_2 \in A_2$.

Lemma 3.1. Let Ω_A be the set of states y for which there is a partition $\{B_1,\{j\},B_2\}$ with B_1 of order s and B_2 of order k-s-1 and $\Pi_{i_1} \succ \Pi_{i_2}$ for all

$i_1 \in B_1$ and $i_2 \in B_2$. If (3.3) holds and $y \in \Omega_A$ then procedures in $C^*_{s,k}$ sample optimally from this point on.

Proof: In a state $y \in \Omega_A$, the stopping criterion for $S^*_{s,k}$ is satisfied as soon as a population with least successes out of n, amongst those whose labels are not in B_2, can be identified. This is equivalent to identifying a population with most failures out of n within that group. Hence, if the $(1-p_{[i]})$ are sufficiently small, it is optimal from this point to sample as a procedure in $\bar{C}^*_{1,s+1}$ would sample if the labels "success" and "failure" were interchanged. Now the procedures in $C^*_{s,k}$ sample in this way after a state $y \in \Omega_A$ has been reached and hence they sample optimally from state y. The condition on $\underset{\sim}{\bar{p}}$ is obtained from Result 3.2: When selecting the population with most failures out of s+1 populations with ordered failure probabilities $q_{[1]} \leq \cdots \leq q_{[s+1]}$, the condition for $\bar{C}^*_{1,s+1}$, with "success" and "failure" labels interchanged, to be optimal is

$$q_{[s+1]}\{1 + \textstyle\sum_{i=1}^{s-1} \Pi_{j=1}^{i} (1-q_{[j]})\} \leq \Pi_{j=1}^{s} (1-q_{[j]}). \tag{3.7}$$

This condition is most difficult to satisfy when the q_i's are large, so the worst case is when the remaining populations have success probabilities $p_{[1]},\ldots,p_{[s+1]}$ substituting $q_{[i]} = 1-p_{[s+2-i]}$ $(i = 1,\ldots,s+1)$ into (3.7) gives the condition (3.3). □

Remark 3.4. A similar result holds for states y where there is a partition $\{B_1, \{j\}, B_2\}$ with B_1 of order s-1, B_2 of order k-s and $\Pi_{i_1} \succ \Pi_{i_2}$ for all $i_1 \in B_1$ and $i_2 \in B_2$. In this situation it is necessary to choose a population with most successes out of k-s+1 populations. The ensuing result is not needed in the proof; the condition on $\underset{\sim}{\bar{p}}$ for optimality of $C^*_{s,k}$ in this case is implied by (3.3).

Proof of Theorem 3.1. Consider a fixed value of $\underset{\sim}{\bar{p}}$ satisfying (3.3). Let P be a procedure in $C_{s,k}$ with a non-randomized sampling rule, that minimizes $E\{N|\underset{\sim}{\bar{p}}\}$--such a procedure can be shown to exist by a backwards induction argument. By Lemma 3.1, P can be modified to agree with a procedure in $C^*_{s,k}$ on Ω_A, with no increase in $E\{N|\underset{\sim}{\bar{p}}\}$; call the modified procedure P'.

Suppose now that there is a state $\bar{y}$ such that the experiment continues past $\bar{y}$ and, when in state $\bar{y}$, P' does not sample a population in the set W, defined by (3.2). Take i such that $\Pi_i \in W$, then there is an ordering of the form (3.1) with $\Pi_i = \Pi_{\{s\}}$. Thus, for state $\bar{y}$

$$f_{\{j\}} > f_i \quad\text{or}\quad (f_{\{j\}} = f_i \;\text{ and }\; s_{\{j\}} \leq s_i), \quad s+1 \leq j \leq k$$

and

$$s_{\{j\}} > s_i \quad\text{or}\quad (s_{\{j\}} = s_i \;\text{ and }\; f_{\{j\}} \leq f_i), \quad 1 \leq j \leq s-1.$$

P' may be modified, to P'' say, so that for all j, P'' samples from Π_i in preference to Π_j if $s_i = s_j$ and $f_i = f_j$; this may be done in such a way that $E\{N|\underset{\sim}{\bar{p}}\}$ remains unchanged and P'' still agrees with a procedure in $C^*_{s,k}$ on Ω_A (for further details see K-J, Lemma 3.2).

We now show that if P'' is used, a further observation on Π_i is inevitable after being in state $\bar{y}$. Suppose not, then on termination without a further observation on Π_i we must have $\Pi_{\{1\}},\ldots,\Pi_{\{s\}} \succ \Pi_{\{s+1\}},\ldots,\Pi_{\{k\}}$ with $\Pi_{\{s\}} = \Pi_i$ (here, $\Pi_{\{1\}},\ldots,\Pi_{\{k\}}$ still denote the ordering constructed for state $\bar{y}$). It follows that the final observation must have been a failure on one of $\Pi_{\{s+1\}},\ldots,\Pi_{\{k\}}$. But this gives a contradiction as the penultimate state is in Ω_A and P'' would then sample from Π_i.

Thus, P'' can be modified, with no increase in $E\{N|\underset{\sim}{\bar{p}}\}$, so that the next observation is taken from Π_i when in state $\bar{y}$ (for further details see K-J, Lemma 3.1). In fact, P'' can be modified gradually, without increasing $E\{N|\bar{p}\}$, to give a procedure in $C^*_{s,k}$. Since $E\{N|\underset{\sim}{\bar{p}}\}$ for the resulting procedure is no greater than $E\{N|\underset{\sim}{\bar{p}}\}$ for an optimal procedure, the procedure is itself optimal; hence all procedures in $C^*_{s,k}$, and also $P^*_{s,k}$, are optimal.

The sufficiency of either of the conditions (3.4) or (3.5) for condition (3.3) is proved in K-J, Theorem 6.2. □

4. MINIMIZING THE EXPECTED NUMBER OF OBSERVATIONS FROM INFERIOR POPULATIONS

In the two population problem an experimenter may wish to minimize the expected number of observations from the inferior population; a natural generalization of this goal for $k \geq 3$ is to minimize $E\{\sum_{i=1}^{k} \lambda_i N_{(i)}|\underset{\sim}{\bar{p}}\}$ where $\lambda_1 \geq \lambda_2 \geq \cdots \geq \lambda_k \geq 0$ and $\underset{\sim}{\lambda} = (\lambda_1,\ldots,\lambda_k)$ may possibly be a function of $\underset{\sim}{\bar{p}}$. For notational ease we shall assume in this Section that the $p_{[i]}$'s are distinct. We say that a procecure is "fully optimal" for a particular $\bar{p}$ if it minimizes $E\{\sum_{i=1}^{k} \lambda_i N_{(i)}|\underset{\sim}{\bar{p}}\}$ for all $\underset{\sim}{\lambda}$ satisfying $\lambda_1 \geq \lambda_2 \geq \cdots \geq \lambda_k \geq 0$. An equivalent condition is that the procedure minimizes $E\{\sum_{i=1}^{r} N_{(i)}|\underset{\sim}{\bar{p}}\}$ for all r $(1 \leq r \leq k)$. The following theorem shows that $P^*_{s,k}$ is fully optimal if all of the $p_{[i]}$'s are sufficiently large.

<u>Theorem 4.1</u>. Amongst all procedures in $C_{s,k}$, $P^*_{s,k}$ and the procedures in $C^*_{s,k}$ minimize $E\{\sum_{i=1}^{r} N_{(i)}|\underset{\sim}{\bar{p}}\}$ for all r $(1 \leq r \leq k)$ if

$$\frac{p_{[1]}^{\,s}}{1-p_{[1]}} \geq s+1. \tag{4.1}$$

Remark 4.1. Substituting s=1 into (4.1) gives the sufficient condition for $P^*_{1,k}$ to be fully optimal: $p_{[1]} \geq 2/3$. This is a weaker result than K-J, Theorem 6.3 which gives the sufficient condition $p_{[1]}+p_{[2]} \geq 1$.

We first prove the following lemma, which will be used later in its dual version.

Lemma 4.1. Amongst all procedures in $C_{1,k}$, those in $\overline{C}^*_{1,k}$ minimize $E\{\sum_{i=k-q+1}^{k} N_{(i)}|\underset{\sim}{p}\}$ for all q $(1 \leq q \leq k)$ whenever

$$\frac{(1-p_{[k]})^{k-1}}{p_{[k]}} \geq k. \tag{4.2}$$

Note that this lemma concerns the minimization of the expected number of observations from populations with high p_i's--the need for such a result will become apparent later. The proof of Lemma 4.1 requires a second lemma.

Lemma 4.2. Let $\overline{y}$ be a state for which there exists a unique population (say Π_1) with $s_1 + f_i \geq n-1$ for $i \neq 1$. Thus, a single success from Π_1 terminates the experiment with the selection of Π_1. If (4.2) holds, there is an optimal procedure in $C_{1,k}$ for minimizing $E\{\sum_{i=k-q+1}^{k} N_{(i)}|\underset{\sim}{p}\}$ which does not take the next observation from Π_1.

Proof: The proof is similar to that of K-J, Lemma 6.2. We consider first states $\overline{y}$ in which no populations have been eliminated and $f_j = f_1$ for all $j \neq 1$. In this case $f_j = n-s_1-1$ $(1 \leq j \leq k)$, n-1 observations have been taken from Π_1 and $n-(r_j+1)$ observations, say, have been taken from Π_j $(j \neq 1)$ where $r_j \geq 1$. Without loss of generality suppose $1 \leq r_2 \leq r_3 \leq \cdots \leq r_k$. Suppose further that P is a Bayes optimal procedure for a given $\underset{\sim}{p}$ and that in state $\overline{y}$, P does take the next observation from Π_1. If this observation is a success the experiment stops; if not it can be seen that P must sample from $\Pi_2,\ldots,\Pi_k$ in order, in each case sampling from Π_j until it is eliminated by the occurrence of a failure or until an additional (r_j+1) successes are obtained from Π_j and the experiment stops with Π_j selected as best. We define the procedure P' as follows: Sample as under P but omit the initial observation from Π_1; if r_j successes are obtained from Π_j take an observation from Π_1 and then proceed again as under P behaving as if the observation from Π_1 had been taken when in state $\overline{y}$. Let $N^{\overline{y}}_{(i)}$ denote the total number of further observations from the population associated with $p_{[i]}$, after being in state $\bar{y}$. It can be shown that for fixed $\bar{p} = (p_{[1]},\ldots,p_{[k]})$

$$
\begin{aligned}
&E_{p}\{\textstyle\sum_{i=k-q+1}^{k} N_{(i)}^{\bar{y}} | \bar{p}\} - E_{p'}\{\textstyle\sum_{i=k-q+1}^{k} N_{(i)}^{\bar{y}} | \bar{p}\} \\
&= E[(1-p_2^{r_2})(1-p_3^{r_3}) \ldots (1-p_k^{r_k})\, \delta(p_1 \geq p_{[k-q+1]}) \\
&\quad -p_1\{\frac{1-p_2^{r_2}}{1-p_2}\, \delta(p_2 \geq p_{[k-q+1]}) + (1-p_2^{r_2})\, \frac{1-p_3^{r_3}}{1-p_3}\, \delta(p_3 \geq p_{[k-q+1]}) + \ldots \\
&\quad + (1-p_2^{r_2}) \ldots (1-p_{k-1}^{r_{k-1}})\, \frac{1-p_k^{r_k}}{1-p_k}\, \delta(p_k \geq p_{[k-q+1]})\}]
\end{aligned}
$$

where the expectation on the right hand side is over the posterior distribution of p, given the observations up to state $\bar{y}$. $\delta(\cdot)$ is the indicator variable taking the value 1 if the event occurs and 0 otherwise. Now

$$
\begin{aligned}
&E\{(1-p_2^{r_2})(1-p_3^{r_3}) \ldots (1-p_k^{r_k})\, \delta(p_1 \geq p_{[k-q+1]})\} \\
&\geq (1-p_{[k]}^{r_2})(1-p_{[k-1]}^{r_3}) \ldots (1-p_{[2]}^{r_k})\, P\{p_1 \geq p_{[k-q+1]}\} \\
&\geq (1-p_{[k]}^{r_2})(1-p_{[k-1]}^{r_3}) \ldots (1-p_{[2]}^{r_k}) \cdot \frac{q}{k}\,.
\end{aligned}
$$

Also $(1-p_i^{r_i})/(1-p_i) \leq (1-p_{[k]}^{r_k})/(1-p_{[k]})$ for all i $(1 \leq i \leq k)$. Thus

$$
\begin{aligned}
&E[-p_1\{\frac{1-p_2^{r_2}}{1-p_2}\, \delta(p_2 \geq p_{[k-q+1]}) + \ldots \\
&\qquad + (1-p_2^{r_2}) \ldots (1-p_{k-1}^{r_{k-1}})\, \frac{1-p_k^{r_k}}{1-p_k}\, \delta(p_k \geq p_{[k-q+1]})\}] \\
&\geq -p_{[k]}\, \frac{1-p_{[k]}^{r_k}}{1-p_{[k]}}\, E\{\delta(p_2 \geq p_{[k-q+1]}) + \ldots + \delta(p_k \geq p_{[k-q+1]})\} \\
&\geq -p_{[k]}\, \frac{1-p_{[k]}^{r_k}}{1-p_{[k]}}\, q.
\end{aligned}
$$

And therefore $E_{p'}\{\sum_{i=k-q+1}^{k} N_{(i)}^{\bar{y}} | \bar{p}\} \leq E_{p}\{\sum_{i=k-q+1}^{k} N_{(i)}^{\bar{y}} | \bar{p}\}$ if

$$
(1-p_{[k]}^{r_2})(1-p_{[k-1]}^{r_3}) \ldots (1-p_{[2]}^{r_k}) \geq p_{[k]}\, \frac{1-p_{[k]}^{r_k}}{1-p_{[k]}}\, k.
$$

Since $1-p_{[2]}^{r_k} \geq 1-p_{[k]}^{r_k}$, a sufficient condition for this inequality to hold is $(1-p_{[k]})^{k-1}/p_{[k]} \geq k$. Thus if (4.2) holds there is an optimal procedure, P', which does not take the next observation from Π_1, as required. The proof for other states $\bar{y}$ follows as in K-J, Lemma 6.2. □

The proof of Lemma 4.1 follows from Lemma 4.2 in the same way that K-J, Theorem 6.2 follows from K-J, Lemma 6.2.

Lemma 4.3. As in Lemma 3.1, let Ω_A be the set of states y for which there is a partition $\{B_1,\{j\},B_2\}$ with B_1 of order s, B_2 of order k-s-1 and $\Pi_{i_1} \succ \Pi_{i_2}$ for all $i_1 \in B_1$ and $i_2 \in B_2$. If (4.1) holds and $y \in \Omega_A$ then procedures in $C^*_{s,k}$ sample optimally, from this point on, for minimizing $E\{\sum_{i=1}^{r} N_{(i)} | \underset{\sim}{p}\}$ for all r $(1 \leq r \leq k)$.

Proof: As in Lemma 3.1, the problem reduces to choosing the population with the lowest p_i out of a set of s+1 populations. At the same time we must minimize the expected total number of further observations from populations with $p_i \leq p_{[r]}$. Letting $q_i = 1-p_i$, the problem is then to select from this set of s+1 populations that population with the largest q_i, while minimizing the expected number of observations from populations with $q_i \geq 1-p_{[r]}$. Restricting attention to these s+1 populations, it follows from Lemma 4.1 that procedures in $\bar{C}^*_{1,s+1}$ with the labels "success" and "failure" interchanged are optimal for this problem if $p_{[1]}^{s}/(1-p_{[1]}) \geq$ s+1. But, by definition, these are the procedures in $C^*_{s,s+1}$ and, in this situation, they sample as procedures in $C^*_{s,k}$. This proves the lemma. □

The proof of Theorem 4.1 follows from Lemma 4.3 in exactly the same way as the proof of Theorem 3.1 follows from Lemma 3.1.

5. AN APPROXIMATION FOR $E\{N_{(i)}\}$ UNDER $P^*_{s,k}$

A mean path approximation may be used to obtain approximations for $E\{N_{(i)}\}$ $(1 \leq i \leq k)$; expressions for $E\{N_{(i)}\}$ are calculated under the assumption that the observed proportion of successes on population Π_i is exactly p_i and are accurate up to order $O(\sqrt{n})$ as $n \to \infty$. Using the methods of Jennison (1984) it can be shown that this approximation is in fact accurate up to order $\underset{\sim}{o}(n)$ if $p_{[k-s]} < p_{[k-s+1]} < p_{[k-s+2]}$.

With probability approaching 1 as $n \to \infty$, the experiment ends with n observations on $\Pi_{(k-s+1)}$ and if there are z successes on $\Pi_{(k-s+1)}$, then there are z successes on each of $\Pi_{(k-s+2)},\ldots,\Pi_{(k)}$ and n-z failures on $\Pi_{(1)},\ldots,\Pi_{(k-s)}$. This leads to the approximation

Table 5.1

Approximations for $E\{N_{(i)}\}$ under Three Procedures when k=4, s=2, $\underset{\sim}{\bar{p}} = (0.2, 0.4, 0.6, 0.8)$ and n = 50.

Procedure	$E\{N_{(1)}\}$	$E\{N_{(2)}\}$	$E\{N_{(3)}\}$	$E\{N_{(4)}\}$
$\mathcal{P}^*_{2,4}$	25.0	33.3	50.0	37.5
$(\mathcal{R}^*_{1,4}, \mathcal{S}^*_{2,4}, \mathcal{T}^*_{2,4})$	25.0	33.3	50.0	50.0
Single-stage procedure	50.0	50.0	50.0	50.0

$$E\{N_{(i)}\} \simeq \begin{cases} \dfrac{np_{[k-s+1]}}{p_{[i]}} & i = k-s+2,\ldots,k \\ n & i = k-s+1 \\ \dfrac{n(1-p_{[k-s+1]})}{1-p_{[i]}} & i = 1,\ldots,k-s \end{cases}$$

which is accurate up to order $\underset{\sim}{o}(n)$ as $n \to \infty$, if $p_{[k-s]} < p_{[k-s+1]} < p_{[k-s+2]}$ and up to order $O(\sqrt{n})$ otherwise.

By comparison, the single-stage procedure always takes n observations from each population. Alternatively if the procedure $(\mathcal{R}^*_{1,k}, \mathcal{S}^*_{s,k}, \mathcal{T}^*_{s,k})$ is used,

$$E\{N_{(i)}\} \simeq \begin{cases} n & i = k-s+1,\ldots,k \\ \dfrac{n(1-p_{[k-s+1]})}{1-p_{[i]}} & i = 1,\ldots,k-s. \end{cases}$$

We illustrate the relative efficiency of $\mathcal{P}^*_{s,k}$ with an example: Table 5.1 compares the above approximations for $E\{N_{(i)}\}$ when k = 4, s = 2, $\underset{\sim}{\bar{p}}$ = (0.2, 0.4, 0.6, 0.8) and n = 50.

SUMMARY

We consider the problem of selecting the s out of k Bernoulli populations with the largest single-trial success probabilities. Bechhofer and Kulkarni (1982) proposed a class of sequential procedures for this problem which take no more than n observations from any of the k populations and which achieve the same probability of correct selection as does a single-stage procedure taking exactly n observations from each population. We describe a procedure which minimizes the expected total number of observations to terminate sampling, amongst procedures in this class, when all of the success probabilities are sufficiently high. A conjugate procedure minimizes the expected sample size when all of the success probabilities are low.

REFERENCES

Bechhofer, R. E. (1954). A single-sample multiple decision procedure for ranking means of normal populations with known variances. Ann. Math. Statist. 25, 16-39.

Bechhofer, R. E. and Kulkarni, R. V. (1982). Closed adaptive sequential procedures for selecting the best of $k \geq 2$ Bernoulli populations. Proceedings of the Third Purdue Symposium on Statistical Decision Theory and Related Topics (ed. by S. S. Gupta and J. Berger), New York, Academic Press, I, 61-108.

Bofinger, E. (1978). Selection using "play-the-winner" sampling for non-negative observations. Comm. Statist. A - Theory Methods A7(15), 1489-1499.

Jennison, C. (1983a). Equal probability of correct selection for Bernoulli selection procedures. Comm. Statist. - Theory Methods 12(24), 2887-2896.

Jennison, C. (1984). On the expected sample size for the Bechhofer-Kulkarni Bernoulli selection procedure. Accepted for publication in Comm. Statist. C - Sequential Anal.

Kulkarni, R. V. and Jennison, C. (1983). Optimal properties of the Bechhofer-Kulkarni Bernoulli selection procedure. Technical Report No. 600, School of Operations Research and Industrial Engineering, Cornell University.

Sobel, M. and Huyett, M. J. (1957). Selecting the best one of several binomial populations. Bell System Tech. J. 36, 537-576.

ASYMPTOTICALLY OPTIMAL ALLOCATION OF TREATMENTS IN SEQUENTIAL EXPERIMENTS*

T.L. Lai

Department of Mathematical Statistics
Columbia University
New York, New York

Herbert Robbins

Department of Mathematical Statistics
Columbia University
New York, New York

1. INTRODUCTION

Let $\Pi_1,\ldots,\Pi_k$ be statistical populations (treatments, manufacturing processes, etc.) specified respectively by univariate density functions $f(x;\theta_1),\ldots,f(x;\theta_k)$ with respect to some measure ν, where $f(\cdot;\cdot)$ is known and the θ_j are unknown parameters belonging to some set Θ. Assume that $\int_{-\infty}^{\infty} |x|\ f(x;\theta)d\nu(x) < \infty$ for all $\theta \in \Theta$. How should we sample $x_1,x_2,\ldots$ sequentially from the k populations if our objective is to achieve the greatest possible expected value of the sum $S_n = x_1 + \ldots + x_n$ in the long run (as $n \to \infty$)? Starting with Robbins (1952), there has been a considerable literature on the subject, which is often called the "multi-armed bandit problem".

Let $\mu(\theta) = \int_{-\infty}^{\infty} xf(x;\theta)d\nu(x)$, and define

$$\mu^* = \max\{\mu(\theta_1),\ldots,\mu(\theta_k)\} = \mu(\theta_\ell) \text{ for some } \ell \in \{1,\ldots,k\}. \tag{1}$$

An <u>adaptive allocation rule</u> ϕ consists of a sequence of random variables $\phi_1,\phi_2,\ldots$ taking values in the set $\{1,\ldots,k\}$ and such that the event $\{\phi_n=j\}$ ("sample from Π_j at stage n") belongs to the σ-field F_{n-1} generated by $\phi_1,x_1,\ldots,\phi_{n-1},x_{n-1}$. For $n \leq \infty$, let

* Research supported by the National Science Foundation and the National Institutes of Health.

$$T_n(j) = \sum_{i=1}^{n} I_{\{\phi_i = j\}} \tag{2}$$

denote the number of times that the rule ϕ samples from Π_j through stage n. Then

$$ES_n = \sum_{j=1}^{k} \mu(\theta_j)\, ET_n(j) \tag{3}$$

(cf. Lai and Robbins, 1984b), and maximizing ES_n is equivalent to minimizing the "regret"

$$R_n(\theta_1,\ldots,\theta_k) = n\mu^* - ES_n = \sum_{j:\mu(\theta_j)<\mu^*} (\mu^*-\mu(\theta_j))ET_n(j) . \tag{4}$$

While a rule ϕ that minimizes the regret uniformly for all parameter configurations $\underset{\sim}{\theta} = (\theta_1,\ldots,\theta_k)$ does not exist except in trivial cases, we showed (1984b) that under certain assumptions there are rules that are asymptotically optimal, in the sense of attaining the asymptotic lower bound for the regret given by

<u>Theorem 1</u>. Define the Kullback-Leibler number

$$I(\theta,\lambda) = \int_{-\infty}^{\infty} [\log(f(x;\theta)/f(x;\lambda))]\, f(x;\theta)d\nu(x). \tag{5}$$

Assume that

$$0 < I(\theta,\lambda) < \infty \quad \text{whenever} \quad \mu(\lambda) > \mu(\theta), \tag{6}$$

and that for every $\varepsilon > 0$ and $\theta,\lambda \in \Theta$ such that $\mu(\lambda) > \mu(\theta)$, there exists $\delta > 0$ such that

$$|I(\theta,\lambda) - I(\theta,\lambda')| < \varepsilon \text{ if } \mu(\lambda) \le \mu(\lambda') \le \mu(\lambda) + \delta. \tag{7}$$

Moreover, assume that

$$\forall\lambda \in \Theta \text{ and } \forall\delta > 0,\ \exists\lambda' \in \Theta \text{ such that } \mu(\lambda) < \mu(\lambda') < \mu(\lambda) + \delta. \tag{8}$$

Let ϕ be any allocation rule such that for every fixed $\underset{\sim}{\theta} = (\theta_1,\ldots,\theta_k)$

$$R_n(\underset{\sim}{\theta}) = o(n^r) \quad \text{for every} \quad r > 0. \tag{9}$$

Then for every $\underset{\sim}{\theta}$ such that the $\mu(\theta_j)$ are not all equal

$$\liminf_{n\to\infty} R_n(\underset{\sim}{\theta})/\log n \geq \sum_{j:\mu(\theta_j)<\mu^*} (\mu^* - \mu(\theta_j))/I(\theta_j,\theta_\ell), \tag{10}$$

where μ^* and θ_ℓ are defined by (1).

For example, suppose that the k populations have normal distributions with unknown means $\theta_1,\ldots,\theta_k$ and variance 1. Then $\mu(\theta) = \theta$, $\Theta = (-\infty,\infty)$, $I(\theta,\lambda) = \frac{1}{2}(\theta-\lambda)^2$, and conditions (6), (7), (8) are satisfied. For normal populations, as well as for several other families of distributions, we showed (1984b) how to construct a rule ϕ such that

$$R_n(\underset{\sim}{\theta}) \sim \{\sum_{j:\mu(\theta_j)<\mu^*} (\mu^* - \mu(\theta_j))/I(\theta_j,\theta_\ell)\} \log n \tag{11}$$

for every fixed $\underset{\sim}{\theta} = (\theta_1,\ldots,\theta_k)$ such that the $\mu(\theta_j)$ are not all equal. Under the assumptions of Theorem 1, allocation rules that satisfy (11) whenever the $\mu(\theta_j)$ are not all equal were called _asymptotically efficient_ in Lai and Robbins (1984b).

In this paper we consider two different kinds of _partial_ knowledge about the k populations and show that one can use such knowledge to reduce the order of magnitude of the regret given by (11). One kind of partial knowledge about the k populations is considered in Section 2, where we show in particular that it is possible to construct a rule whose regret is _bounded_ in n if there is prior knowledge of a number γ that lies between μ^* and $\max\{\mu(\theta_j) : \mu(\theta_j) < \mu^*\}$. This generalizes a result of Feldman (1962) and of Lai and Robbins (1984a) on bounded regret in the two-armed bandit problem in which we have the prior information that $(\theta_1,\theta_2) \in \{(a,b),(b,a)\}$, where a and b are two given parameter values.

In Section 3 we consider another kind of partial knowledge about the k populations where, contrary to (8), $\{\mu(\theta): \theta \in \Theta\}$ is a countable set of real numbers. For example, suppose it is known that the mean of each population is a positive integer (cf. Hammersley, 1950; Robbins, 1970). This consideration leads us to provide a generalization of Theorem 1 applicable to such cases. By extending the methods of Lai and Robbins (1984b), we also obtain for these situations asymptotically optimal allocation rules that have logarithmic regret at certain parameter configurations and bounded regret at other parameter configurations.

2. BOUNDED REGRET WHEN A SEPARATING VALUE BETWEEN THE LARGEST AND THE OTHER MEANS IS KNOWN

Suppose that we know a number γ such that

$$\mu^* > \gamma > \max\{\mu(\theta_j) : \mu(\theta_j) < \mu^*\}. \tag{12}$$

For $j = 1,\dots,k$, sample Π_j at stage j. After n ($\geq k$) observations, let $j_n \in \{1,\dots,k\}$ be such that

$$\sum_{i=1}^{T_n(j_n)} (Y_{j_n,i}-\gamma) = \max_{1\leq j\leq k} \sum_{i=1}^{T_n(j)} (Y_{ji}-\gamma), \tag{13}$$

where $Y_{j1}, Y_{j2}, \dots$ denote successive observations from Π_j, and at stage $n+1$ samples from Π_{j_n}. This allocation rule will be denoted by $\phi(\gamma)$, and the following theorem shows that it has bounded regret.

Theorem 2. Assume that the k populations have finite second moments and that γ satisfies (12). Then for the rule $\phi(\gamma)$, $ET_\infty(j) < \infty$ for every j such that $\mu(\theta_j) < \mu^*$.

Proof. Fix any j such that $\mu(\theta_j) < \mu^*$ and define

$$\tau = \sup\{n \geq 1: \sum_{i=1}^{n} (Y_{ji}-\gamma) + \sup_{m\geq 1} \sum_{i=1}^{m} (\gamma-Y_{\ell i}) \geq 0\} \quad (\sup \emptyset = 0).$$

Then $T_\infty(j) \leq \tau+1$, and therefore it suffices to show that $E\tau < \infty$. Since $Y_{\ell 1}, Y_{\ell 2}, \dots$ are i.i.d. with $EY_{\ell 1}^2 < \infty$ and $E(\gamma-Y_{\ell 1}) = \gamma-\mu^* < 0$, we have $E[\sup_{m\geq 1}(\sum_{i=1}^{m}(\gamma-Y_{\ell i}))^+] < \infty$ (cf. Chow and Lai, 1975). Let $0 < \varepsilon < \gamma-\mu(\theta_j)$, and define $T_\varepsilon = \sup\{n \geq 1: \sum_{i=1}^{n}(Y_{ji}-\gamma) > -\varepsilon n\}$. Since $Y_{j1}, Y_{j2}, \dots$ are i.i.d. with $EY_{j1}^2 < \infty$ and $E(Y_{j1}-\gamma) = \mu(\theta_j)-\gamma < -\varepsilon$, $ET_\varepsilon < \infty$ (cf. Chow and Lai, 1975). On the event $\{\tau > T_\varepsilon\}$ we have $\varepsilon\tau \leq \sup_{m\geq 1}(\sum_{i=1}^{m}(\gamma-Y_{\ell i}))^+$. The proof is completed by the inequality

$$E\tau \leq ET_\varepsilon + \varepsilon^{-1} E[\sup_{m\geq 1}(\sum_{i=1}^{m}(\gamma-Y_{\ell i}))^+] < \infty. \qquad \square$$

We now extend the preceding argument to the cases where (i) μ^* is known, and (ii) $\max\{\mu(\theta_j): \mu(\theta_j) < \mu^*\}$ is known. First suppose that we know μ^*. For $j = 1,\dots,k$, sample from Π_j at stage j. After n ($\geq k$) observations, let Π_{j_n} denote a population that has the largest value of

$$\sum_{i=1}^{T_n(j)} (Y_{ji}-\mu^*) + \delta T_n^\alpha(j), \tag{14}$$

where $\delta > 0$, $\frac{1}{2} < \alpha < 1$, and sample from Π_{j_n} at stage $n+1$. This allocation rule will be denoted by $\bar{\phi}$, and the following theorem shows that it has bounded regret.

Theorem 3. Assume that the k populations have finite second moments and that $E(Y_{\ell 1}^{-})^{(\alpha+1)/\alpha} < \infty$. Then for the rule $\bar{\phi}$, $ET_\infty(j) < \infty$ for every j such that $\mu(\theta_j) < \mu^*$.

Proof. Fix any j such that $\mu(\theta_j) < \mu^*$ and define

$$\tilde{\tau} = \sup\{n \geq 1: \sum_{i=1}^{n} (Y_{ji} - \mu^*) + \delta n^\alpha + \sup_{m \geq 1} [\sum_{i=1}^{m} (\mu^* - Y_{\ell i}) - \delta m^\alpha] \geq 0\}.$$

Then $T_\infty(j) \leq \tilde{\tau}+1$. Since $EY_{\ell 1}^2 < \infty$, $E(\mu^* - Y_{\ell 1}) = 0$, and $E(Y_{\ell 1}^{-})^{(\alpha+1)/\alpha} < \infty$ with $\alpha > 1/2$, we have (cf. Chow and Lai, 1975)

$$E\{\sup_{m \geq 1} [\sum_{i=1}^{m} (\mu^* - Y_{\ell i}) - \delta m^\alpha]^+\} < \infty.$$

An argument similar to that in the proof of Theorem 2 then completes the proof. □

Now consider the case where we know $\lambda = \max\{\mu(\theta_j): \mu(\theta_j) < \mu^*\}$. For $j = 1, \ldots, k$, sample from Π_j at stage j. After n ($\geq k$) observations, let Π_{j_n} denote a population that has the largest value of

$$\sum_{i=1}^{T_n(j)} (Y_{ji} - \lambda) - \delta T_n^\alpha(j), \tag{15}$$

where $\delta > 0$, $1/2 < \alpha < 1$, and sample from Π_{j_n} at stage n+1. This allocation rule will be denoted by $\underline{\phi}$, and the following theorem shows that it has bounded regret.

Theorem 4. Assume that the k populations have finite second moments and that $E(Y_{\ell 1}^{-})^{2/\alpha} < \infty$ and $E(Y_{j1}^{+})^{2/\alpha} < \infty$ for all j such that $\mu(\theta_j) = \lambda$. Then for the rule $\underline{\phi}$, $ET_\infty(j) < \infty$ for every j such that $\mu(\theta_j) < \mu^*$.

Proof. First consider the case where $\mu(\theta_j) = \lambda$. We note that

$$P\{\sum_{i=1}^{n} (Y_{ji} - \lambda) - \delta n^\alpha + \max_{m \leq n/4} [\sum_{i=1}^{m} (\lambda - Y_{\ell i}) + \delta m^\alpha] \geq 0\}$$

$$\leq P\{\sum_{i=1}^{n} (Y_{ji} - \lambda) \geq \frac{1}{4} \delta n^\alpha\} + P\{\max_{m \leq n/4} \sum_{i=1}^{m} (\mu^* - Y_{\ell i}) \geq \frac{1}{4} \delta n^\alpha\}. \tag{16}$$

Moreover, letting $\lambda < \gamma < \mu^*$, we have for all large n

$$P\{\sum_{i=1}^{n} (Y_{ji} - \lambda) - \delta n^\alpha + \sup_{m \geq n/4} [\sum_{i=1}^{m} (\lambda - Y_{\ell i}) + \delta m^\alpha] \geq 0\}$$

$$\leq P\{\sum_{i=1}^{n} (Y_{ji} - \lambda) \geq \delta n^\alpha\} + P\{\gamma \geq \inf_{m \geq n/4} m^{-1} \sum_{i=1}^{m} Y_{\ell i}\}. \tag{17}$$

Since $EY_{j1} = \lambda$, $EY_{\ell 1} = \mu^*$, $EY_{j1}^2 < \infty$, $EY_{\ell 1}^2 < \infty$, $E(Y_{j1}^+)^{2/\alpha} < \infty$, and $E(Y_{\ell 1}^-)^{2/\alpha} < \infty$, it follows that (cf. Chow and Lai, 1975)

$$\sum_{n=1}^{\infty} P\{\sum_{i=1}^{n}(Y_{ji}-\lambda) \geq \frac{1}{4}\delta n^{\alpha}\} < \infty, \qquad \sum_{n=1}^{\infty} P\{\gamma \geq \inf_{m \geq n/4} m^{-1} \sum_{i=1}^{m} Y_{\ell i}\} < \infty,$$

$$\sum_{n=1}^{\infty} P\{\max_{m \leq n/4} \sum_{i=1}^{m}(\mu^*-Y_{\ell i}) \geq \frac{1}{4}\delta n^{\alpha}\} < \infty. \tag{18}$$

By (16), (17), and (18),

$$\sum_{n=1}^{\infty} P\{\sum_{i=1}^{n}(Y_{ji}-\lambda) - \delta n^{\alpha} \geq \sum_{i=1}^{m}(Y_{\ell i}-\lambda) - \delta m^{\alpha} \text{ for some } m \geq 1\} < \infty,$$

and therefore $ET_{\infty}(j) < \infty$.

For the case $\mu(\theta_j) < \lambda$, we can use an argument similar to the proof of Theorem 2 to show that $ET_{\infty}(j) < \infty$. □

To see the basic idea behind the allocation rule $\phi(\gamma)$ of Theorem 2, consider the example where the distributions of the k populations belong to the exponential family with densities

$$f(x;\theta) = e^{\theta x - \psi(\theta)} \tag{19}$$

with respect to some nondegenerate measure ν. Here $\mu(\theta) = \psi'(\theta)$ is increasing in θ, and

$$I(\theta,\lambda) = (\theta-\lambda)\psi'(\theta) - (\psi(\theta) - \psi(\lambda)). \tag{20}$$

First consider the case $k = 2$ and assume that $(\theta_1,\theta_2) \in \{(a,b), (b,a)\}$, where $a > b$. As shown by Feldman (1962), the Bayes procedure with equal prior probabilities for (a,b) and (b,a) samples from Π_1 or Π_2 with probability 1/2 at the first stage, and samples at stage n+1 from Π_1 or Π_2 according as $\Lambda_n > 1$ or $\Lambda_n < 1$, with indifference between Π_1 and Π_2 when $\Lambda_n = 1$, where

$$\Lambda_n = \{\prod_{i=1}^{T_n(1)} f(Y_{1i};a)\cdot \prod_{i=1}^{T_n(2)} f(Y_{2i};b)\}/\{\prod_{i=1}^{T_n(1)} f(Y_{1i};b)\cdot \prod_{i=1}^{T_n(2)} f(Y_{2i};a)\}$$

$$= \exp\{(a-b)(\sum_{i=1}^{T_n(1)} Y_{1i} - \sum_{i=1}^{T_n(2)} Y_{2i}) - (\psi(a) - \psi(b))(T_n(1) - T_n(2))\}.$$

Thus, at stage n+1, the rule chooses the population that has the larger value of $\sum_{i=1}^{T_n(1)}(Y_{1i}-\gamma)$ and $\sum_{i=1}^{T_n(2)}(Y_{2i}-\gamma)$, where

$$\gamma = (\psi(a) - \psi(b))/(a-b). \tag{21}$$

Since ψ is convex,

$$\mu(a) = \psi'(a) > \gamma > \psi'(b) = \mu(b).$$

The Bayes rule is therefore the "sample-from-the-leader" rule $\phi(\gamma)$, where "leader" is defined by the criterion (13) with γ given by (21).

Now consider the case of k populations in which it is known that one population has the density function (19) with parameter $\theta = a$ and the other (k-1) populations have the same density function (19) but with $\theta = b$. This corresponds to the so-called "slippage configuration" in the selection and ranking literature (cf. Bechhofer et al., 1968). In analogy with the case k = 2 discussed above, we define the "leader" at the conclusion of stage n to be the population that attains

$$\max_{1 \le j \le k} \{ \prod_{i=1}^{T_n(j)} f(Y_{ji};a) \cdot \prod_{\nu \ne j} \prod_{i=1}^{T_n(\nu)} f(Y_{\nu i};b)\},$$

or equivalently, the population that has the largest $\sum_{i=1}^{T_n(j)} (Y_{ji}-\gamma)$, where γ is defined in (21).

More generally, suppose it is known that the k populations have density functions $f(x;\theta_1),\ldots,f(x;\theta_k)$, where f is defined in (19), and that $\theta_j \in \Theta = \{a,b\}$ for $j = 1,\ldots,k$. We again define the "leader" at the conclusion of stage n to be the population that has the largest $\sum_{i=1}^{T_n(j)} (Y_{ji}-\gamma)$, where γ is defined in (21). By Theorem 2, the "sample-from-the-leader" rule $\phi(\gamma)$ has bounded regret:

$$\sup_n R_n(\underset{\sim}{\theta}) < \infty \text{ for all } \underset{\sim}{\theta} = (\theta_1,\ldots,\theta_k) \in \Theta^k, \tag{22}$$

with regret 0 is all the θ_j are equal.

3. A GENERALIZATION OF THEOREM 1 AND ASYMPTOTICALLY OPTIMAL ALLOCATION RULES FOR GENERAL PARAMETER SPACES

We begin by considering the following generalization of Theorem 1.

Theorem 5. Assume that $0 < I(\theta,\lambda) < \infty$ whenever $\mu(\lambda) > \mu(\theta)$, where $I(\theta,\lambda)$ is defined in (5). Let ϕ be an allocation rule whose regret satisfies

(9) for every fixed $\underset{\sim}{\theta} = (\theta_1,\ldots,\theta_k)$. Then for all $\theta_1,\ldots,\theta_k$ and $\lambda \in \Theta$ such that $\mu(\lambda) > \mu^* = \max_j \mu(\theta_j)$,

$$\liminf_{n\to\infty} R_n(\underset{\sim}{\theta})/\log n \geq \sum_{j:\mu(\theta_j)<\mu^*} (\mu^*-\mu(\theta_j))/I(\theta_j,\lambda). \tag{23}$$

<u>Proof</u>. Fix any j such that $\mu(\theta_j) < \mu^*$. It suffices to show that

$$\liminf_{n\to\infty} E_{\underset{\sim}{\theta}} T_n(j)/\log n \geq 1/I(\theta_j,\lambda). \tag{24}$$

Define $\underset{\sim}{\beta} = (\beta_1,\ldots,\beta_k)$ by putting

$$\beta_j = \lambda \text{ and } \beta_i = \theta_i \text{ for } i \neq j.$$

Then since $R_n(\underset{\sim}{\beta}) = o(n^r)$, and since $\mu(\beta_j) = \mu(\lambda) > \mu^* = \max_{i\neq j} \mu(\beta_i)$,

$$\sum_{i\neq j} E_{\underset{\sim}{\beta}} T_n(i) = o(n^r) \quad \text{for every} \quad r > 0.$$

An argument similar to the proof of Theorem 2 of Lai and Robbins (1984b) can then be applied to establish (24). □

Suppose that the distributions of the populations Π_j $(j = 1,\ldots,k)$ belong to the exponential family (19) with parameters $\theta_j \in \Theta$, where Θ is a given subset of the natural parameter space. We note that

$$\mu(\theta) \text{ is an increasing function of } \theta, \tag{25}$$

and that for every fixed θ

$$I(\theta,\lambda) \text{ is an increasing function of } \lambda > \theta. \tag{26}$$

Given $\underset{\sim}{\theta} = (\theta_1,\ldots,\theta_k) \in \Theta^k$, define

$$\theta^* = \max_j \theta_j, \; \overline{\theta} = \inf\{\lambda \in \Theta: \lambda > \theta^*\} \quad (\inf \emptyset = \infty). \tag{27}$$

It will be shown that allocation rules can be constructed such that for every fixed $\underset{\sim}{\theta}$ whose components θ_j are not all equal

$$R_n(\underset{\sim}{\theta}) \sim \{\sum_{j:\theta_j<\theta^*} (\mu^*-\mu(\theta_j))/I(\theta_j,\overline{\theta})\} \log n \quad \text{if } \overline{\theta} < \infty, \tag{28a}$$

and

$$\sup_n R_n(\underset{\sim}{\theta}) < \infty \quad \text{if } \overline{\theta} = \infty. \tag{28b}$$

In view of Theorem 5, such allocation rules are asymptotically optimal in the sense of attaining the asymptotic lower bound (23) for the regret.

For a simple example, consider the case where Θ has three elements $a > b > c$. By Theorem 5, if ϕ is an allocation rule such that for every $\underset{\sim}{\theta} = (\theta_1, \ldots, \theta_k) \in \Theta^k$

$$R_n(\underset{\sim}{\theta}) = o(n^r) \quad \text{for every } r > 0, \tag{29}$$

then for every $\underset{\sim}{\theta} \in \Theta^k$ such that $\max_j \theta_j = b$,

$$\liminf_{n \to \infty} R_n(\underset{\sim}{\theta})/\log n \geq \#_c[\mu(b)-\mu(c)]/I(c,a), \tag{30}$$

where $\#_c$ denotes the number of θ_j equal to c. Moreover, as will be shown below, an allocation rule ϕ^* can be constructed such that

$$\begin{aligned} &\sup_n R_n(\underset{\sim}{\theta}) < \infty \text{ if } \max_j \theta_j = a, \text{ and} \\ &R_n(\underset{\sim}{\theta}) \sim \#_c[\mu(b)-\mu(c)](\log n)/I(c,a) \text{ if } \max_j \theta_j = b. \end{aligned} \tag{31}$$

If Θ satisfies the continuum-type assumption (8) of Theorem 1, then $\overline{\theta} = \theta^*$, and we showed (1984b) that certain "confidence sequences" for θ_j can be used to construct asymptotically optimal allocation rules satisfying

$$R_n(\underset{\sim}{\theta}) \sim \{\sum_{j:\theta_j<\theta^*} (\mu^*-\mu(\theta_j))/I(\theta_j,\theta^*)\} \log n, \tag{32}$$

which is a special case of (28a) since $\overline{\theta} = \theta^*$. If (8) fails to hold, as is the case when Θ is discrete, then some modifications of the approach in Lai and Robbins (1984b) are needed to construct allocations rules that satisfy (28a) and (28b).

The first modification involves the confidence sequences for θ_j. To fix the ideas, let $X_1, X_2, \ldots$ be i.i.d random variables with a common density function $f(x;\theta)$ given by (19). It will be shown in Lemma 1 below how Borel functions $g_{ni}: R^i \to R \cup \{\infty\}$ $(n = 1,2,\ldots;\ i = 1,\ldots,n)$ can be constructed such that for every θ and every $0 < \gamma < 1$

$$\begin{aligned} &P_\theta\{\theta < g_{ni}(X_1,\ldots,X_i) \text{ for all } 1 \leq i \leq n \text{ and all } \gamma m \leq n \leq m\} \\ &= 1 - O(m^{-1}(\log m)^{-p}) \text{ for some } p > 1, \end{aligned} \tag{33}$$

$$\sum_{i=1}^{n} P_\theta\{g_{ni}(X_1,\ldots,X_i) \geq \lambda\} \sim (\log n)/I(\theta,\lambda) \text{ whenever } \lambda > \theta, \tag{34}$$

and

$$g_{ni} \text{ is nondecreasing in } n\ (\geq i) \text{ for every fixed } i = 1,2,\ldots \tag{35}$$

In Lai and Robbins (1984b), where allocation rules satisfying (32) were constructed, we used confidence sequences that satisfy instead of (33) the weaker condition

$$P_\theta\{r \le g_{ni}(X_1,\ldots,X_i) \text{ for all } i \le n\} = 1 - o(n^{-1}) \text{ for every } r < 0. \tag{36}$$

However, to construct allocation rules satisfying (28a), which is not the same as (32) when Θ is discrete, we need the stronger condition (33).

Lemma 1. Let $X_1, X_2, \ldots$ be i.i.d. with a common density function $f(x;\theta)$ given by (19). Let $S_n = X_1+\ldots+X_n$, and let $h(\cdot)$ be a positive continuous function on $(-\infty,0)$ such that $\int_{-\infty}^0 h(t)dt = 1$. Assume that the natural parameter space is not bounded below, and define for $a \ge 1$

$$U(i,S_i;a) = \inf\{\theta: \int_{-\infty}^0 \exp\{tS_i - i(\psi(\theta+t)-\psi(\theta))\}h(t)\,dt \ge a\} \tag{37}$$

$(\inf \emptyset = \infty)$. Then for all θ and all $\lambda > \theta$,

(i) $P_\theta\{\theta < U(i,S_i;a) \text{ for all } i \ge 1\} \ge 1-a^{-1}$,

(ii) $\sum_{1\le i\le a} P_\theta\{U(i,S_i;a) \ge \lambda\} \sim (\log a)/I(\theta,\lambda)$ as $a \to \infty$.

Proof. Let $Z_i(\theta,t) = \exp\{tS_i - i(\psi(\theta+t)-\psi(\theta))\}$. For fixed $t < 0$, $Z_i(\theta,t)$ is a continuous increasing function of θ, since ψ is convex. Hence we have the equivalence

$$\theta \ge U(i,S_i;a) \iff \int_{-\infty}^0 Z_i(\theta,t)\,h(t)dt \ge a$$

$$\iff \int_{-\infty}^0 \{\prod_{r=1}^i f(X_r;\theta+t)/\prod_{r=1}^i f(X_r;\theta)\}\,h(t)dt \ge a.$$

Since $\{\int_{-\infty}^0 \{\prod_{r=1}^i f(X_r;\theta+t)/\prod_{r=1}^i f(X_r;\theta)\}\,h(t)dt,\ i \ge 1\}$ is a nonnegative martingale with mean 1 under P_θ,

$$P_\theta\{\theta \ge U(i,S_i;a) \text{ for some } i \ge 1\}$$

$$\le P_\theta\{\int_{-\infty}^0 \{\prod_{r=1}^i f(X_r;\theta+t)/\prod_{r=1}^i f(X_r;\theta)\}\,h(t)dt \ge a \text{ for some } i \ge 1\}$$

$$\le a^{-1}$$

(cf. Pollak and Siegmund, 1975), and (i) follows.

To prove (ii), let $\lambda > \theta$ and define

$$N_a = \sum_{i=1}^{\infty} I_{\{U(i,S_i;a)\geq\lambda\}}, \qquad L_a = \sup\{i: \int_{-\infty}^{0} Z_i(\lambda,t)\,h(t)dt \leq a\},$$
$$T_a = \inf\{i: \int_{-\infty}^{0} Z_i(\lambda,t)\,h(t)dt \geq a\}.$$

We note that

$$T_a - 1 \leq N_a \leq L_a. \tag{38}$$

Since $h(\cdot)$ is positive and continuous on $(-\infty,0)$ and since $\theta = \lambda+t$ for some $t < 0$, it follows from Theorem 1 of Pollak and Siegmund (1975) that $E_\theta T_a \sim (\log a)/I(\theta,\lambda)$ as $a \to \infty$. Moreover, an argument similar to the proof of Theorem 1 of Pollak and Siegmund (1975) also shows that $E_\theta L_a \sim (\log a)/I(\theta,\lambda)$ and that

$$\sum_{i\geq a} P_\theta\{U(i,S_i;a) \geq \lambda\} = o(1) \text{ as } a \to \infty. \tag{39}$$

Hence it follows from (38) that

$$E_\theta N_a \sim (\log a)/I(\theta,\lambda) \text{ as } a \to \infty. \tag{40}$$

From (39) and (40), (ii) follows.

Let b_n be a nondecreasing sequence of positive numbers such that

$$b_n \sim (\log n)^p \text{ as } n \to \infty \text{ for some } p > 1. \tag{41}$$

Define

$$g_{ni}(X_1,\ldots,X_i) = U(i,S_i;nb_n), \tag{42}$$

where $U(i,S_i;a)$ is given by (37). Clearly (35) holds, and by Lemma 1(ii), condition (34) is also satisfied. Moreover, by Lemma 1(i),

$$P_\theta\{\theta < g_{ni}(X_1,\ldots,X_i) \text{ for all } i \leq n \text{ and all } \gamma m \leq n \leq m\}$$
$$\geq P_\theta\{\theta < U(i,S_i;[\gamma m]b_{[\gamma m]}) \text{ for all } i \geq 1\} \geq 1 - ([\gamma m]b_{[\gamma m]})^{-1},$$

and therefore (33) holds.

The maximum likelihood estimate of θ based on $X_1,\ldots,X_i$ can be written as

$$h_i(X_1,\ldots,X_i) = \mu^{-1}((X_1+\ldots+X_i)/i),$$

where $\mu(\theta) = \psi'(\theta)$ is increasing in θ. A standard argument shows that h_i satisfies the following condition: for all θ, $0 < \gamma < 1$, $0 < \beta < 1/2$, and $\delta > 0$,

$$P_\theta\{\max_{\gamma n \le i \le n} |h_i(X_1,\ldots,X_i) - \theta| > \delta n^{-\beta}\}$$
$$= o(n^{-1}(\log n)^{-q}) \text{ with } q > 1. \tag{43}$$

This property of the estimator of θ will be useful in proving the asymptotic optimality of the allocation rule described below.

In general, without assuming the exponential family of distributions, suppose that there exist Borel functions $h_i: R^i \to R$ and $g_{ni}: R^i \to R \cup \{\infty\}$ $(n = 1,2,\ldots;\ i = 1,\ldots,n)$ such that for every $\theta \in \Theta$, conditions (33), (34), (35), and (43) are satisfied. We now modify the procedure proposed in Lai and Robbins (1984b) to obtain an allocation rule satisfying (28a) and (28b). For $j = 1,\ldots,k$, let $T_n(j)$ denote the number of times that the rule samples from Π_j through stage n, and let $Y_{j1},\ldots,Y_{j,T_n(j)}$ denote the successive observations from Π_j through stage n. Define

$$\hat\theta_n(j) = h_{T_n(j)}(Y_{j1},\ldots,Y_{j,T_n(j)}),$$
$$U_n(j) = g_{n,T_n(j)}(Y_{j1},\ldots,Y_{j,T_n(j)}). \tag{44}$$

Let $\delta > 0$, $0 < \beta < 1/2$, and $0 < \rho < 1/k$. At stage $j \in \{1,\ldots,k\}$, the rule takes one observation from Π_j. Suppose that the rule has taken $n \ge k$ observations. Since $T_n(1) + \ldots + T_n(k) = n$, we can choose $j_n \in \{1,\ldots,k\}$ such that

$$\hat\theta_n(j_n) = \max\{\hat\theta_n(j): T_n(j) \ge \rho n\}. \tag{45}$$

Define

$$\overline{\theta}_n = \inf\{\theta \in \Theta: \theta > \hat\theta_n(j_n) + \delta n^{-\beta}\} \quad (\inf \emptyset = \infty). \tag{46}$$

At stage $n+1 = km+j$, where m is a positive integer and $j \in \{1,\ldots,k\}$, we sample from Π_j if

$$\overline{\theta}_n < \infty \text{ and } \overline{\theta}_n \le U_n(j), \tag{47}$$

and from Π_{j_n} otherwise. This allocation rule will be denoted by ϕ^*.

The allocation rule ϕ^* is similar to that proposed in Lai and Robbins (1984b) where $\hat{\theta}_n(j_n)$ was used instead of $\overline{\theta}_n$ in (47). While $\hat{\theta}_n(j_n)$ can be regarded as an estimate of θ^*, we can regard $\overline{\theta}_n$ as an estimate of $\overline{\theta}$. The following heuristic argument explains why we use $\overline{\theta}_n$ here instead of $\hat{\theta}_n(j_n)$. At the conclusion of stage n, Π_{j_n} is the apparently superior population. Suppose that its parameter θ_{j_n} were known. For an apparently inferior population Π_j, we can be reasonably confident that it is indeed not better than Π_{j_n} if the upper confidence bound for θ_j falls below $\inf\{\theta \in \Theta: \theta > \theta_{j_n}\}$. Since ϕ has sampled from Π_{j_n} in at least ρn of the n trials, $\hat{\theta}_n(j_n)$ provides a relatively good estimate of θ_{j_n}. Therefore in ignorance of θ_{j_n}, we replace $\inf\{\theta \in \Theta : \theta > \theta_{j_n}\}$ by $\overline{\theta}_n = \inf\{\theta \in \Theta: \theta > \hat{\theta}_n(j_n) + \delta n^{-\beta}\}$ and do not sample from Π_j at stage n+1 if $U_n(j) < \overline{\theta}_n$.

The asymptotic optimality of the rule ϕ^* is established in

<u>Theorem 6</u>. Assume that $0 < I(\theta,\lambda) < \infty$ whenever $\lambda > \theta$, and that $I(\theta,\lambda)$ satisfies (26). Moreover, assume that (25) holds for $\mu(\theta)$. Suppose that the functions g_{ni} and h_i satisfy conditions (33), (34), (35), and (43) for every $\theta \in \Theta$. Then the regret of ϕ^* satisfies (28a) and (28b) for every $\underset{\sim}{\theta}$ whose components θ_j are not all equal.

<u>Proof</u>. It suffices to show that for every fixed j such that $\theta_j < \theta^*$,

$$E_{\underset{\sim}{\theta}}T_n(j) \leq \left(\frac{1}{I(\theta_j,\overline{\theta})} + o(1)\right) \log n \quad \text{if } \overline{\theta} < \infty, \tag{48}$$

and

$$\sup_n E_{\underset{\sim}{\theta}}T_n(j) < \infty \quad \text{if } \overline{\theta} = \infty. \tag{49}$$

This implies that $\sup_n R_n(\underset{\sim}{\theta}) < \infty$ if $\overline{\theta} = \infty$, and in the case $\overline{\theta} < \infty$ that

$$\limsup_{n \to \infty} R_n(\underset{\sim}{\theta})/\log n \leq \sum_{j:\theta_j<\theta^*} (\mu^*-\mu(\theta_j))/I(\theta_j,\overline{\theta}).$$

On the other hand, in view of (25) and (26), we can apply Theorem 5 in the case $\overline{\theta} < \infty$ to obtain that

$$\liminf_{n \to \infty} R_n(\underset{\sim}{\theta})/\log n \geq \sum_{j:\theta_j<\theta^*} (\mu^*-\mu(\theta_j))/I(\theta_j,\overline{\theta}),$$

establishing (28a).

To prove (48) and (49), let $L = \{1 \leq \ell \leq k: \theta_\ell = \theta^*\}$. Using the notation #A to denote the number of elements of a set A, we note that for any fixed j such that $\theta_j < \theta^*$

$$T_n(j) \le \#\ \{1 \le n \le N-1: j_n \in L,\ |\hat{\theta}_n(j_n) - \theta^*| \le \delta n^{-\beta}, \text{ and}$$

$$\phi^* \text{ samples from } \Pi_j \text{ at stage } n+1\} + 1$$

$$+ \#\ \{1 \le n \le N-1: j_n \in L \text{ and } |\hat{\theta}_n(j_n) - \theta^*| > \delta n^{-\beta}\}$$

$$+ \#\ \{1 \le n \le N-1: j_n \notin L\}. \tag{50}$$

From the definitions of $\overline{\theta}_n$ and $\overline{\theta}$, we have the implications

$$|\hat{\theta}_n(j_n) - \theta^*| \le \delta n^{-\beta} \Rightarrow \hat{\theta}_n(j_n) + \delta n^{-\beta} \ge \theta^*$$

$$\Rightarrow \overline{\theta}_n \ge \overline{\theta} \text{ if } \overline{\theta} < \infty \text{ and } \overline{\theta}_n = \infty \text{ if } \overline{\theta} = \infty. \tag{51}$$

Hence if $\overline{\theta} = \infty$, then

$$\#\ \{1 \le n \le N-1: j_n \in L,\ |\hat{\theta}_n(j_n) - \theta^*| \le \delta n^{-\beta}, \text{ and}$$

$$\phi^* \text{ samples from } \Pi_j \text{ at stage } n+1\} = 0.$$

On the other hand, if $\overline{\theta} < \infty$, then from (51) and the definition of ϕ^*, it can be shown by an argument similar to that of the proof of Theorem 3 of Lai and Robbins (1984b) that

$$E_{\underset{\sim}{\theta}}(\#\{2 \le m \le N: j_{m-1} \in L,\ |\hat{\theta}_{m-1}(j_{m-1}) - \theta^*| \le \delta(m-1)^{-\beta}, \text{ and}$$

$$\phi^* \text{ samples from } \Pi_j \text{ at stage } m\})$$

$$\le 1 + E_{\underset{\sim}{\theta}}(\#\{1 \le i \le N: g_{ni}(Y_{j1},\ldots,Y_{ji}) \ge \overline{\theta}\})$$

$$\sim (\log N)/I(\theta_j,\overline{\theta}), \quad \text{by (34)}.$$

Since $T_n(j_n) \ge \rho n$ by (45), it follows that

$$\lim_{n\to\infty} E_{\underset{\sim}{\theta}}(\#\{1 \le n \le N-1: j_n \in L \text{ and } |\hat{\theta}_n(j_n) - \theta^*| > \delta n^{-\beta}\})$$

$$\le \sum_{n=1}^{\infty} \sum_{\ell \in L} P_{\underset{\sim}{\theta}}\{\max_{\rho n \le i \le n} |h_i(Y_{\ell 1},\ldots,Y_{\ell i}) - \theta^*| > \delta n^{-\beta}\}$$

$$< \infty, \text{ by (43)}.$$

Hence, (48) and (49) will follow from (50) if we can show that

$$\sum_{n=1}^{\infty} P_{\underset{\sim}{\theta}}\{j_n \notin L\} < \infty. \tag{52}$$

We now modify the proof of Lemma 1 of Lai and Robbins (1984b) to show that (52) indeed holds. Choose an integer $c > (1-k\rho)^{-1}$ and define for $r = 0,1,\ldots,$

$$A_r = \bigcap_{1\le j\le k} \{\max_{\rho c^{r-1}\le n\le c^{r+1}} |h_n(Y_{j1},\ldots,Y_{jn})-\theta_j| \le \delta c^{-\beta(r+1)}\},$$

$$B_r = \bigcap_{\ell\in L} \{g_{ni}(Y_{\ell 1},\ldots,Y_{\ell i}) > \theta^* \text{ for all } 1 \le i \le n \text{ and } c^{r-1} \le n \le c^{r+1}\}.$$

Then by (33) and (43),

$$P_{\underset{\sim}{\theta}}(\bar{A}_r) = O(c^{-r}r^{-q}),\ P_{\underset{\sim}{\theta}}(\bar{B}_r) = O(c^{-r}r^{-p}) \tag{53}$$

with $p > 1$ and $q > 1$, where $\bar{A}$ denotes the complement of an event A. We now show that for $r \ge r_0$ (sufficiently large)

$$\text{on } A_r \cap B_r,\ \ j_n \in L \text{ for all } c^r \le n \le c^{r+1}. \tag{54}$$

Consider the stage $n+1 = km+\ell$ with $\ell \in L$ and $c^{r-1} \le n < c^{r+1}$. If $j_n \in L$, then ϕ^* samples from either Π_{j_n} or Π_ℓ at stage n+1. Now assume that $j_n \notin L$. Then on $A_r \cap B_r$ with r sufficiently large, $\hat{\theta}_n(j_n) + \delta n^{-\beta} < \theta^*$, and therefore

$$\bar{\theta}_n \le \theta^* < g_{n,T_n(\ell)}(Y_{\ell 1},\ldots,Y_{\ell,T_n(\ell)}) = U_n(\ell),$$

so ϕ^* samples from Π_ℓ. An argument similar to that of the proof of Lemma 1 of Lai and Robbins (1984b) then proves (54).

From (53) and (54) we obtain that for $r \ge r_0$

$$\sum_{c^r\le n\le c^{r+1}} P_{\underset{\sim}{\theta}}\{j_n \notin L\} \le \sum_{c^r\le n\le c^{r+1}} \{P_{\underset{\sim}{\theta}}(\bar{A}_r) + P_{\underset{\sim}{\theta}}(\bar{B}_r)\}$$

$$= O(r^{-q}) + O(r^{-p}).$$

Since $p > 1$ and $q > 1$, the desired conclusion (52) follows.

SUMMARY

Given k statistical populations with finite means, we consider the problem of sampling $x_1, x_2, \ldots$ sequentially from these populations so as to achieve the greatest possible expected value of the sum $S_n = x_1 + \ldots + x_n$. We describe a general method for constructing asymptotically optimal allocation rules that have logarithmic regret at certain parameter configurations and bounded regret at other parameter configurations.

REFERENCES

Bechhofer, R. E., Kiefer, J., and Sobel, M. (1968). Sequential Identification and Ranking Procedures. Univ. Chicago Press.

Chow, Y. S. and Lai, T. L. (1975). Some one-sided theorems on the tail distribution of sample sums with applications to the last time and largest excess of boundary crossings. Trans. Amer. Math. Soc. 208, 51-72.

Feldman, D. (1962). Contributions to the two-armed bandit problem. Ann. Math. Statist. 33, 847-856.

Hammersley, J. M. (1950). On estimating restricted parameters. J. Roy. Statist. Soc. Ser. B 12, 192-240.

Lai, T. L. and Robbins, H. (1984a). Optimal sequential sampling from two populations. Proc. Natl. Acad. Sci. USA, in press.

Lai, T. L. and Robbins, H. (1984b). Asymptotically efficient adaptive allocation rules. Adv. Appl. Math., in press.

Pollak, M. and Siegmund, D. (1975). Approximations to the expected sample size of certain sequential tests. Ann. Statist. 3, 1267-1282.

Robbins, H. (1952). Some aspects of the sequential design of experiments. Bull. Amer. Math. Soc. 55, 527-535.

Robbins, H. (1970). Sequential estimation of an integer mean. Scientists at Work. Almqvist and Wiksells, Uppsala, 205-210.

A NONPARAMETRIC ANALYSIS OF URBAN, RURAL, AND INTERSTATE TRAFFIC FATALITY RATES

Thomas J. Lorenzen

Mathematics Department
General Motors Research
Laboratories
Warren, Michigan

Gary C. McDonald

Mathematics Department
General Motors Research
Laboratories
Warren, Michigan

1. INTRODUCTION

In an earlier study [McDonald (1979)] nonparametric subset selection rules were applied to state traffic fatality rates for the years 1960 to 1976. New England states and Middle Atlantic states comprised most of the subset asserted, with a 90% confidence level, to contain the state with the smallest fatality rate; whereas, Southern states, Southwestern states, and Rocky Mountain states generally comprised the subset for the state with the largest fatality rate. In this article, the hypothesis that the particular separation of states is attributable to the differences among states in the urban, rural, and interstate mileage mix is considered.

The urban/rural/interstate mix factor is certainly not the only factor that could account for the observed differences in the state fatality rates. Recht (1977) discusses in some detail many factors that affect the trend in fatal accidents. Although his discussion centers on the national long-term trend in motor-vehicle fatalities, many of the same factors could account for state-to-state differences in fatality rates over time. Some of these factors are fraction of youthful drivers, law enforcement, sizes and types of vehicles, traffic speed, and condition of roads. A few of these factors have been quantitatively evaluated using national data or data from a single state in time series. Enüstün and Yang (1975) investigated the 55 mph speed limit effect on accidents in Michigan. Peltzman (1975) uses national data to relate motor-vehicle fatality rates to alcohol consumption, vehicle speed, earned income, and several other factors to infer the effect of highway safety regulations pursuant to the vehicle safety act of 1966.

We have chosen to look at the urban/rural/interstate mileage mix for three reasons; because of its plausibility in accounting for the results of McDonald (1979) and Gupta and Hsu (1980), because data is available to adjust for this factor, and because there is a sparsity of analyses concerned with this factor. Chatfield (1973) establishes the plausibility by showing that sections of a highway system with higher travel densities typically have lower fatal accident rates.

Section 2 of this article describes and graphically displays the data. Both the direct and indirect adjustments are used to account for a difference in the urban, rural, and interstate fatality experience. Similar adjustments have been made for U.S. mortality data by Duffy and Carroll (1967) and the methodology is described therein. Section 3 contains the assumptions and techniques of the data analysis and provides some graphical displays of the results of the analysis. Section 4 comments on the statistical techniques employed in the analysis, and section 5 brings together our conclusions.

2. DESCRIPTION OF DATA AND DATA ADJUSTMENT

The data of interest in this study is the motor-vehicle traffic fatality rates for each of the 48 contiguous states for the years 1967 through 1975. By definition the motor-vehicle traffic fatality rate is the number of motor-vehicle deaths per 100,000,000 (10^8) vehicle miles. The motor-vehicle fatalities are recorded for the state in which the accident occurs and the number of vehicle miles are estimated by the individual states.

Since 1967 states have applied a uniform procedure in estimating vehicle miles travelled. This procedure takes into account gallons of gasoline sold (using gas tax revenues), the size and mix of automobiles (using registration data), fuel economy estimates, traffic density counters (both permanent and temporary), traffic counts, and survey data. Although no estimate of the accuracy of the method has been developed, the procedure at least is applied consistently. Prior to 1967, no systematic procedure was required.

On the other hand, one would think that the fatality count would be very accurate. Every accident resulting in a death must be reported to the police who report to the state authorities. In practice, however, the count is not without ambiguities. Should deaths resulting from accidents on farms and on private property be counted? What should be done about deaths that occur some time after an accident but are a direct result of an accident (such as coma victims)? Do we wait a fixed amount of time or estimate whether or not the victim will survive?

At least five sources are available for the above figures. Accident Facts (1969-1977) contain the number of deaths and the fatality rates by state for the years 1967 through 1975. Automobile Facts and Figures (1969-1975), its replacement MVMA Motor Vehicle Facts and Figures (1976-1977), and Highway Statistics (1967-1976) contain vehicle miles by state for the years 1967 through 1975. Highway Statistics Summary to 1975 (1975) contains vehicle miles, deaths, and fatality rates up to 1975. Fatal and Injury Accident Rates on Federal-Aid and other Highway Systems (1967-1976) contain vehicle miles, deaths and the fatality rates by states for all of the years of interest.

A comparison of the tabulations in these sources is somewhat disappointing. No two of the sources are in perfect agreement. The number of deaths reported differed by as much as 121 (1164 compared to 1285), the vehicle mileage by as much as 928 million miles (15,950 compared to 15,022) and the fatality rate by an equivalent amount. Most of the discrepancies occurred during the early years (1967-1972) with the later years agreeing quite closely.

Some of the observed differences can be traced to differences in the definition of fatality and the use of preliminary estimates of vehicle miles. But even when the definitions of fatality are the same and no preliminary estimates are used, differences can occur. These are due primarily to reporting errors and updated estimates.

The most accurate set of data is contained in Highway Statistics Summary to 1975 (1975) (not only because this source contains all of the updates but also because the states reviewed and corrected all reporting errors during 1973). Unfortunately, the statistics contained in Highway Statistics Summary to 1975 (1975) are not detailed sufficiently for the analyses pursued here.

To overcome this difficulty, a complete breakdown of the data used to generate Highway Statistics Summary to 1975 (1975) was obtained from the U.S. Department of Transportation. For this data, a fatality is defined as a death occurring within one year of an accident involving a motor-vehicle on a trafficway. No preliminary estimates are used and all rates are rounded to two decimal places. Subsequent to 1975, some states failed to supply data in sufficient detail to enable an analysis similar to the one used in this paper.

Table 1 contains the motor-vehicle traffic fatality rates for each of the 48 contiguous states (Alaska, Hawaii and Washington, D.C. excluded) for the years 1967 through 1975. These are, by definition, the number of fatalities per 100,000,000 vehicle miles. All values are rounded to two decimal places.

Tables for rural and urban fatality rates are given in Lorenzen and McDonald (1984). Generally speaking, rural rates are greater than urban rates. To more clearly display this relationship, the urban and rural rates are plotted in Figure 11.1 for the year 1975. The line equating urban and rural rates is included for reference. For this year the correlation coefficient between the two rates is 0.254. Over the 9 years of interest, the lowest correlation is 0.135 occurring in 1968 and the highest is 0.351 in 1972.

If both the urban and rural rates are smaller for one state compared to another (for example, Massachusetts compared to Pennsylvania in Figure 11.1) then a straightforward ordering of the two states is possible. In this case the state with the lower rates is "better" than the other. However, when two states are ordered differently depending on the two fatality rates (for example, New Mexico and Idaho in Figure 11.1) then we truly have a bivariate problem and a weighting of the two rates is required to completely order the states.

A method to reduce the bivariate problem to a univariate problem is to adjust the fatality rates for the urban/rural mix. Two adjustment methods (commonly used on census data) can be used, the direct method and the indirect method. With the direct method the actual rural and urban rates of each state are weighted by the same urban/rural mix (typically the national average mix).

The indirect method is slightly more complex. With it the <u>national</u> fatality rate is multiplied by the ratio of the state's actual fatality rate divided by the state's expected fatality rate calculated with the state's actual urban/rural mix and some specified (usually national) urban and rural fatality rates. Detailed references on the direct and the indirect method are Woolsey (1959) and Duffy and Carroll (1967).

The two adjustment methods, in practice, give similar results. (The correlations between these two methods for the individual years 1967 through 1975 range from 0.930 to 0.982). However, each method has specific properties. With the direct method the urban/rural mix must be specified. The <u>actual</u> adjusted rates will depend on the mix chosen. Fortunately the <u>relative</u> values are approximately the same as long as an extreme mix is not used. Thus, the use of the national urban/rural mix is recommended.

Table 1

Motor Vehicle Traffic Fatalities per 10^8 Vehicle Miles

State	67	68	69	70	Year 71	72	73	74	75
Alabama	7.26	7.04	7.08	5.65	5.91	5.49	5.13	4.08	3.93
Arizona	6.22	6.92	6.52	6.29	5.70	5.49	5.96	4.77	4.19
Arkansas	6.58	6.84	5.53	5.45	5.72	5.93	5.00	3.91	4.06
California	4.70	4.58	4.55	4.19	3.78	3.95	3.78	3.15	3.16
Colorado	5.52	6.07	5.28	5.21	4.45	4.64	4.17	3.81	3.56
Connecticut	3.18	2.96	2.61	2.75	2.88	2.62	2.81	2.21	2.18
Delaware	4.96	5.86	4.46	5.14	3.65	3.85	3.67	3.17	3.45
Florida	5.42	6.05	5.68	5.19	5.01	4.59	4.49	3.91	3.31
Georgia	6.62	6.87	6.38	6.10	5.72	5.69	5.41	4.40	3.53
Idaho	6.74	7.12	7.57	6.86	6.59	6.60	6.40	5.78	4.84
Illinois	5.00	4.79	4.70	4.24	4.18	3.80	3.90	3.39	3.42
Indiana	5.73	5.14	5.47	4.79	4.70	4.23	4.14	3.33	3.03
Iowa	6.06	5.51	4.70	5.33	4.64	4.64	4.18	3.60	3.43
Kansas	5.39	5.16	5.92	4.91	4.89	4.53	4.04	3.41	3.34
Kentucky	6.18	5.95	5.75	5.36	4.76	4.75	4.58	3.33	3.57
Louisiana	8.31	8.48	7.21	6.15	6.27	6.22	5.99	4.42	4.62
Maine	5.00	4.15	4.55	4.52	4.21	3.85	3.56	3.23	3.27
Maryland	4.53	4.64	4.10	3.85	3.57	3.45	3.22	3.07	2.74
Massachusetts	3.97	3.86	3.54	3.47	3.24	3.37	3.45	3.40	3.04
Michigan	4.71	4.98	4.89	4.10	3.87	3.91	3.78	3.36	3.11
Minnesota	5.16	5.31	4.75	4.41	4.38	4.14	4.07	3.47	3.03
Mississippi	9.40	7.41	7.56	7.72	7.74	7.03	6.42	4.62	4.23
Missouri	5.53	6.00	5.98	5.69	5.23	5.04	4.72	3.52	3.50
Montana	7.82	6.96	7.64	6.53	6.46	7.35	5.68	5.13	5.21
Nebraska	5.11	5.05	4.59	4.29	4.95	4.49	3.88	3.55	3.35
Nevada	5.91	7.30	7.50	7.43	7.42	6.66	6.24	5.15	4.98
New Hampshire	4.38	4.68	4.50	4.42	4.43	3.55	2.77	3.27	2.85
New Jersey	3.28	3.58	3.25	3.09	3.05	2.79	2.81	2.35	2.23
New Mexico	7.35	7.43	7.97	7.62	6.70	6.63	6.74	5.72	5.73
New York	4.83	4.88	5.39	5.22	5.04	4.91	4.56	4.03	3.77
North Carolina	7.08	7.16	6.51	5.98	5.88	5.83	5.29	4.53	4.17
North Dakota	5.84	5.58	4.98	4.65	5.74	5.05	4.84	3.71	3.75
Ohio	5.07	4.84	4.98	4.60	3.90	3.86	3.66	3.01	2.82
Oklahoma	5.78	4.96	5.33	4.72	4.51	4.21	3.68	3.49	3.35
Oregon	5.68	5.39	5.61	5.09	4.84	4.84	3.98	4.41	3.59
Pennsylvania	4.48	4.24	4.35	3.98	3.74	3.48	3.61	3.37	3.27
Rhode Island	2.62	3.26	2.97	2.96	2.46	2.25	2.39	1.77	1.98
South Carolina	6.98	7.03	6.42	6.24	5.76	5.64	4.73	4.36	3.98
South Dakota	5.34	5.92	6.79	5.08	5.44	5.84	5.56	4.50	3.91
Tennessee	6.94	6.33	7.01	6.65	5.04	4.80	4.44	4.05	3.52
Texas	5.71	5.60	5.37	5.23	5.00	4.82	4.56	3.87	4.05
Utah	5.21	5.20	5.31	5.48	5.15	5.48	4.96	3.06	3.45
Vermont	5.83	5.79	5.68	4.61	4.94	4.67	4.64	4.10	4.35
Virginia	5.40	4.96	4.84	4.33	3.99	3.84	3.52	3.12	2.97
Washington	4.93	4.89	4.17	4.22	4.01	3.82	3.31	3.37	3.21
West Virginia	6.97	6.31	6.17	6.55	5.82	5.45	4.68	4.44	4.60
Wisconsin	5.48	5.25	4.78	4.59	4.37	4.24	4.03	3.26	3.29
Wyoming	5.54	5.86	7.62	6.37	5.19	5.85	5.55	5.65	5.78

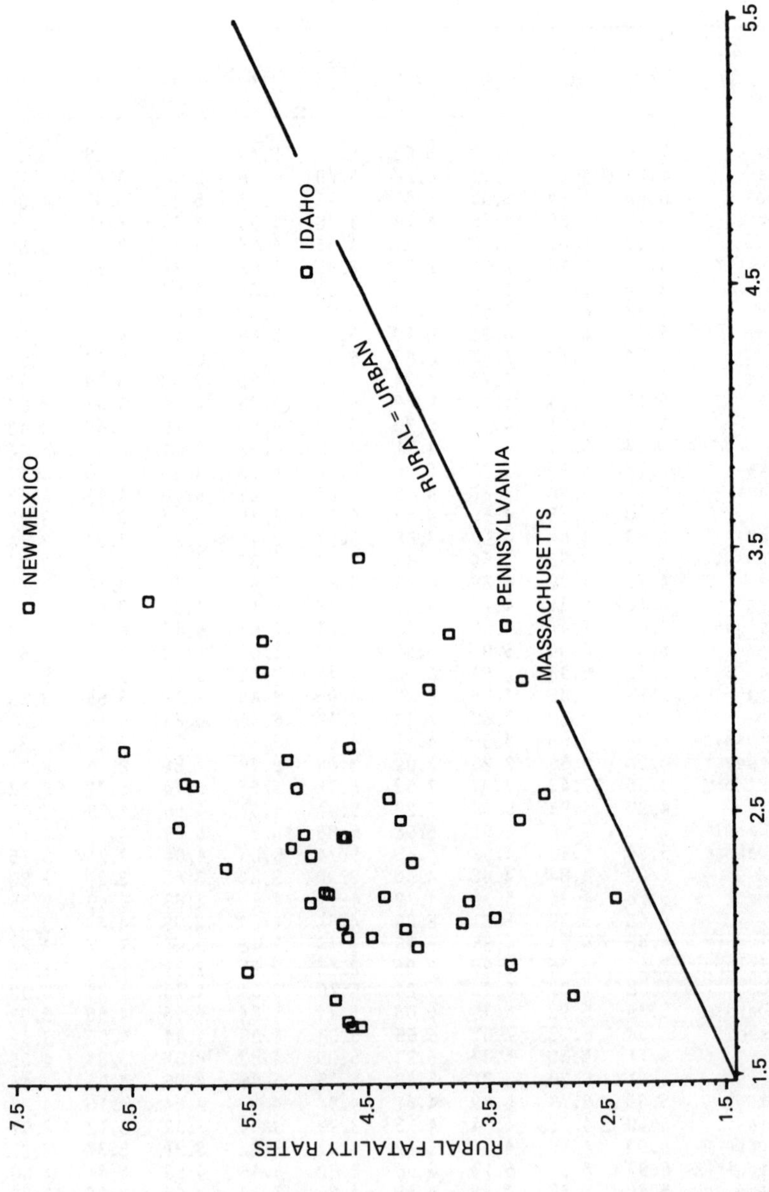

FIGURE 11.1 Urban and rural traffic fatality rates for 1975.

Table 2

Fatality Rates Indirectly Adjusted for Urban/Rural/Interstate Mileages

State	67	68	69	70	Year 71	72	73	74	75
Alabama	7.07	6.91	7.02	5.72	5.95	5.59	5.22	4.23	4.07
Arizona	6.87	7.48	7.20	6.83	6.19	6.15	6.52	5.22	4.68
Arkansas	5.99	6.19	5.07	4.99	5.27	5.48	4.63	3.65	3.80
California	5.59	5.32	5.30	4.87	4.40	4.65	4.47	3.77	3.84
Colorado	5.71	6.37	5.44	5.43	4.71	4.95	4.45	4.03	3.78
Connecticut	4.04	3.78	3.35	3.53	3.72	3.37	3.63	2.86	2.89
Delaware	4.80	5.73	4.40	5.10	3.92	4.17	3.98	3.23	3.44
Florida	5.29	6.34	6.08	5.60	5.33	5.02	4.95	4.30	3.62
Georgia	6.57	6.87	6.40	6.16	5.74	5.75	5.47	4.42	3.60
Idaho	5.98	6.38	6.79	6.25	6.04	6.07	5.88	5.37	4.60
Illinois	5.54	5.38	5.33	4.89	4.79	4.37	4.60	3.99	4.05
Indiana	5.78	4.98	5.33	4.66	4.67	4.35	4.33	3.50	3.19
Iowa	5.61	5.23	4.49	5.09	4.45	4.47	4.03	3.51	3.34
Kansas	4.99	4.85	5.61	4.64	4.64	4.32	3.86	3.26	3.21
Kentucky	5.67	5.48	5.30	4.96	4.48	4.52	4.37	3.17	3.43
Louisiana	8.04	8.16	6.92	5.92	6.08	6.09	5.85	4.31	4.50
Maine	4.40	3.64	4.01	3.99	3.73	3.42	3.19	2.87	2.91
Maryland	4.52	4.65	4.17	3.96	3.69	3.58	3.38	3.20	2.87
Massachusetts	4.78	4.67	4.27	4.22	3.97	4.15	4.47	4.46	4.01
Michigan	4.86	5.15	5.14	4.41	4.20	4.28	4.19	3.72	3.46
Minnesota	4.87	5.03	4.55	4.27	4.23	4.14	4.08	3.48	3.21
Mississippi	8.42	6.65	6.80	7.02	7.09	6.49	5.92	4.25	3.91
Missouri	5.72	6.21	6.15	5.88	5.45	5.12	4.93	3.64	3.75
Montana	6.91	6.13	6.75	5.79	5.75	6.60	5.08	4.68	4.74
Nebraska	4.62	4.55	4.16	3.96	4.57	4.40	3.81	3.48	3.30
Nevada	6.18	7.59	7.84	7.79	7.73	7.03	6.49	5.48	5.30
New Hampshire	4.02	4.33	4.19	4.13	4.12	3.33	2.59	3.11	2.72
New Jersey	3.86	4.22	3.74	3.56	3.54	3.47	3.48	2.92	2.79
New Mexico	7.33	7.44	8.04	7.67	6.75	6.71	6.92	5.88	5.89
New York	5.30	5.39	6.08	6.00	5.63	5.63	5.32	4.71	4.33
North Carolina	6.38	6.49	5.93	5.46	5.42	5.37	4.91	4.22	3.92
North Dakota	4.93	4.72	4.25	3.99	4.92	4.35	4.19	3.22	3.24
Ohio	5.34	5.19	5.37	4.97	4.18	4.20	4.02	3.31	3.05
Oklahoma	5.62	4.91	5.28	4.68	4.50	4.27	3.78	3.61	3.48
Oregon	5.60	5.31	5.58	5.10	4.61	4.60	3.83	4.22	3.40
Pennsylvania	4.67	4.42	4.56	4.16	3.92	3.65	3.81	3.59	3.46
Rhode Island	3.39	4.29	3.95	3.94	3.27	3.01	3.19	2.35	2.60
South Carolina	6.19	6.26	5.78	5.82	5.40	5.27	4.42	4.11	3.76
South Dakota	4.57	5.09	5.83	4.44	4.79	5.14	4.91	3.96	3.47
Tennessee	6.93	6.32	7.06	6.82	5.24	5.03	4.67	4.29	3.71
Texas	6.07	5.98	5.74	5.61	5.43	5.30	5.03	4.27	4.46
Utah	5.54	5.59	5.74	5.92	5.58	5.88	5.34	3.36	3.94
Vermont	5.15	5.10	4.99	4.04	4.31	4.11	4.09	3.63	3.82
Virginia	5.18	4.75	4.71	4.23	4.05	3.88	3.60	3.20	3.03
Washington	5.05	5.11	4.37	4.72	4.46	4.20	3.65	3.70	3.54
West Virginia	6.28	5.68	5.56	6.11	5.45	5.02	4.32	4.09	4.23
Wisconsin	5.41	5.24	4.79	4.46	4.24	4.30	4.15	3.34	3.35
Wyoming	5.17	5.42	7.05	5.93	4.85	5.49	5.24	5.34	5.47

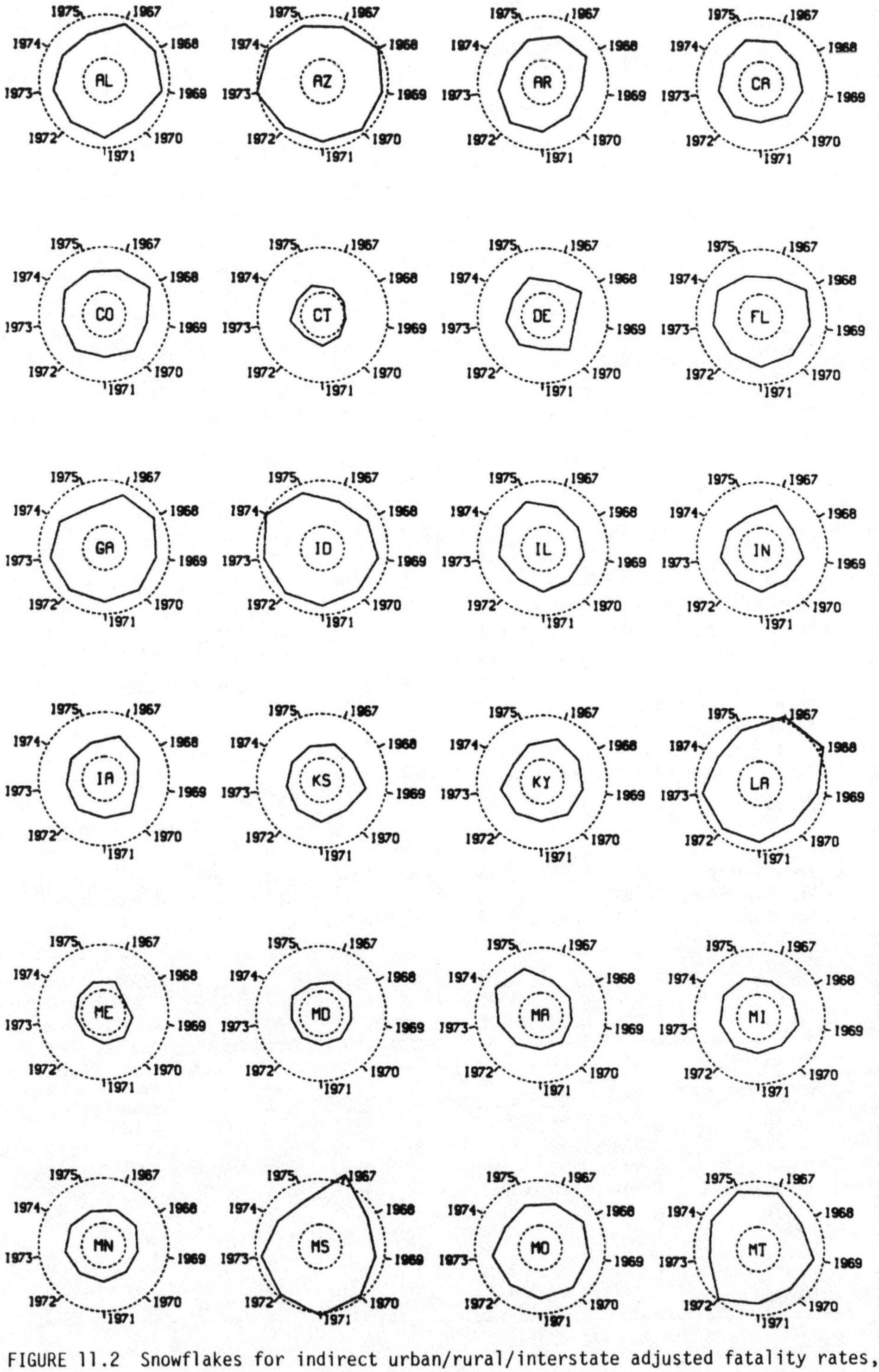

FIGURE 11.2 Snowflakes for indirect urban/rural/interstate adjusted fatality rates, 48 states and 9 years. States are normalized within each year. The dotted lines

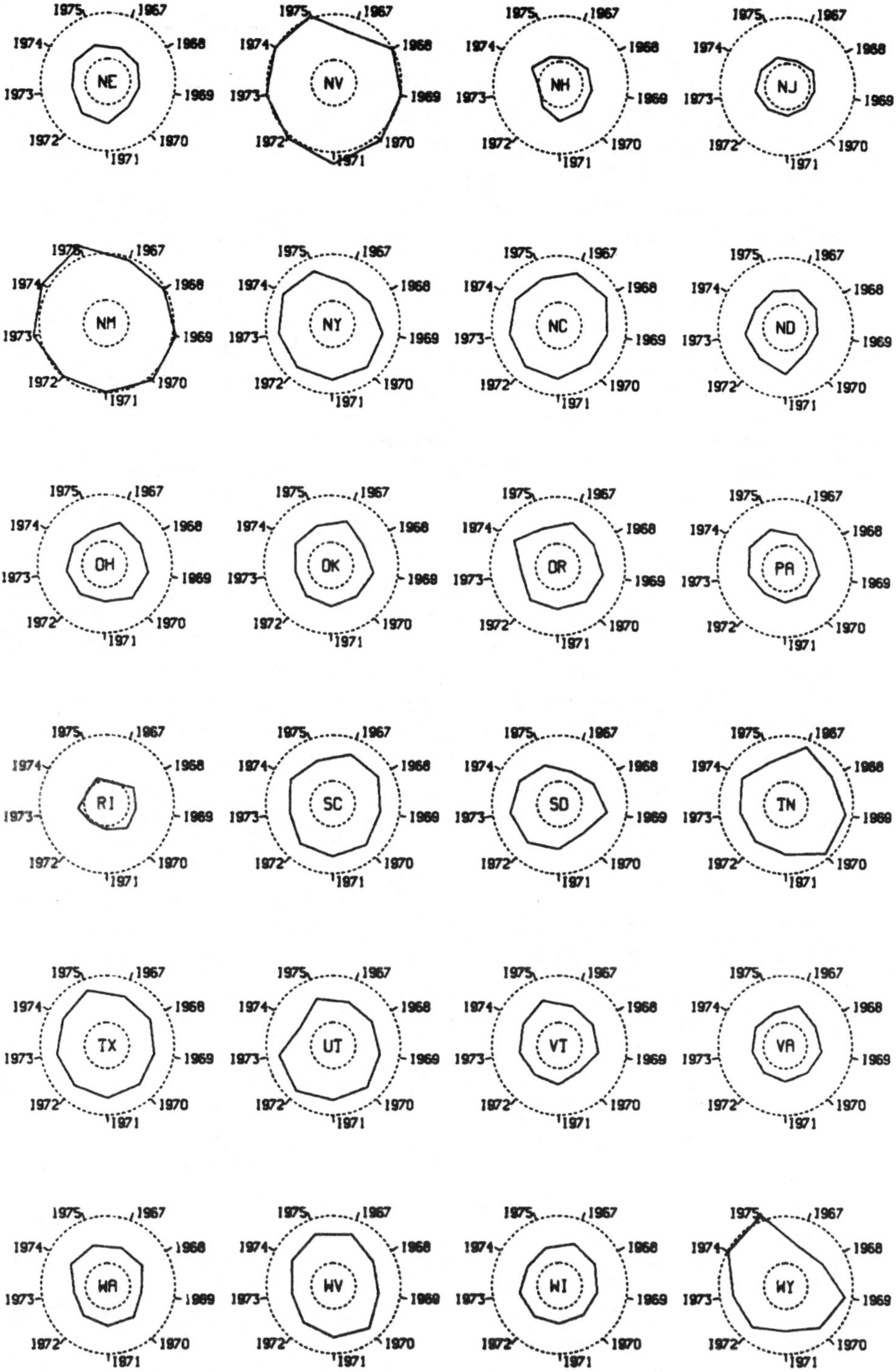

represent two standard deviations above and below the average.

With the indirect method, a similar statement is true. The actual adjusted rates depend strongly on the specified urban and rural fatality rates. Thus, the use of the national urban and rural fatality rates is recommended.

The appropriate method to use depends on the purpose of the adjustment. If two states with the same urban and rural fatality rates are to have the same adjusted rate, then the direct adjusted method is preferable. If two states with the same unadjusted rate and the same urban/rural mix should have the same adjusted rate, then the indirect method is preferable. With missing data, only the indirect method can be used.

Lorenzen and McDonald (1984) contains tables of fatality rates on urban roads, on rural roads, and after direct and indirect adjusting for urban/rural differences. Table 2 gives traffic fatality rates indirectly adjusted for urban/rural/interstate differences. Direct adjustment is not possible as some states had years with no interstate travel. To facilitate visual comparisons, the indirect adjusted values are displayed in Figure 11.2 using the multivariate technique known as snowflakes. Here each snowflake represents a state, the length of each ray represents the adjusted fatality rate (relative to the average for that year), and the rays in order represent the years 1967 through 1975. The outer (inner) dashed circle represents two standard deviations above (below) the average. For example, the state of New Mexico is uniformly two standard deviations above the average of all states for the years 1967 through 1975. New Jersey is uniformly two standard deviations below the average. The snowflake for Wyoming reveals a steady increase (relative to the average) in the adjusted fatality rates for the years 1967 through 1975. These snowflakes were produced by an in-house program but can be constructed using the STARS option in the package of Welsch (1976).

3. ANALYSIS AND SELECTION

3.1. Mathematical Model: Interactions and Transformations

Our basic model assumption is that the transformed fatality rates for the ith state and jth year may be represented as

$$Y_{ij} = \mu + \theta_i + \beta_j + \varepsilon_{ij}, \tag{3.1}$$

where Y_{ij} is the fatality rate under consideration (rural, direct adjusted, etc.) μ is an overall mean effect, θ_i is an effect due to the state $(i = 1,\ldots,k)$ and β_j is an effect due to the year $(j = 1,\ldots,n)$. The error term, ε_{ij}, may have any (not necessarily normal) continuous distribution. To reasonably assume the model form of (3.1) the fatality rates are transformed to $X_{ij} = \ell n(Y_{ij}-c)$, where Y_{ij} is the appropriate fatality rate and c is a rate specific constant. This transformation is chosen to yield the Tukey test for nonadditivity [Tukey (1949)] insignificant (at $\alpha = .25$) for the unadjusted and adjusted rates. The six tail areas corresponding to the calculated F-statistics for $c = 0.62$ range from .25 to .86. Rate specific c values can be determined so that all tail areas exceed 0.50.

3.2. Selection Rules

In this section a selection rule for choosing a subset of states, guaranteed with a specified probability to include the state with the smallest effect on fatality rates, will be described. A similar rule will be used to determine a subset to include the state with the largest effect on the fatality rates. These selection rules are based on ranks and utilize only the stochastic ordering of the distribution of fatality rates as implied by the model (3.1). Although (3.1) is stated for transformed rates, the transformation is monotone and since these selection procedures are based on ranks the analysis can be carried out with the original data--the rates themselves.

Let R_{ij} denote the rank of observation x_{ij} among $x_{1j},\ldots,x_{kj}$; i.e., if there are exactly r of the observations x_{mj}, $m = 1,\ldots,k$, less than x_{ij} then $R_{ij} = r + 1$. These ranks are well-defined since the error ε_{ij} is assumed to have a continuous distribution. The variables R_{ij} take integer values from 1 to k inclusive. Our selection procedures are based on the quantities $T_i = \sum_{j=1}^{n} R_{ij}$, the sum of ranks associated with the <u>ith</u> state, $i = 1,\ldots,k$.

Let $\theta_{[i]}$ denote the <u>ith</u> smallest state effect so that $\theta_{[1]} \leq \theta_{[2]} \leq \cdots \leq \theta_{[k]}$. That (unknown) state characterized by $\theta_{[1]}$ will be called the best, and that characterized by $\theta_{[k]}$ called the worst. The selection rule for choosing a subset to contain the best state is denoted by R_B and that to contain the worst state by R_W. These rules can be stated as:

R_B: Choose the ith state iff $T_i \le \min_{1 \le j \le k} T_j + b$,

and

R_W: Choose the ith state iff $T_i \ge \max_{1 \ge j \ge k} T_j - b$.

The constant b (≥ 0) is chosen as small as possible to yield a specified probabilistic guarantee for the resultant subset.

If a selection rule chooses a subset which actually contains the best state (or worst if that is the goal) then a correct selection occurs. Ideally the constant b used for R_B and R_W is chosen so that the probability of a correct selection is no less than a specified P^* ($k^{-1} < P^* < 1$) whatever the configuration of the true (unknown) θ_i's might be. For the applications in this article the $b \equiv b(n,k,P^*)$ is determined from an asymptotic approximation; specifically,

$$b \doteq h[nk(k+1)/6]^{1/2},$$

where h satisfies

$$\int_{-\infty}^{\infty} [\Phi(x+h2^{1/2})]^{k-1}\phi(x)dx = P^*.$$

The quantities $\Phi(\cdot)$ and $\phi(\cdot)$ refer to the cumulative distribution function and probability density function, respectively, of a standard normal variate. The h quantity appearing in this expression has been tabulated by Gupta (1963); Gupta *et al*. (1973); and Milton (1963).

Unfortunately, this choice of b does not provide the ideal probabilistic guarantee as desired. To mathematically guarantee a lower P^* bound on the correct selection probability the inference space of the θ_i's has to be restricted to a slippage space. This restriction is discussed in detail by McDonald (1972,1979). As described in Section 4, simulation studies with usual error distributions (e.g., normal) indicate that such a restriction may not be necessary. Although these simulations do not provide the global mathematical assurance, we feel the evidence is sufficient to warrant use of these rules. Other nonparametric rules which do provide the mathematical assurance are described in the above mentioned references, but generally result in a larger chosen subset.

For the applications in this article, $k = 48$ states and $n = 9$ years, values of h can be obtained from Gupta (1963) Table I and corresponding b-values:

P^*	.75	.90	.95	.99
b	124.44	153.01	170.29	202.96

.

Table 3

Rank Sums and Subset Selections for Traffic Fatality Rates, P* = 0.90.

State	Unadjusted	Urban	Rural	Urban/rural direct adjusted	Urban/rural indirect adjusted	Interstate indirect adjusted
Alabama	342.5 †	361.0 †	351.5 †	367.0 †	369.0 †	355.0 †
Arizona	348.5 †	336.0 †	379.0 †	384.0 †	387.5 †	403.0 †
Arkansas	314.0 †	268.5 †	261.5 †	274.0 †	277.0 †	262.0
California	91.0 *	215.0	200.0	208.0	214.5	220.0
Colorado	225.5	220.5	256.5 †	252.0	253.5	261.0
Connecticut	16.5 *	72.5 *	38.0 *	24.0 *	22.0 *	29.0 *
Delaware	135.5 *	264.0 †	116.0 *	170.0 *	168.0 *	141.0 *
Florida	233.0	188.5	335.0 †	303.0 †	296.5 †	283.0 †
Georgia	324.5 †	202.5	346.5 †	336.5 †	345.5 †	350.0 †
Idaho	399.0 †	271.5 †	348.0 †	367.0 †	373.0 †	375.0 †
Illinois	124.0 *	184.5	266.5 †	250.5	233.0	236.5
Indiana	177.0	196.5	163.0 *	170.0 *	168.5 *	181.0
Iowa	216.0	123.0 *	194.0	170.5	182.0	175.0
Kansas	196.0	166.5 *	165.5 *	156.0 *	162.0 *	148.5 *
Kentucky	247.5	182.5	194.5	193.0	201.0	192.5
Louisiana	386.0 †	352.5 †	382.0 †	397.0 †	395.0 †	385.5 †
Maine	103.0 *	71.5 *	73.5 *	43.0 *	39.0 *	37.0 *
Maryland	50.0 *	263.5 †	44.5 *	71.5 *	59.0 *	49.0 *
Massachusetts	60.0 *	310.5 †	21.0 *	62.0 *	156.0 *	162.5 *
Michigan	113.5 *	87.5 *	251.5 †	175.5	162.0 *	157.0 *
Minnesota	149.5 *	180.0 *	153.0 *	154.0 *	139.5 *	120.0 *
Mississippi	410.0 †	340.0 †	382.5 †	393.0 †	388.5 †	382.0 †
Missouri	269.0 †	280.0 †	256.5 †	285.5 †	275.5 †	290.5 †
Montana	399.0 †	149.0 *	364.0 †	311.5 †	360.5 †	360.0 †
Nebraska	161.5 *	75.0 *	152.5 *	102.5 *	113.0 *	106.5 *
Nevada	395.0 †	359.0 †	404.0 †	402.5 †	410.0 †	414.0 †
New Hampshire	81.0 *	213.0	47.0 *	66.5 *	44.0 *	44.0 *
New Jersey	27.5 *	171.0 *	17.0 *	17.0 *	33.0 *	28.0 *
New Mexico	420.0 †	413.0 †	404.0 †	421.0 †	421.0 †	423.0 †
New York	225.0	240.5	361.0 †	353.0 †	335.0 †	323.5 †
North Carolina	357.0 †	165.0 *	350.0 †	304.0 †	336.5 †	313.0 †
North Dakota	256.5	97.5 *	155.0 *	112.5 *	129.5 *	128.5 *
Ohio	105.5 *	245.0	117.5 *	137.0 *	143.0 *	150.5 *
Oklahoma	178.0	280.5 †	128.5 *	195.0	173.0	167.0 *
Oregon	239.0	220.5	202.5	216.5	209.0	213.0
Pennsylvania	79.0 *	257.5	55.0 *	104.5 *	101.0 *	101.5 *
Rhode Island	12.0 *	27.5 *	199.0	125.0 *	16.0 *	17.5 *
South Carolina	337.0 †	317.0 †	271.5 †	315.0 †	310.0 †	291.0 †
South Dakota	299.0 †	203.0	200.5	205.5	206.5	210.0
Tennessee	299.5 †	311.5 †	310.0 †	317.5 †	318.0 †	324.0 †
Texas	253.5	295.5 †	302.0 †	315.5 †	309.5 †	312.5 †
Utah	221.5	341.5 †	153.5 *	259.0	237.5	290.5 †
Vermont	260.5	139.0 *	151.0 *	138.5 *	138.5 *	153.0 *
Virginia	107.5 *	129.5 *	103.5 *	90.0 *	88.0 *	94.5 *
Washington	95.5 *	188.5	142.0 *	139.5 *	145.5 *	150.5 *
West Virginia	338.0 †	268.5 †	292.0 †	295.0 †	300.0 †	293.0 †
Wisconsin	156.0 *	39.0 *	258.0 †	152.0 *	168.5 *	155.0 *
Wyoming	348.5 †	298.5 †	263.0 †	281.0 †	270.0 †	324.0 †

*Member of the subset chosen to contain the smallest fatality rate.
†Member of the subset chosen to contain the largest fatality rate.

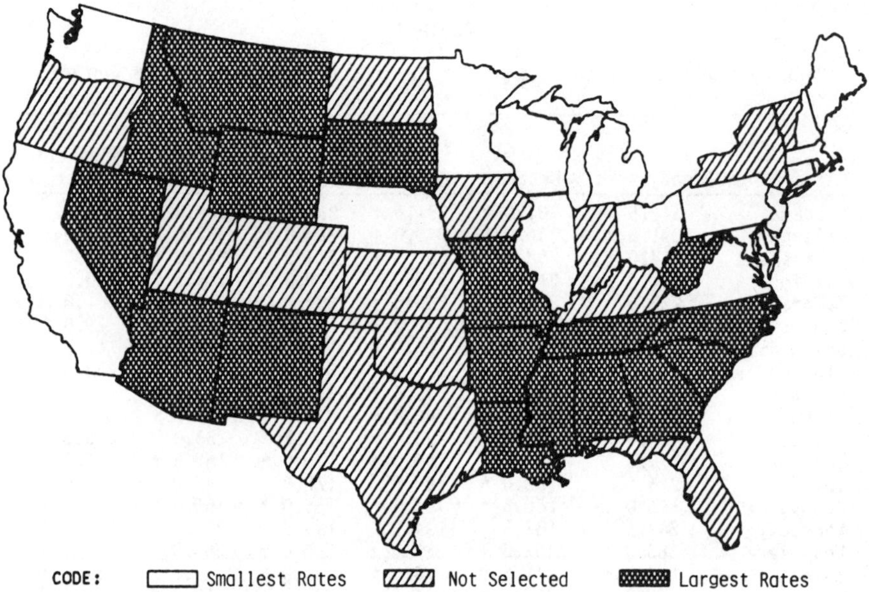

FIGURE 11.3 Selected states for unadjusted traffic fatality rates ($P^* = .90$).

3.3. Selection of States

The results of applying the selection procedures R_B and R_W to the fatality rates appearing in Tables 1 and 2 as well as the urban, rural, direct, and indirect adjusted urban/rural, and indirect adjusted urban/rural/interstate rates are given in Table 3 with the corresponding state rank sums. The lower bound on the probability of a correct selection is fixed at $P^* = .90$ and thus $b = 153.01$ for the selection rules. For the unadjusted rates the minimum state rank sum is 12 (Rhode Island) and R_B chooses all those states for which the rank sum is no greater than $12 + 153.01 = 165.01$. For the same rates the maximum rank sum is 420 (New Mexico) and so R_W chooses all those states for which the rank sum is no less than $420 - 153.01 = 266.99$. The selection rules apply similarly for the other five fatality rates. Results are indicated in Table 3. The selected states using unadjusted and urban/rural/interstate adjusted rates are geographically displayed in Figures 11.3 and 11.4, respectively.

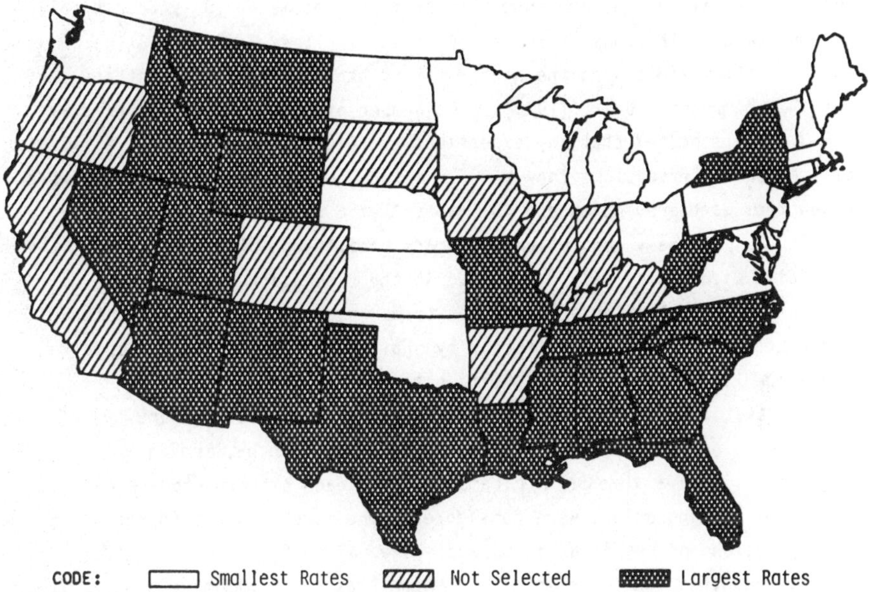

FIGURE 11.4 Selected states for indirect urban/rural/interstate adjusted rates.

4. JUSTIFICATION OF THE SELECTION PROCEDURE USED

As mentioned in the previous section, to mathematically guarantee a lower P^* bound on the probability of correct selection, P(CS), for the rank sum procedure, the inference space on the θ_i's must be restricted to a slippage configuration. In fact, a counterexample given by McDonald (1972) shows that, even for $k = 3$ populations, the minimal P(CS) for unrestricted θ_i need not occur when the θ_i are all equal. Thus the choice of b given in the previous section and used for the selection procedure does not guarantee mathematically an unqualified lower P^* bound on P(CS).

To determine operating characteristics of the selection rule, R_W (equivalent to those for R_B), simulation studies were performed. The purpose of these studies was to check the usual error distributions under various parameter configurations to verify that P(CS) is no less than P^* and is minimized when the θ_i are all equal (which gives $P(CS) = P^*$).

The error distributions are chosen so that the parameter θ stochastically orders the distributions; that is, $\theta_1 < \theta_2$ implies $F_{\theta_1}(x) \geq F_{\theta_2}(x)$.

All simulations used the procedures in the 1980 update of Edition 7 of the IMSL computer library (1979). The number of populations was fixed at $k = 3$ in the belief that any exceptions to P(CS) for arbitrary k will also appear as exceptions for $k = 3$. A total of $n = 8$ observations were taken from each population and b was set at 6 giving an exact P* of .9068. To precisely control the accuracy of all probability estimates, the simultion size was set at $N = 5000$. If the simulation estimated the probability of selecting a population to be p, the standard deviation of the estimate is given approximately by $[p(1-p)/5000]^{1/2}$. This standard deviation is bounded above by .00707 obtained for $p = 0.5$.

To test location parameters the following distributions were simulated: Normal, Logistic, Double Exponential, Cauchy, and the Exponential with a threshold parameter. For each distribution three parameter configurations were considered. One is called a slippage configuration and has location parameters equal to 0, 0, and δ for $\delta \geq 0$. For this configuration it is known that the minimum value of P(CS) is P* and is achieved at $\delta = 0$. The second can be called a reverse slippage configuration and has parameters at 0, δ, and δ for $\delta \geq 0$. This is the configuration leading to the McDonald (1972) counterexample involving uniform variates over disjoint support intervals. The final configuration is called equal spacing and has parameters at 0, δ, and 2δ for $\delta \geq 0$.

The chosen distributions cover a wide spectrum of tail densities and one is non-symmetric. In all cases P(CS) was a strictly increasing function of δ indicating that the minimal P(CS) occurs at $\delta = 0$ (at least for the chosen distributions and parameter configurations). A plot of the probability of selecting each population for the Cauchy distribution under equal spacing is given in Figure 11.5. Plots of the other distributions give the same qualitative relationship.

Next distributions ordered by a scale parameter were considered. Two representative distributions were chosen--the exponential distribution, and the gamma distribution. For both distributions, three scale configurations were considered, a slippage configuration (scale parameters = 1, 1, δ for $\delta \geq 1$), a reverse slippage configuration (scale parameters = 1, δ, δ for $\delta \geq 1$), and an equal power configuration (scale parameters = 1, δ, δ^2 for $\delta \geq 1$). Again $k = 3$ populations, $n = 8$ observations, $b = 6$, P* = .9068, and $N = 5000$ simulations were run. In all cases P(CS) was

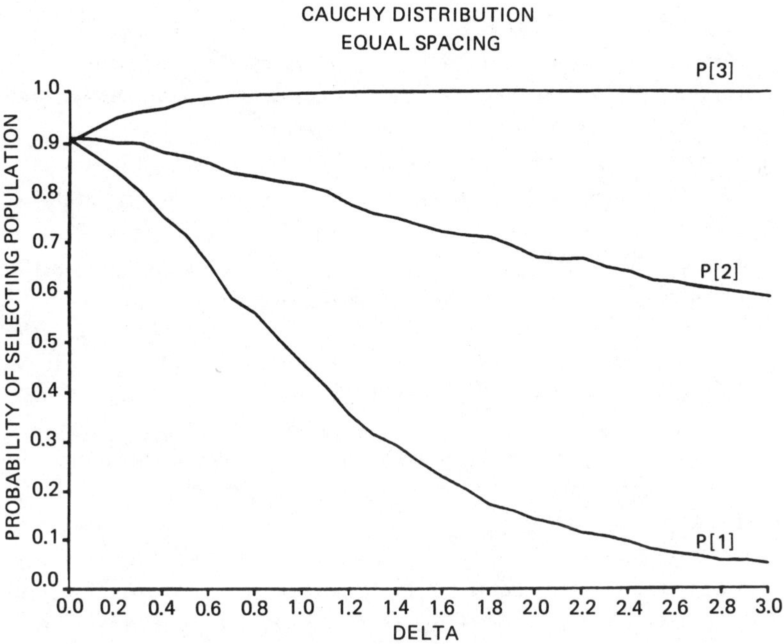

FIGURE 11.5 Probability of choosing population as a function of δ where $\theta_1 = \theta$, $\theta_2 = \delta$, and $\theta_3 = 2\delta$. ($P^* = .90$).

strictly increasing in δ showing the minimal P(CS) again equals P^* and occurs at $\delta = 1$.

Finally, the counterexample itself was considered. This distribution is a mixture of two distinct uniform random variables described by its cumulative distribution function: $F(x) = 0$ for $x \leq -c$, $F(x) = (x+c)/(2c)$ for $-c \leq x \leq 0$, $F(x) = 1/2$ for $0 \leq x \leq 1$, $F(x) = 1/2 + (x-1)/(2a)$ for $1 \leq x \leq 1+a$, and $F(x) = 1$ for $x \geq 1+a$. The counterexample is established for P^* near 1, c large, $a < 1$, and the reverse slippage configuration with slippage equal to 1. The simulations used the reverse slippage configuration with slippage equal to 1, $a = 0.5$, and increasing c. Two different specifications were considered, $k = 3$, $n = 8$, $b = 6$, $P^* = .9068$, and $k = 3$, $n = 50$, $b = 25.58$, $P^* = .99$. While P(CS) decreased as c increased, it remained above the nominal P^* value for both examples. Going back to the McDonald (1972) paper, using the normal approximation (appropriate for n large), the exact formula for P(CS), and the tables of Milton (1963) show that the nominal P^* must be greater than .9992 for the inequality leading to the counterexample to hold even for the largest value of c. Thus it is not surprising that the simulation did not uncover a P(CS) less than P* and suggests that the minimum value of P(CS) equals P* for reasonable values of P*.

These simulations indicate the minimal P(CS) actually equals P* for the usual location or scale distributions and, even with unusual distributions, for reasonable values of P*. This suggests that the practical inference corresponding to the selection procedure of the previous section need not be restricted to slippage configurations.

5. CONCLUSIONS

The rank sum procedure has been the primary tool used in the selection and anaylsis of motor vehicle traffic fatality rates. Monte Carlo simulations with a wide variety of statistical distributions and parameter configurations have indicated that, without further assumptions, the probability of correct selection (i.e. selecting the state with the smallest or largest mean fatality rate) is bounded below by a known tabulated (P*) value.

The broad conclusions of these analyses are that the Southern, Southwestern, and Rocky Mountain states generally have large motor-vehicle traffic fatality rates and are contained in the selected subset which, with 90% confidence, includes the state with the largest mean fatality rate; correspondingly, the New England, Middle Atlantic, and Midwestern states generally have small rates and comprise most of the subset including the state with the smallest rate. These conclusions evolve from analyzing the unadjusted fatality rates as well as the direct and indirect adjusted rates which are normalized, in some sense, to account for the difference among states in the urban, rural, and interstate mileage mix. Although the broad geographic pattern persists after these adjustments, the statistical inferences concerning a few individual states do change. Some of these changes are noted in the next several paragraphs. The sizes of the selected subsets using adjusted rates are larger than those obtained using unadjusted rates.

Focusing on the rural fatality rate in Table 3, most of the Southern, Southwestern, and Rocky Mountain states are in the subset for the largest fatality rate. Utah is an interesting exception. While all of its bordering states are included in the subset for the largest rural rate, it is included in the subset for the smallest rate. New York is a similar exception in the opposite direction. In considering the urban rates,

Massachusetts appears to be an isolated member in the subset for the largest rate. Its neighboring states are either not selected for a subset or belong to that subset chosen for the smallest fatality rate.

In comparing the subsets selected for the *largest* unadjusted and indirect urban/rural/interstate adjusted rates, fifteen states are common to the two subsets. Using unadjusted rates, Arkansas and South Dakota are chosen; whereas the direct adjustment excludes Arkansas and South Dakota, but does include Florida, New York, and Utah. Except for Utah, these latter states are included in part because of their relatively large rural fatality rates.

The subsets selected for the *smallest* unadjusted and direct adjusted fatality rates contain fifteen states in common. In addition the unadjusted rates lead to the selection of California and Illinois. In using the indirect adjusted rates these two states are not selected; however, Kansas, North Dakota, Oklahoma, and Vermont are chosen.

Comparing the subsets chosen with direct urban/rural adjusted rates to those chosen with indirect urban/rural adjusted rates shows a difference in inference for one state--Michigan. Indirect rates lead to the inclusion of Michigan in the subset for smallest fatality rate, whereas the direct rates do not lead to the inclusion.

SUMMARY

In a recent article by McDonald (*Technometrics* *21* (1979) 515-523) subset selection rules were applied to state traffic fatality rates. New England and Middle Atlantic states comprised most of the subset asserted to contain the state with the smallest fatality rate. Southern, Southwestern, and Rocky Mountain states generally comprised the subset corresponding to a largest fatality rate. In this article, additional properties of the selection procedure are determined and the possibility that the particular separation of states is attributable to differences among state urban, rural, and interstate mileage mix is considered.

Two new traffic fatality indices are constructed--a "direct" and an "indirect" adjusted rate--to normalize for the urban/rural/interstate mileage differences between states. Applying the nonparametric selection procedure to those new indices indicates the broad geographic conclusions of the earlier study generally hold. Inferences for a few states do change and are noted. Extensive Monte Carlo simulations quantifying the selection probabilities for the statistical procedure are summarized.

ACKNOWLEDGMENTS

The authors gratefully acknowledge the assistance of Ms. Mary Beth Hatton for computer assistance on the simulation and analysis of variance studies, and Mr. Harold Gugel for the computer graphics implementation of the snowflake method used for Figure 11.2.

REFERENCES

Automobile Manufacturers Association. Automobile Facts and Figures. Annual 1969 through 1975. Automobile Manufacturers Assoc., Inc., Detroit, Michigan 48202.

Chatfield, B.V. (1973). Fatal accidents and travel density. Highway Research Record No. 469, 40-51.

Duffy, E.A. and Carroll, R.E. (1967). United States Metropolitan Mortality, 1959-1961, PHS Publication No. 999-AP-39, U.S. Public Health Service, National Center for Air Pollution Control.

Enüstün, N. and Yang, A.H. (1975). The 55-mph limit: efect on accidents. Traffic Engineering, 45, 22-25.

Gupta, S.S. (1963). Probability integrals of multivariate normal and multivariate t., Ann. Math. Statist., 34, 792-828.

Gupta, S.S. and Hsu, J.C. (1980). Subset selection procedures with application to motor vehicle fatality data in a two-way layout. Technometrics, 22, 543-546.

Gupta, S.S., Nagel, K., and Panchapakesan, S. (1973). On the order statistics from equally correlated normal random variables, Biometrika, 60, 403-413.

IMSL Library 1 Reference Manual (1979). International Mathematical and Statistical Libraries, Inc., Sixth Floor, GNB Building, 7500 Bellaire Blvd., Houston, TX 77036.

Lorenzen, T.J. and McDonald, G.C. (1984). A nonparametric analysis of urban, rural, and interstate traffic fatality rates. General Motors Research Publication, GMR-3341R, General Motors Research Laboratories, Warren, Michigan 48090.

McDonald, G.C. (1972). Some multiple comparison selection procedures based on ranks. Sankhyā, A, 34, 53-64.

McDonald, G.C. (1979). Nonparametric selection procedures applied to state traffic fatality rates. Technometrics, 21, 515-523.

Milton, R.C. (1963). Tables of the equally correlated multivariate normal probability integral. Tech. Report No. 27, Department of Statistics, University of Minnesota 55455.

Motor Vehicle Manufacturers Association. MVMA Motor Vehicle Facts and Figures. Annual 1976 and 1977. Motor Vehicle Manufacturers Assoc., Detroit, Michigan 48202.

National Safety Council. Accident Facts. Annual 1969 through 1977. National Safety Council, 444 N. Michigan Ave., Chicago, Illinois 60611.

Peltzman, S. (1975). The effects of automobile safety regulation. Journal of Political Economy, 83, 677-725.

Recht, J.L. (1977). The long-term trend in fatal accidents. Traffic Safety, 77, 22-24 and 34-36.

Tukey, J.W. (1949). One degree of freedom for non-additivity. Biometrics, 5, 232-242.

U.S. Department of Transportation, Federal Highway Administration. Fatal and Injury Accident Rates on Federal-Aid and Other Highway Systems. Annual 1967 through 1976. U.S. Government Printing Office, Washington, D.C. 20402.

U.S. Department of Transportation, Federal Highway Administration. Highway Statistics. Annual 1967 through 1976. U.S. Government Printing Office, Washington, D.C. 20402.

U.S. Department of Transportation, Federal Highway Administration. Highway Statistics Summary to 1975. U.S. Government Printing Office, Washington, D.C. 20402.

Welsch, R.E. (1976). Graphics for data analysis. Computers and Graphics, 2, 31-37.

Woolsey, T.D. (1959). Adjusted death rates and other indices of mortality. Techniques of Vital Statistics Reprint of Chapters I-IV of Vital Statistics in the United States, (F.E. Linder and R.D. Grove, ed.). National Office of Vital Statistics, Washington, D.C. 20402. Stock number 1722-00236.

TWO-STAGE SELECTION PROCEDURES BASED ON TESTS*

Klaus J. Miescke

Department of Mathematics,
Statistics, and Computer Science
University of Illinois at Chicago
Chicago, Illinois

1. INTRODUCTION

The following procedure, which is used in certain clinical studies, may serve as a motivation for the considerations in this paper. Suppose that k new treatments have been developed with the purpose of replacing the standard treatment with the best new one, provided that it is actually an improvement on the standard treatment. At a preliminary stage each new treatment is applied several times and screened out if it is not considered to be significantly better than the standard treatment. Hereby, judgment is gained through suitable statistical tests at a fixed level of α_0. If all k new treatments are eliminated the standard treatment will not be replaced. If exactly one new treatment is not eliminated this will be taken as a replacement. In all other cases, the remaining treatments are further examined in a follow-up study through additional applications, and finally that one which appears to be the best will be used as a replacement of the standard treatment. The natural questions of how to choose the tests in the first stage and the terminal decision in the second stage are the topic of this paper.

Let $\pi_1,\ldots,\pi_k$ be k populations associated with unknown parameters $\theta_1,\ldots,\theta_k \in \Omega \subseteq \mathbb{R}$. Let $\theta_0 \in \Omega$ be a control value which may be known or

* This research was supported by the Office of Naval Research Contract N00014-75-C-0455 at Purdue University.

unknown. In the latter case, assume that there is also a control population, π_0. A population π_i is considered to be better than π_0 if $\theta_i > \theta_0$, $i = 1,\ldots,k$. The goal is to determine, in two stages, whether there is any population better than the control and, in the affirmative, which one is associated with the largest parameter. Assume that samples $\underline{X}_i = \{X_{ij}\}_{j=1,\ldots,n_i}$, $i = 0,1,\ldots,k$, and $\underline{Y}_i = \{Y_{ij}\}_{j=1,\ldots,m_i}$, $i = 1,\ldots,k$, of fixed predetermined sample sizes $n_0,\ldots,n_k$ and $m_1,\ldots,m_k$, respectively, can be drawn from the populations $\pi_0,\pi_1,\ldots,\pi_k$ at the first and at the second stage, respectively, which are mutually independent. Let $\{f_\theta\}_{\theta \in \Omega}$ be a given family of densities with respect to μ, the Lebesgue measure on $\mathbb{R}$ or the counting measure on any lattice in $\mathbb{R}$, and assume that for every $i \in \{0,1,\ldots,k\}$ all observations from π_i have a common distribution with density f_{θ_i}. Later on, after Theorem 1 has been proved, we will make the additional assumption that for every sample $\underline{Z}$ of size n from one population there exists a sufficient statistic $T_n(\underline{Z})$ such that the family of joint densities has nondecreasing likelihood ratios in T_n. For notational convenience let them be denoted below by $U_i = T_{n_i}(\underline{X}_i)$, $i = 0,1,\ldots,k$, $V_i = T_{m_i}(\underline{Y}_i)$, and $W_i = T_{n_i+m_i}(\underline{X}_i,\underline{Y}_i)$, $i = 1,\ldots,k$.

To simplify the presentation the case of a known control value θ_0 will be considered first. Before we define a natural class of two-stage procedures in a concise way, let us briefly describe how these procedures will be typically applied. For every testing problem H_i: $\theta_i \leq \theta_0$ versus K_i: $\theta_i > \theta_0$ the experimenter chooses a test based on $\underline{X}_i$ with a fixed level α_0 and another test based on $(\underline{X}_i,\underline{Y}_i)$ with a variable level α. At <u>Stage 1</u> he discards all populations which are not significant at level α_0 under the first set of tests. If none (exactly one) is left, he decides that none (this one) is better than the control (and is the best population). If more than one population survives he proceeds to Stage 2. At <u>Stage 2</u>, he draws additional samples $\underline{Y}_i$ from those populations which have been selected at Stage 1 and makes a final decision in favor of that population among the selected ones which has the smallest p-value (i.e. is most significant) based on the respective $(\underline{X}_i,\underline{Y}_i)$ under the associated second test.

If these tests are upper level tests, which for simplicity may be non-randomized for a moment to fix ideas, based on some real-valued statistics $\tilde{U}_i$ and $\tilde{W}_i$, say, $i = 1,\ldots,k$, then the procedure considered above can be equivalently described as follows: At Stage 1 all π_i's are selected with $\tilde{U}_i \geq c_i$ (where c_i is the $1-\alpha_0$ quantile of $\tilde{U}_i$ under $\theta_i = \theta_0$), and a final decision is made in terms of the largest $\tilde{W}_j$ among the selected π_j's. The truncated versions of such procedures (i.e. which perform Stage 1 only) have been studied by several authors, see, for example, Gupta and Sobel (1958) and Lehmann (1961). For further references see Gupta and Panchapakesan (1979), Chapter 20. Some preliminary results concerning two-stage procedures of the type described above in the case of $n_1 = \ldots = n_k$ and $m_1 = \ldots = m_k$ can be found in Gupta and Miescke (1982), which include a comparison with the one-stage analog by Bechhofer and Turnbull (1978).

Intuitively, one would expect that procedures based on better (more powerful) tests should have better performance characteristics. The purpose of this paper is to focus on that point and to clarify to what extent such a presumption can be confirmed. In Section 2 techniques from the theory of testing hypotheses will be used to derive some basic formulas for the relevant performance characteristics. They will serve as a basis for various considerations throughout the remainder of this paper. The central problem, however, of how to maximize the probability of a correct selection has to be attacked in a different approach. This will be done in Section 3. Under the assumption of monotone likelihood ratios it can be shown that the use of the respective uniformly most powerful test at Stage 2 is optimum in this respect, provided that $n_1+m_1 = \ldots = n_k+m_k$. Techniques used in the proof are mainly conditional arguments (given the outcomes at Stage 1) and the decreasing in transposition (DT) property by Hollander, Proschan and Sethuraman (1977). Analogous results concerning the choice of the tests at Stage 1 do not hold true. However, several strong justifications can be given for the use of the respective uniformly most powerful tests at Stage 1, partly under the more restrictive assumption of an underlying strongly unimodal (log-concave) exponential family. Finally, in Section 4, several applications with illustrations in the normal case will be given.

2. TWO-STAGE SELECTION PROCEDURES AND THEIR BASIC PROPERTIES

To begin with, let us point out that several definitions given in Miescke (1979) will be relevant in the sequel and thus are briefly recalled below. First of all, tests may be randomized ones taking values in [0,1]. This typically occurs in discrete cases or in continuous cases where nonparametric (rank) tests are under concern. Thus significance statements as well as p-values are understood to be based on additional randomization schemes as are used in Miescke (1979). To be more specific, let $\underline{A} = (A_1,\ldots,A_k)$ and $\underline{B} = (B_1,\ldots,B_k)$ be the randomization schemes for the first and the second stage, respectively. The A_i's and B_j's are independent random variables which are uniformly distributed in [0,1]. They are used to break ties in the case of their occurrence among the p-values of the corresponding tests. Note that the $\underline{X}_i$'s, $\underline{Y}_j$'s, $\underline{A}$, and $\underline{B}$ altogether are assumed to be mutually independent.

The Class $\mathcal{D}$ of Two-stage Procedures

For $i = 1,\ldots,k$, let $\phi_i = \{\phi_{i,\alpha}\}_{\alpha\in[0,1]}$ be a test for H_i versus K_i which is right continuous and monotone in α and is standardized at $\theta_i = \theta_0$, i.e. which satisfies $E_{\theta_0}(\phi_{i,\alpha}(\underline{X}_i)) = \alpha$ for every $\alpha \in [0,1]$. Moreover, assume that $\phi_{i,1} \equiv 1$ outside of the support of the distribution of $\underline{X}_i$ at $\theta_i = \theta_0$. These assumptions are made for technical reasons and are in fact not very restrictive as has been shown in Miescke (1979). Let $\underline{\phi}_{\alpha_0} = (\phi_{1,\alpha_0},\ldots,\phi_{k,\alpha_0})$ where from now on $0 < \alpha_0 < 1$ is fixed. Analogously, let $\psi_i = \{\psi_{i,\alpha}\}_{\alpha\in[0,1]}$ be such a test for H_i versus K_i based on $(\underline{X}_i,\underline{Y}_i)$. Let $\underline{\psi} = (\psi_1,\ldots,\psi_k)$. Let $\mathcal{D}$ be the class of all procedures of the following type $(\underline{\phi}_{\alpha_0},\underline{\psi})$:

Stage 1: Select π_i if $p_{\phi_i}(\underline{X}_i,A_i)$, the p-value of $\underline{X}_i$ under ϕ_i, is smaller than α_0, $i = 1,\ldots,k$. If none (exactly one) of the populations is selected, stop and decide none is better than π_0 (this one is better than π_0 and is the best population). Otherwise proceed to Stage 2.

Stage 2: Among the selected populations decide finally in favor of that π_j which has the smallest p-value $p_{\psi_j}(\underline{X}_j,\underline{Y}_j,B_j)$ under ψ_j.

The following result will prove to be useful in various aspects, except for the important question of how to optimize the component $\underline{\psi}$ in $(\underline{\phi}_{\alpha_0},\underline{\psi})$.

Theorem 1. Let $(\underline{\phi}_{\alpha_0},\underline{\psi}) \in \mathcal{D}$. For notational convenience, let $E_i = E_{\theta_i}(\phi_{i,\alpha_0}(\underline{X}_i))$ and $F_i(\alpha) = E_{\theta_i}(\phi_{i,\alpha_0}(\underline{X}_i)\psi_{i,\alpha}(\underline{X}_i,\underline{Y}_i))$, $\alpha \in [0,1]$, $i = 1,\ldots,k$, $\underline{\theta} \in \Omega^k$. Then for every non-empty $D \subseteq \{1,\ldots,k\}$ and $\underline{\theta} \in \Omega^k$,

$$P_{\underline{\theta}}\{\text{final decision falls into } D\} = \int_0^1 \prod_{j\notin D} [1-F_j(\alpha)]\, d(1-\prod_{i\in D}[1-F_i(\alpha)]), \tag{1}$$

$$P_{\underline{\theta}}\{\text{final decision is in favor of } \pi_i\} = \int_0^1 \prod_{\substack{j=1\\ j\neq i}}^{k} [1-F_j(\alpha)]\, dF_i(\alpha), \quad i = 1,\ldots,k, \tag{2}$$

$$P_{\underline{\theta}}\{\text{final decision is made at Stage 1 in favor of } \pi_i\} = \prod_{\substack{j=1 \\ j \neq i}}^{k} [1-E_j]\, E_i, \quad i = 1,\ldots,k, \tag{3}$$

$$P_{\underline{\theta}}\{\text{final decision is in favor of the control}\} = \prod_{j=1}^{k} [1-E_j]. \tag{4}$$

Proof: It has been shown in Miescke (1979) [cf. (2.3) - (2.5) loc. cit.] that the distribution function of each p-value appearing in $(\underline{\phi}_{\alpha_0},\underline{\psi})$ equals to the power function of the corresponding test, which is a continuous function of $\alpha \in [0,1]$ at every fixed parameter point, and which at $\alpha = 1$ assumes the value 1.

Let now D be a non-empty subset of $\{1,\ldots,k\}$. For $j = 1,\ldots,k$, let $p^*_{\psi_j}(\underline{X}_j,\underline{Y}_j,B_j)$ be equal to $p_{\psi_j}(\underline{X}_j,\underline{Y}_j,B_j)$ if $p_{\phi_j}(\underline{X}_j,A_j) \leq \alpha_0$, and let it be equal to 1 otherwise. Then it is easy to see that the l.h.s. of (1) is equal to

$$P_{\underline{\theta}}\{\min_{i \in D}\{p^*_{\psi_i}(\underline{X}_i,\underline{Y}_i,B_i)\} < \min_{j \notin D}\{p^*_{\psi_j}(\underline{X}_j,\underline{Y}_j,B_j)\}\}.$$

Since for $i \in \{1,\ldots,k\}$ and $\theta_i \in \Omega$, $P_{\theta_i}\{p^*_{\psi_i}(\underline{X}_i,\underline{Y}_i,B_i) \leq \alpha\}$ is equal to $F_i(\alpha)$ if $0 \leq \alpha < 1$, and is equal to 1 if $\alpha = 1$, (1) follows by standard arguments. (2) is a special case of (1) which was stated only because of its relevance in later applications. The verification of (3) and (4) is straightforward and therefore the proof is omitted.

Remark 1. It will be shown in Section 3 that under the assumption of monotone likelihood ratios (MLR), every $(\underline{\phi}_{\alpha_0},\underline{\psi}) \in \mathcal{D}$ is dominated by $(\underline{\phi}_{\alpha_0},\underline{\psi}^*)$ if $n_1 + m_1 = \ldots = n_k + m_k$, where $\underline{\psi}^*$ consists of the uniformly most powerful (UMP) tests. Hereby other techniques than those provided by Theorem 1 will be used since one has to apply conditional arguments. There is, however, a particular situation where (1) and (2) can be used for a similar purpose. Suppose that the data of Stage 1 are not available but the information on which populations have been significant is at hand. Then one has to use tests at Stage 2 which depend only on the $\underline{Y}_i$'s from the selected populations. In this case every $F_i(\alpha)$ factorizes into the product of the two power functions of ϕ_{i,α_0} and $\psi_{i,\alpha}$, respectively, and therefore (1) and (2) are completely determined through these power functions. For example, if $D_+(\underline{\theta}) = \{i|\theta_i > \theta_0, i \in \{1,\ldots,k\}\}$ is not empty, then (1) for $D = D_+(\underline{\theta})$ is maximized by the procedure which uses the UMP-tests at both stages. This is true even if the $n_i + m_i$'s are not assumed to be all equal.

Since these and related results in such a special case, however, are considered to be of less statistical importance they will not be discussed in further detail.

In the case of an unknown control parameter θ_0 some obvious changes have to be made. First of all the tests ϕ_{i,α_0} depend now on $(\underline{X}_i,\underline{X}_0)$, $i = 1,\dots,k$, whereas $\underline{\psi}$ remains the same as before. Let $\mathcal{D}'$ denote the class of two-stage procedures of the type $(\underline{\phi}_{\alpha_0},\underline{\psi})$ in this case. The analog of Theorem 1 for $\mathcal{D}'$ can be attained by replacing the right hand sides of (1) - (4) by their integrals with respect to the distribution of $\underline{X}_0$. If not explicitly stated otherwise, the results to be derived in the sequel for $\mathcal{D}$ have analogous counterparts for $\mathcal{D}'$ which will not be formulated or proved for brevity because of the close similarities.

3. OPTIMALITY RESULTS FOR MLR-FAMILIES

In the following let $D_+(\underline{\theta}) = \{i|\theta_i > \theta_0, i \in \{1,\dots,k\}\}$ be the "good" populations and $D_-(\underline{\theta}) = \{1,\dots,k\}\backslash D_+(\underline{\theta})$ be the "bad" ones, $\underline{\theta} \in \Omega^k$. Also let us partition the parameter space Ω^k into $\Omega_-^k = \{\underline{\theta} \in \Omega^k|\theta_i \le \theta_0, i = 1,\dots,k\}$ and its complement Ω_+^k, say. A procedure is said to make a correct selection (CS) at $\underline{\theta} \in \Omega_-^k$ if all populations are eliminated at Stage 1, and it is said to make a correct selection at $\underline{\theta} \in \Omega_+^k$ if a final decision is made in favor of a population with the largest θ-value. Let the goal be now to find a procedure in $\mathcal{D}$ which has a large probability of a correct selection (PCS) on Ω^k. From now on we assume that the family $\{f_\theta\}_{\theta\in\Omega}$ has the MLR-property as specified in Section 1. Then the following partial solution to our problem can be given.

Theorem 2. Let $(\underline{\phi}_{\alpha_0},\underline{\psi}) \in \mathcal{D}$. If $n_1 + m_1 = \dots = n_k + m_k$ and $\psi_1 = \dots = \psi_k$ then for all $\underline{\theta} \in \Omega^k$,

$$\sum_\sigma P_{\sigma(\underline{\theta})}\{CS \text{ under } (\underline{\phi}_{\alpha_0},\underline{\psi})\} \le \sum_\sigma P_{\sigma(\underline{\theta})}\{CS \text{ under } (\underline{\phi}_{\alpha_0},\underline{\psi}^*)\}\ , \tag{5}$$

where ψ^* consists of the UMP-tests for H_i versus K_i, $i = 1,\dots,k$, which in this case are all identical, and where the summation is with respect to all k! permutations of $(1,\dots,k)$. $\sigma(\underline{\theta}) = (\theta_{\sigma(1)},\dots,\theta_{\sigma(k)})$.

Proof: Only an outline of the proof will be given since it follows by similar decision theoretic arguments as have been used previously in Gupta and Miescke (1984).

Under the assumptions stated above, the associated decision function of $(\underline{\phi}_{\alpha_0},\underline{\psi})$ which determines final selections at Stage 2 is permutation invariant. The loss function which is implicitly employed is zero if a correct selection is made, and is one, otherwise. Its component which is associated with final selections at Stage 2 is permutation invariant and favors selections of populations with large parameters.

Let now $\underline{\theta} \in \Omega_+^k$ be fixed and, in a Bayes approach, assume that the unknown parameter vector is random and has a prior distribution which gives equal mass 1/k! to all permutations $\sigma(\underline{\theta})$ of $\underline{\theta}$. Then the posterior distribution of the parameter vector, given $W_i = w_i$, $i = 1,\ldots,k$, has the decreasing in transposition (DT) property. This is a multivariate extension of the MLR-property which was studied in detail by Hollander, Proschan, and Sethuraman (1977). From this fact and the properties of the loss function stated above it follows that the optimal final decision at Stage 2 is the natural one which is made with respect to the non-eliminated population with the largest W_i, where ties are broken at random. Clearly this is equivalent to selecting the non-eliminated population with the smallest p-value under test ψ^*. The proof is now completed by noting that (5) gives a comparison of the corresponding Bayes risks, where of course at all $\underline{\theta} \in \Omega_-^k$ the probabilities of a correct selection are the same for both procedures.

Corollary 1. Under the assumptions of Theorem 2, suppose further that $n_1 = \ldots = n_k$ and $\phi_{1,\alpha_0} = \ldots = \phi_{k,\alpha_0}$ holds. Then for all $\underline{\theta} \in \Omega^k$

$$(6)\quad P_{\underline{\theta}}\{CS \text{ under } (\underline{\phi}_{\alpha_0}, \underline{\psi})\} \le P_{\underline{\theta}}\{CS \text{ under } (\underline{\phi}_{\alpha_0}, \underline{\psi}^*)\}.$$

Proof: Since both procedures considered here are completely permutation invariant, and since also the 0-1 loss function employed is permutation invariant, their risk functions are symmetric functions of $\underline{\theta}$ Ω^k. Therefore all summands on the l.h.s. of (5) coincide and the same holds for the summands on the r.h.s. of (5).

Remark 2. The proof of Theorem 2 actually applies more generally to the following situation. No matter how the populations are eliminated at Stage 1, if $n_1 + m_1 = \ldots = n_k + m_k$, and if only permutation invariant final decision functions are admitted at Stage 2, then every Bayes procedure w.r.t. any symmetric prior employs the natural rule at Stage 2. Of course, also from a non-Bayesian point of view, (5) is an intuitively appealing criterion. It simply reflects the lack of knowledge of how the sample sizes are associated with the k ordered populations parameters. Since there is not even an approximately similar result available in the case of unequal $n_i + m_i$'s, it is strongly recommended to "repair" the design of every experiment with unequal n_i's by choosing the m_i's appropriately to get equal overall sample sizes. Let us assume from now on that $n_1 + m_1 = \dots = n_k + m_k = N$, say, holds.

Actually, in various selection problems authors have chosen their designs such that the statistics on which the natural final decision rule is based have joint distributions with the DT property. To mention a few relevant examples, Bechhofer (1954), Bechhofer, Dunnett and Sobel (1954), and Dudewicz and Dalal (1975) have done so to be able to implement their procedures at certain specified performance requirements. It may now be added that exactly in these designs the employed natural final decisions are optimal in terms of the risk or the PCS, respectively, uniformly on all parameter configurations.

Lack of the DT property in distributions of statistics used for final decisions, in cases where this property could be attained in principle, should be considered as pathological designs. Even in the simplest case of a one-stage selection procedure serious difficulties arise, as may be illustrated by the following problem which was emphasized by Bechhofer (1982).

Example. Let $\overline{X}_1,\dots,\overline{X}_k$ be independent sample means with unknown expectations $\theta_1,\dots,\theta_k$ and with known but different variances $q_1,\dots,q_k$ generated from k normal populations. For the problem of selecting a population with the largest θ-value no procedure exists which has a largest PCS, uniformly in $\underline{\theta} \in \mathbb{R}^k$. At least for $k = 2$ the natural rule which selects the population with the largest sample mean can be seen to be Bayes with respect to independent priors $N(0,q_1)$ and $N(0,q_2)$, respectively. It follows that it is also admissible under the 0-1 loss function provided that the loss is zero (or constant) whenever $\theta_1 = \theta_2$. This is so because of the continuity of the risk function at all $\underline{\theta}$ with $\theta_1 \neq \theta_2$. However, if $k \geq 3$ the natural rule cannot be Bayes with respect to any (multivariate) normal prior. The question of whether it is admissible is still open. On the other hand, for every normal prior with expectation $(0,\dots,0)$ for which the posterior distribution has the DT-property (which leads to a simple solution) the Bayes rule selects in terms of the largest $\overline{X}_i/q_i$, $i = 1,\dots,k$.

In the remainder of this paper only procedures of the type $(\underline{\phi}_{\alpha_0},\underline{\psi}^*) \in \mathcal{D}$ will be considered. This is justified in view of the assumption $n_1 + m_1 = \dots = n_k + m_k = N$ and of Theorem 2. The following result, which generalizes Theorem 3 of Gupta and Miescke (1982), can be used to find least favorable parameter configurations (LFC) of such procedures on suitable subspaces of Ω^k.

Theorem 3. Let $(\underline{\phi}_{\alpha_0},\underline{\psi}^*) \in \mathcal{D}$ where for every $i \in \{1,\ldots,k\}$ the power function of ϕ_{i,α_0} is nondecreasing in θ_i. Then the performance characteristics considered in Theorem 1 have the following monotonicity properties.
(1) is nondecreasing in θ_i, $i \in D$ and nonincreasing in θ_j, $j \notin D$. (2) and (3) are nondecreasing in θ_i and nonincreasing in θ_j, $j \neq i$. (4) is nonincreasing in $\theta_1,\ldots,\theta_k$.

Proof: The assertions concerning (3) and (4) are obviously true. To prove those concerning (1) and (2) note that for $i \in \{1,\ldots,k\}$, $\underline{\theta} \in \Omega^k$, and $\alpha \in [0,1]$,

$$F_i(\alpha) = E_{\theta_i}(\phi_{i,\alpha_0}(\underline{X}_i))\, E_{\theta_i}\{\psi^*_{i,\alpha}(\underline{X}_i,\underline{Y}_i)|p_{\phi_i}(\underline{X}_i,A_i) \le \alpha_0\}. \tag{7}$$

The first factor on the r.h.s. of (7) is nondecreasing in θ_i according to the assumptions made above. The second factor can be seen to have the same property by applying Theorem 1 of Simons (1980) which guarantees that for every sample from a MLR-family, conditionally on any proportion of the information in the sample which one might choose to extract, likelihood ratios are still stochastically nondecreasing. Since $\psi^*_{i,\alpha}(\underline{X}_i,\underline{Y}_i)$ is a nondecreasing function of W_i, the proof is completed by noting that A_i could be ignored since the arguments apply to every situation $A_i = a_i$, where $a_i \in [0,1]$ is held fixed.

Corollary 2. Under the assumptions of Theorem 3, let $n_1 = \ldots = n_k$ and $\phi_{1,\alpha_0} = \ldots = \phi_{k,\alpha_0}$. Then for every $\underline{\theta} \in \Omega^k$ with $\theta_1 \le \ldots \le \theta_k$, the probability of a final decision in favor of population π_i is nondecreasing in $i \in \{1,\ldots,k\}$.

Proof: Given the assumptions above it can be seen from the proof of Theorem 3 that for $\theta_1 \le \ldots \le \theta_k$,

$$F_1(\alpha) \le F_2(\alpha) \le \ldots \le F_k(\alpha), \quad \alpha \in [0,1]. \tag{8}$$

Therefore the assertion follows from (2) by the same technique which was used in (IV) of Miescke (1979).

Theorem 4. Let $(\underline{\phi}_{\alpha_0},\underline{\psi}^*) \in \mathcal{D}$ where $\underline{\phi}_{\alpha_0}$ consists of consistent tests. Then for increasing sample sizes n_i and $m_i = N - n_i$, $i = 1,\ldots,k$, the probability of a correct selection tends to one at all $\underline{\theta} \in \Omega^k_+$ and at all $\underline{\theta} \in \Omega^k_-$ with $\theta_i < \theta_0$, $i = 1,\ldots,k$.

Proof: If $\underline{\theta} \in \Omega^k_+$ with exactly one coordinate greater than θ_0, or if $\underline{\theta} \in \Omega^k_-$ with $\theta_i < \theta_0$, $i = 1,\ldots,k$, the assertion follows immediately from (3) and

(4), respectively. For all other $\underline{\theta} \in \Omega_+^k$, the probability that all populations π_i with $\theta_i > \theta_0$ will not be eliminated at Stage 1 tends to one. Moreover, $P_{\underline{\theta}}\{W_i > W_j$ for all $j \neq i\}$ also tends to one if θ_i is the unique maximum of $\theta_1,\ldots,\theta_k$. This can be seen as follows. Selecting in terms of the largest W_j is equivalent to selecting in terms of the smallest p-value under tests $\tilde{\psi}_j$ which are essentially the same tests as ψ_j^* but now standardized at θ_i. By Theorem 2 of Miescke (1979) it follows that

$$P_{\underline{\theta}}\{W_i > W_j \text{ for all } j \neq i\} = \int_0^1 \prod_{j\neq i} [1-E_{\theta_j}(\tilde{\psi}_{j,\alpha}(\underline{X}_j,\underline{Y}_j))]d\alpha, \tag{9}$$

which now can be seen to tend to one if N tends to infinity. If, however, all the good populations are not eliminated at Stage 1 and $W_i > W_j$, for all $j \neq i$, then a correct selection is made. This completes the proof in the given parameter configuration. The case of more than one best population can be treated similarly.

Focussing now on the first component $\underline{\phi}_{\alpha_0}$ in the procedures $(\underline{\phi}_{\alpha_0},\underline{\psi}^*) \in \mathcal{D}$, the natural choice is, of course, $\underline{\phi}_{\alpha_0}^*$ which consists of the corresponding UMP-tests for H_i versus K_i based on $\underline{X}_i$, or more precisely, based on U_i, $i = 1,\ldots,k$. Even though such a choice cannot be justified (not even in the case of $n_1 = \ldots = n_k$) by an overall improvement on the PCS, several strong reasons can be quoted in support of choosing $(\underline{\phi}_{\alpha_0}^*,\underline{\psi}^*)$. First, of course, all results derived hitherto hold for this procedure. Second, the following can be stated.

Theorem 5. Among all procedures in $\mathcal{D}$, $(\underline{\phi}_{\alpha_0}^*,\underline{\psi}^*)$ maximizes

$$P_{\underline{\theta}}\{CS\} \text{ at every } \underline{\theta} \in \Omega_-^k, \tag{10}$$

$$P_{\underline{\theta}}\{CS \text{ at Stage 1}\} \text{ at every } \underline{\theta} \in \Omega_+^k \text{ with exactly one } \theta_i > \theta_0, \tag{11}$$

$$E_{\underline{\theta}} \text{ (number of good populations selected at Stage 1) at every } \underline{\theta} \in \Omega_+^k, \tag{12}$$

$$E_{\underline{\theta}} \text{ (number of bad populations eliminated at Stage 1) at every } \underline{\theta} \in \Omega^k. \tag{13}$$

Proof: (10) follows from (4) and (11) follows from (3). All arguments are standard and are based on the well known properties of the power functions of the UMP-tests ϕ_{i,α_0}^*, $i = 1,\ldots,k$. Therefore, no further details will be given.

Third, all permutation invariant procedures $(\underline{\phi}_{\alpha_0},\underline{\psi}^*)$, except $(\underline{\phi}_{\alpha_0}^*,\underline{\psi}^*)$ itself, can be modified, without changing the sizes of the selected subsets of populations at Stage 1, in a specific way which leads to an improvement on the PCS, provided that the family $\{f_\theta\}_{\theta\in\Omega}$ is a strongly unimodal exponential family. It should be noted that the modified procedure is also based on tests but is no longer a member of the class $\mathcal{D}$. More precisely, from Corollary 2 in Gupta and Miescke (1984) the following result can be derived.

Theorem 6. Let $\{f_\theta\}_{\theta\in\Omega}$ be a strongly unimodal (i.e. log-concave) exponential family. Then every $(\underline{\phi}_{\alpha_0},\underline{\psi}^*) \in \mathcal{D}$ with $n_1 = \ldots = n_k$ and $\phi_{1,\alpha_0} = \ldots = \phi_{k,\alpha_0}$ can be improved by simply replacing the selected populations at Stage 1 by the same number of populations, but now by those which are associated with the largest U_i's, where ties are broken at random. Then for all $\underline{\theta} \in \Omega^k$,

$$P_{\underline{\theta}}\{\text{CS under } (\underline{\phi}_{\alpha_0},\underline{\psi}^*)\} \leq P_{\underline{\theta}}\{\text{CS under the modified procedure}\}. \tag{14}$$

4. APPLICATIONS WITH ILLUSTRATIONS IN THE NORMAL CASE

In applications the procedure $(\underline{\phi}^*_{\alpha_0},\underline{\psi}^*)$ usually will be implemented as to meet certain performance requirements. This will be described in this section and will be illustrated by the example of k normal populations $N(\theta_i,\sigma_i^2)$, $i = 1,\ldots,k$, where U_i and W_i are the corresponding sample means, $i = 1,\ldots,k$. Here the procedure can be considered to be the two-stage analog of the one-stage procedure by Bechhofer and Turnbull (1978). At first consider the basic requirement

$$\inf\{P_{\underline{\theta}}\{\text{CS under } (\underline{\phi}^*_{\alpha_0},\underline{\psi}^*)\} \,|\, \underline{\theta} \in \Omega^k_-\} = P^*_0 , \tag{15}$$

where P^*_0 is a predetermined constant. In view of (4) this can be accomplished by choosing α_0 to satisfy

$$(1 - \alpha_0)^k = P^*_0 . \tag{16}$$

Then in the normal case the procedure is of the following form. At Stage 1, population π_i is selected if $n_i^{1/2}(U_i - \theta_0)/\sigma_i \geq \Phi^{-1}(1 - \alpha_0)$, $i = 1,\ldots,k$, where Φ denotes the c.d.f. of N(0,1). And at Stage 2, a final decision is made in terms of the largest W_j from the selected populations.

Since (15) actually involves only the properties of the procedure at the first stage, the P^*_0-condition can be attained by employing techniques used for one-stage procedures. Thus (15) can be solved by taking recourse to relevant papers in this area. For further details see Gupta and Panchapakesan (1979).

A second requirements will typically employ the indifference zone approach which is due to Bechhofer (1954). Let $\Delta > 0$ be fixed and let $\Omega^k_\Delta = \{\underline{\theta} \in \Omega^k \mid \theta_i + \Delta \geq \theta_0, \theta_1, \ldots, \theta_{i-1}, \theta_{i+1}, \ldots, \theta_k$ for some $i\}$. Now consider the requirement

$$\inf\{P_{\underline{\theta}}\{CS \text{ under } (\underline{\phi}^*_{\alpha_0}, \underline{\psi}^*)\} \mid \underline{\theta} \in \Omega^k_\Delta\} = P^*_1 , \tag{17}$$

where P^*_1 is a second predetermined constant. Even though Theorem 3 can be used to find the LFC, it is technically too difficult to attain (17) exactly. Therefore, the following conservative approach, which over-protects the experimenter with respect to (17), is recommended and is easy to perform. Let $\underline{\theta} \in \Omega^k_\Delta$ with $\theta_i = \max\{\theta_1, \ldots, \theta_k\}$, say. If population π_i is selected at Stage 1 and W_i is the unique maximum of $W_1, \ldots, W_k$ then a correct selection is made. Therefore if the following two conditions are fulfilled, and β_1 and β_2 are chosen to meet $\beta_1 + \beta_2 - 1 = P^*_1$ then by Bonferroni's inequality it follows that the l.h.s. of (17) is not smaller than P^*_1. The conditions are

$$E_{\theta_0+\Delta}(\phi^*_{j,\alpha_0}(\underline{X}_j)) \geq \beta_1, \quad j = 1, \ldots, k, \text{ and} \tag{18}$$

$$\inf\{P_{(\theta+\Delta,\theta,\ldots,\theta)}\{W_1 > W_2, \ldots, W_k\} \mid \theta \in \Omega\} \geq \beta_2 . \tag{19}$$

In the normal case it is well known (cf. Tamhane and Bechhofer (1979)) that Slepian's inequality leads to better results than Bonferroni's inequality. Use of the former allows us to choose β_1 and β_2 according to the condition $\beta_1\beta_2 = P^*_1$ which is preferable since $\beta_1 + \beta_2 - 1 < \beta_1\beta_2$ for $0 < \beta_1, \beta_2 < 1$.

To meet (18), standard techniques from the theory of testing hypotheses can be used. And (19) can be attained by using results of single-stage selection procedures in the indifference zone approach due to Bechhofer (1954).

SUMMARY

Suppose that k new treatments have been developed with the purpose of replacing the standard treatment with the best new one, provided that it is actually an improvement on the standard treatment. In a parametric approach, mainly under the assumption of MLR, procedures are considered which, at a first stage, screen out inferior treatments through statistical tests at a common level of significance α_0. If none (exactly one) is not eliminated, none (this one) will be used as a replacement. Otherwise, if more than one treatment overcomes this screening process, that one of the non-eliminated treatments will be chosen as the replacement which is judged to be the best, after additional data have been observed from the selected treatments. Topics of this paper are the questions of how to choose the terminal decision at the second stage and the tests at the first stage, respectively, and how to implement the appropriate procedures at certain pre-specified performance criteria.

ACKNOWLEDGMENT

The author wishes to express his thanks to a referee for several helpful comments which have led to an improvement of the presentation of the material.

REFERENCES

Bechhofer, R. E. (1954). A single sample multiple decision procedure for ranking means of normal populations with known variances. Ann. Math. Statist. 25, 16-39.

Bechhofer, R. E. (1982). Private communication.

Bechhofer, R. E., Dunnett, C. W., and Sobel, M. (1954). A two-sample multiple-decision procedure for ranking means of normal populations with a common unknown variance. Biometrika 41, 170-176.

Bechhofer, R. E. and Turnbull, B. W. (1978). Two (k+1)-decision selection procedures for comparing k normal means with a specified standard. J. Amer. Statist. Assoc. 73, 385-392.

Dudewicz, E. J. and Dalal, S. R. (1975). Allocation of observations in ranking and selection with unequal variances. Sankhyā Ser. B 37, 28-78.

Gupta, S. S. and Miescke, K. J. (1982). On the problem of finding a best population with respect to a control in two stages. In Statistical Decision Theory and Related Topics III (Ed. Gupta, S. S. and Berger, J. O.), Vol. 1, 473-496. Academic Press, New York.

Gupta, S. S. and Miescke, K. J. (1984). Sequential selection procedures - A decision theoretic approach. Ann. Statist. 12, 336-350.

Gupta, S. S. and Panchapakesan, S. (1979). Multiple Decision Procedures: Theory and Methodology of Selecting and Ranking Populations. J. Wiley, New York.

Gupta, S. S. and Sobel, M. (1958). On selecting a subset which contains all populations better than a standard. Ann. Math. Statist. 29, 235-244.

Hollander, M., Proschan, F., and Sethuraman, J. (1977). Functions decreasing in transposition and their applications in ranking problems. Ann. Statist. 5, 722-733.

Lehmann, E. L. (1961). Some model I problems of selection. Ann. Math. Statist. 32, 990-1012.

Miescke, K. J. (1979). Identification and selection procedures based on tests. Ann. Statist. 7, 207-219.

Simons, G. (1980). Extensions of the stochastic ordering property of likelihood ratios. Ann. Statist. 8, 833-839.

Tamhane, A. C. and Bechhofer, R. E. (1979). A two-stage minimax procedure with screening for selecting the largest normal mean (II): An improved PCS lower bound and associated tables. Commun. Statist. - Theor. Meth. A8, 337-358.

DESIGNING EXPERIMENTS FOR SELECTING A NORMAL POPULATION WITH A LARGE MEAN AND A SMALL VARIANCE*

Thomas J. Santner

School of Operations Research
and Industrial Engineering
Cornell University,
Ithaca, New York

Ajit C. Tamhane

Department of Industrial Engineering
and Management Sciences
Northwestern University
Evanston, Illinois

1. INTRODUCTION

Let $\Pi_1,\Pi_2,\ldots,\Pi_p$ be $p \geq 2$ normal populations where observations X_{ij} from Π_i are independently distributed as $N(\mu_i,\sigma_i^2)$ $(i = 1,2,\ldots,p;$ $j = 1,2,\ldots)$. All the means μ_i and variances σ_i^2 are unknown. For this setup Dudewicz and Dalal (1975) and Rinott (1978) studied the problem of selecting the population having the largest mean. However, in many practical situations an experimenter will not be satisfied in selecting the population with the largest mean if it has (nearly) the largest variance. For example, suppose that a manufacturing process is to be selected from $p \geq 2$ possible alternatives. It may be desired to select a process that is highly productive (large mean daily output) as well as highly reliable (small day-to-day variation in the output). If no such process exists then it may be desired to select none.

Motivated by the above consideration, the present paper proposes a new selection goal that is appropriate when both means and variances are important. In §2 we give a mathematical formulation of this new selection goal and an associated probability requirement based on the indifference-zone approach of Bechhofer (1954). This formulation requires the experimenter to specify two integers a and b $(1 \leq a,b \leq p)$ such that any

*This research was performed while the second author was visiting Cornell University in 1982-83 on a sabbatical leave with partial support from the U.S. Army Research Office-Durham Contract DAAG-29-81-K0168.

Π_i with mean μ_i among the a largest ones and variance σ_i^2 among the b smallest ones is regarded as a "good" population. The goal then is to select the population with the largest mean from the set of "good" populations if that set is nonempty and to select no population if it is empty. In §3 we propose different two-stage selection procedures for the two cases: (I) $a + b > p$, in which case the set of "good" populations cannot be empty, and (II) $a + b \leq p$. The probabilities of a correct selection (P(CS)) associated with these two-stage procedures are derived in §4 and their infimums over the appropriate region ("preference-zone") of the parameter space are studied. Based on the expressions for these infimums §5 gives computer programs for calculating the constants required to implement the procedures. The conservatism of certain bounds employed in deriving the infimums on the P(CS) is also studied in this section by Monte Carlo simulation. Some concluding remarks are made in §6.

2. FORMULATION OF THE PROBLEM

Denote $\underset{\sim}{\mu} = (\mu_1,\mu_2,\ldots,\mu_p)$, $\underset{\sim}{\sigma}^2 = (\sigma_1^2,\sigma_2^2,\ldots,\sigma_p^2)$, $\underset{\sim}{\omega} = (\underset{\sim}{\mu},\underset{\sim}{\sigma}^2)$ and the parameter space of $\underset{\sim}{\omega}$ by Ω. Let $\mu_{[1]} \leq \mu_{[2]} \leq \cdots \leq \mu_{[p]}$ and $\sigma_{[1]}^2 \leq \sigma_{[2]}^2 \leq \cdots \leq \sigma_{[p]}^2$ be the ordered means and variances, respectively. For <u>specified</u> integers a and b $(1 \leq a,b \leq p)$ and $\underset{\sim}{\omega} \in \Omega$ define the index sets

$$A = A(\underset{\sim}{\mu};a) = \{i: \mu_i \geq \mu_{[p-a+1]}\}$$

and

$$B = B(\underset{\sim}{\sigma}^2;b) = \{i: \sigma_i^2 \leq \sigma_{[b]}^2\}.$$

Let the index set of "good" populations be denoted by

$$C = C(\underset{\sim}{\omega};a,b) = A(\underset{\sim}{\mu},a) \cap B(\underset{\sim}{\sigma}^2,b)$$

and let $c = \text{card}\ (C)$; clearly $\max(a+b-p,0) \leq c \leq \min(a,b)$.

The experimenter's <u>goal</u> is to select a population associated with $\max\{\mu_i : i \in C\}$ if C is nonempty and select no population if C is empty. If a selection procedure achieves this goal then a <u>correct selection</u> (CS) is said to have been made.

<u>Remark 2.1</u>. If $a + b > p$ then C cannot be empty and $\max\{\mu_i : i \in C\} = \max\{\mu_i : i \in B\}$. In this case the above goal reduces to selecting the population with the largest mean among the ones having the b smallest variances. Note that this goal is independent of the value of a as long as $a + b > p$. This special goal was considered by Handa and Jain (1982).

In the special subcase $b = p$ of $a + b > p$, the goal reduces to selecting the population with the largest mean $\mu_{[p]}$ which was considered by Dudewicz and Dalal (1975) and Rinott (1978). In the subcase $(a,b) = (p,1)$ of $a + b > p$, the goal reduces to selecting the population with the smallest variance $\sigma^2_{[1]}$ which was considered by Bechhofer and Sobel (1954).

To formulate the probability requirement for the above selection goal we adopt the <u>indifference-zone approach</u> of Bechhofer (1954). In this formulation we assume that the experimenter can specify three constants $\{\delta,\theta,\gamma\}$ where $\delta > 0$, $\theta > 1$ and $1/p < \gamma < 1$. The first two constants (together with specified integers a and b) define the following <u>preference-zone</u>:

$$\Omega^* = \bigcup_{i=0}^{p} \Omega^*_i$$

where

$$\Omega^*_0 = \{\underset{\sim}{\omega} \in \Omega: C = \phi,\ \sigma^2_{[b+1]}/\sigma^2_{[b]} \geq \theta,\ \mu_{[p-a+1]} - \mu_{[p-a]} \geq \delta\}$$

and for $1 \leq i \leq p$,

$$\Omega^*_i = \{\underset{\sim}{\omega} \in \Omega: i \in C,\ \sigma^2_{[b+1]}/\sigma^2_{[b]} \geq \theta,\ \mu_i \geq \mu_j + \delta\ \forall\ j \in C,\ j \neq i \ \&\ \forall\ j \notin A\}.$$

The indifference-zone is $\Omega - \Omega^*$.

<u>Remark 2.2</u>. If $a = p$ then the last condition in the definition of Ω^*_0 is vacuous. If $b = p$ then the second condition in the definition of Ω^*_i $(0 \leq i \leq p)$ is vacuous. Also note that if $c > 1$ (resp., $c = 1$) then for $i = 1,\ldots,p$ the last condition in the definition of Ω^*_i is equivalent to $\mu_i \geq \mu_j + \delta\ \ \forall\ j \in C,\ j \neq i$ (resp., $\mu_i \geq \mu_j + \delta\ \forall\ j \notin A$).

Our <u>probability requirement</u> is

$$\inf_{\Omega^*} P_{\underset{\sim}{\omega}}(CS) \geq \gamma. \tag{2.1}$$

An equivalent way of stating (2.1) which will be used in the sequel is as follows: Let d_i denote the decision to select the population Π_i $(1 \leq i \leq p)$ and d_0 denote the decision to select no population. Then the probability requirement is

$$\inf_{\Omega^*_i} P_{\underset{\sim}{\omega}}(d_i) \geq \gamma \qquad (0 \leq i \leq p). \tag{2.2}$$

3. TWO-STAGE SELECTION PROCEDURES P_I AND P_{II}

3.1 Description of the Procedures

Since $\underset{\sim}{\sigma}^2$ is assumed to be unknown and the parameter of primary interest for selection purposes is the mean of a population, it follows from the arguments of Dantzig (1940) and Dudewicz (1971) that except for the case $a = p$, $b = 1$, no single-stage procedure can guarantee (2.2). (For $a = p$, $b = 1$, when the goal is to select the population associated with $\sigma^2_{[1]}$, a single-stage procedure can be used, e.g., the Bechhofer-Sobel (1954) procedure.)

We now propose two-stage procedures P_I and P_{II} for Case I: $a + b > p$ and Case II: $a + b \leq p$, respectively.

<u>Case I $(a + b > p)$: Procedure</u> P_I

<u>Stage 1</u>: For $i = 1,2,\ldots,p$, take a random sample of size $n_i \geq 2$ (fixed in advance) from Π_i and calculate the sample variances

$$S_i^2 = \left[\sum_{j=1}^{n_i} X_{ij}^2 - \left(\sum_{j=1}^{n_i} X_{ij}\right)^2 \Big/ n_i\right] \Big/ (n_i - 1).$$

Let $S^2_{[1]} \leq S^2_{[2]} \leq \cdots \leq S^2_{[k]}$ be the ordered sample variances,

$$\hat{B} = \{i: S_i^2 \leq S^2_{[b]}\}, \tag{3.1}$$

and

$$N_i = \max\{n_i, \langle (S_i h/\delta)^2 \rangle\}, \; i \in \hat{B} \tag{3.2}$$

where $\langle x \rangle$ denotes the smallest integer $\geq x$ and $h > 0$ is a predetermined constant. The choice of h is discussed in Remark 4.3.

<u>Stage 2</u>: For $i \in \hat{B}$ take an additional random sample of size $(N_i - n_i)$ from Π_i and calculate the overall sample mean $\overline{X}_i = \sum_{j=1}^{N_i} X_{ij}/N_i$. Select the population yielding $\max\{\overline{X}_i: i \in \hat{B}\}$.

<u>Case II $(a + b \leq p)$: Procedure</u> P_{II}

<u>Stage 1</u>: The first stage is the same as that of P_I except that N_i must be calculated by (3.2) for all Π_i $(1 \leq i \leq p)$.

<u>Stage 2</u>: For $i = 1,2,\ldots,p$ take an additional random sample of size $(N_i - n_i)$ from Π_i and calculate the overall sample mean $\overline{X}_i$. Let $\overline{X}_{[1]} \leq \overline{X}_{[2]} \leq \cdots \leq \overline{X}_{[p]}$ be the ordered sample means,

$$\hat{A} = \{i: \overline{X}_i \geq \overline{X}_{[p-a+1]}\},$$

$\hat{B}$ be as in (3.1) and $\hat{C} = \hat{A} \cap \hat{B}$. Select no population if $\hat{C} = \phi$; otherwise, select the population yielding $\max\{\overline{X}_i : i \in \hat{C}\}$.

For convenience, we shall hereafter assume that $n_i = n$ for $1 \leq i \leq p$. Thus each S_i^2 is based on the same degrees of freedom $\nu = n - 1$.

3.2 Some Remarks on the Procedures

(i) There are two design constants, n and h, in P_I and P_{II} which must be chosen to guarantee (2.1) for given p and specified $\{a,b,\delta,\theta,\gamma\}$. (For P_I the design constants do not depend on a.) For fixed n, h must be chosen so that the infimum of $P_{\underset{\sim}{\omega}}(CS)$ over Ω^* is $\geq \gamma$. It will be shown that (see Remarks 4.2 and 4.3) that such a finite h exists iff $n > n_0$ where n_0 is a positive integer which depends on p,b,θ and γ.

(ii) It follows from Remark 2.1 that in P_I we do not need to estimate sets A and $C = A \cap B$ while in P_{II} we do. Thus in Case I it suffices to select the population yielding the largest $\overline{X}_i$ among the ones in the estimated set $\hat{B}$. However, in Case II we must estimate set C to determine whether to make decision d_0. Thus it is necessary to sample from all populations in the second stage.

(iii) Ideally one would like to use the data from both the stages to estimate the σ_i^2 and the set B. However, this poses the same technical difficulties that were present in Stein's original (1945) problem.

4. PROBABILITIES OF A CORRECT SELECTION OF P_I AND P_{II}

4.1 Case I (a + b > p): Procedure P_I

Since $C \neq \phi$ in this case, it suffices to consider the infimum of $P_{\underset{\sim}{\omega}}(d_p | P_I)$ over Ω_p^* (say) in order to guarantee (2.2). Our result is given in the following theorem.

Theorem 4.1. Let $\Phi(\cdot)$ and $F_\nu(\cdot)$ denote the cdf's of a N(0,1) and a χ_ν^2 random variable (rv), respectively, where $\nu = n - 1$. Let

$$G(v,w) = G(v,w|\nu,h,\theta) = \int_0^{\theta w} \Phi\left(\frac{h}{\sqrt{\nu/v + \nu/y}}\right) dF_\nu(y). \tag{4.1}$$

Then for given p and specified $\{a,b,\delta,\theta\}$ where $b < p$ and $a + b > p$,

$$\inf_{\Omega_p^*} P_{\underset{\sim}{\omega}}(d_p|P_I) \geq (p-b)\int_0^\infty [\int_0^{\theta w} \{G(v,w)\}^{b-1} dF_\nu(v)]$$

$$\times\{1 - F_\nu(w)\}^{p-b-1} dF_\nu(w). \qquad (4.2)$$

Proof. Fix $\underset{\sim}{\omega} \in \Omega_p^*$ and thus $A = A(\underset{\sim}{\mu};a)$, $B = B(\underset{\sim}{\sigma}^2;b)$ and $C = A \cap B$ with $p \in C$. Consider

$$P_{\underset{\sim}{\omega}}(d_p|P_I) = \sum_Q P_{\underset{\sim}{\omega}}\{\overline{X}_p > \overline{X}_i\ \forall\ i \in Q,\ i \neq p;\ S_i^2 < S_j^2\ \forall\ i \in Q,\ j \notin Q\} \quad (4.3)$$

where the sum is over all subsets Q of $\{1,2,\ldots,p\}$, $p \in Q$ with card(Q) = b. A lower bound on (4.3) is given by the term for $Q = B$, viz.,

$$P_{\underset{\sim}{\omega}}\{\overline{X}_p > \overline{X}_i\ \forall i \in B,\ i \neq p;\ S_i^2 < S_j^2\ \forall i \in B,\ j \notin B\}. \qquad (4.4)$$

This bound is sharp for $\underset{\sim}{\omega} \in \Omega_p^*$; if $\overline{A} \cap \overline{B} = \phi$ (where $\overline{A}$ and $\overline{B}$ denote the complements of A and B, respectively) then we can choose a sequence $\{\underset{\sim}{\omega}_\ell\}$ in Ω_p^* such that $\mu_i \to +\infty\ \forall\ i \in A \cap \overline{B}$ which makes all the terms in (4.3) for $Q \neq B$ approach zero yielding (4.4). If $\overline{A} \cap \overline{B} \neq \phi$ then we can choose the sequence such that, in addition, $\sigma_j^2 \to +\infty\ \forall\ j \in \overline{A} \cap \overline{B}$ thus yielding (4.4).

We shall find the infimum of the lower bound (4.4) which we first write as

$$P_{\underset{\sim}{\omega}}\{Y_{pi} < \frac{\mu_p - \mu_i}{\sqrt{\sigma_p^2/N_p + \sigma_i^2/N_i}}\ \forall\ i \in B,\ i \neq p;\ Z_i < \frac{\sigma_j^2}{\sigma_i^2} Z_j\ \forall\ i \in B,\ j \notin B\}. \quad (4.5)$$

Here

$$Y_{ij} = \frac{\overline{X}_j - \overline{X}_i + \mu_i - \mu_j}{\sqrt{\sigma_i^2/N_i + \sigma_j^2/N_j}}\ (i \neq j,\ 1 \leq i,\ j \leq p),\ Z_i = \frac{\nu S_i^2}{\sigma_i^2}\ (1 \leq i \leq p). (4.6)$$

For fixed $\underset{\sim}{\sigma}^2$, (4.5) is increasing in each $\mu_p - \mu_i$ for $i \in B$, $i \neq p$. So a lower bound on the infimum of (4.5) with respect to (wrt) $\underset{\sim}{\mu}$ is obtained by noting that for $i \in B$, $i \neq p$,

$$\frac{\mu_p - \mu_i}{\sqrt{\sigma_p^2/N_p + \sigma_i^2/N_i}} \geq \frac{\delta}{\sqrt{\sigma_p^2/N_p + \sigma_i^2/N_i}} \quad \text{(by the definition of } \Omega_p^*\text{)}$$

$$\geq \frac{h}{\sqrt{\sigma_p^2/S_p^2 + \sigma_i^2/S_i^2}} \quad \text{(by (3.2))} \tag{4.7}$$

$$= \frac{h}{\sqrt{\nu/Z_p + \nu/Z_i}} \quad \text{(by (4.6)).}$$

Now note that conditional on the $\{Z_i: 1 \leq i \leq p\}$, the $\{Y_{pi}: i \in B, i \neq p\}$ are multivariate normal with zero means, unit variances and nonnegative correlations which depend on the unknown variances. Therefore we apply the Slepian (1962) inequality to obtain a further lower bound on (4.5) by regarding the Y_{pi} as conditionally uncorrelated N(0,1) rv 's. Finally, the infimum of this last bound wrt $\underset{\sim}{\sigma}^2$ is attained when $\sigma_j^2/\sigma_i^2 = \theta \; \forall \; i \in B, j \notin B$. An integral expression for this infimum is obtained by conditioning on $\min_{j \notin B} Z_j = w$, $Z_p = v$, and $Z_i = u_i$ for $i \in B$, $i \neq p$ and noting that the Z_i are iid χ_ν^2 rv's. This expression is

$$(p-b)\int_0^\infty \left[\int_0^{\theta w} \left\{ \prod_{\substack{i \in B \\ i \neq p}} \left[\int_0^{\theta w} \Phi\left(\frac{h}{\sqrt{\nu/u_i + \nu/v}}\right) dF_\nu(u_i)\right] dF_\nu(v) \right\} \times\{1 - F_\nu(w)\}^{p-b-1}\right] dF_\nu(w)$$

which upon substituting (4.1) gives (4.2). □

<u>Remark 4.1</u>. The expression (4.2) is not applicable for $b = p$ because in that case the event $\{S_i^2 \leq S_j^2 \; \forall \; i \in B, j \notin B\}$ is vacuous. Dropping the event $\{\hat{B} = B\}$ in (4.4) yields Rinott's (1978) expression (see his equation (12)) for the infimum of the P(CS). For (a,b) = (p,1), (4.2) reduces to Bechhofer and Sobel's (1954) expression (see their equation (19)) for the infimum of the P(CS).

4.2 Case II $(a + b \leq p)$: Procedure P_{II}

Since C can be empty in this case we must consider both (i) the infimum of $P_{\underset{\sim}{\omega}}(d_0 | P_{II})$ over Ω_0^*, and (ii) the infimum of $P_{\underset{\sim}{\omega}}(d_p | P_{II})$ over Ω_p^* (say). The result for (i) is given in Theorem 4.2 while that for (ii) is given in Theorem 4.3.

Theorem 4.2. Let

$$H(v,w) = H(v,w|\nu,\theta,h) = \int_{w/\theta}^{\infty} \Phi\left(\frac{h}{\sqrt{\nu/v + \nu/y}}\right) dF_{\nu}(y). \tag{4.8}$$

Then for given p and specified $\{a,b,\delta,\theta\}$ where $a + b \leq p$,

$$\inf_{\Omega_0^*} P_{\underset{\sim}{\omega}}(d_0|P_{II}) \geq b \int_0^{\infty} \{H(w,w)\}^a \left[\int_0^w \{H(u,0)\}^a dF_{\nu}(u)\right]^{b-1}$$

$$\times\left[\int_{w/\theta}^{\infty} \{H(v,0)\}^a dF_{\nu}(v)\right]^d dF_{\nu}(w) \tag{4.9}$$

where $d = p - (a + b)$.

Proof. Fix $\underset{\sim}{\omega} \in \Omega_0^*$ and thus $A = A(\underset{\sim}{\mu};a)$, $B = B(\underset{\sim}{\sigma}^2;b)$ and $C = A \cap B = \phi$. Consider

$$P_{\underset{\sim}{\omega}}(d_0|P_{II}) = \sum_{Q,R} P_{\underset{\sim}{\omega}}\{\hat{A} = Q, \hat{B} = R\} \tag{4.10}$$

where the sum is over all sets Q,R which are mutually exclusive subsets of $\{1,2,\ldots,p\}$ with card $(Q) = a$ and card $(R) = b$. As in the proof of Theorem 4.1, a sharp lower bound on (4.10) is

$$P_{\underset{\sim}{\omega}}\{\hat{A} = A, \hat{B} = B\}. \tag{4.11}$$

This lower bound is sharp because for fixed sets A,B we can choose a sequence $\{\underset{\sim}{\omega}_{\ell}\}$ in Ω_0^* such that $\mu_i \to -\infty\ \forall\ i \notin A$ and $\sigma_j^2 \to +\infty\ \forall\ j \notin B$ thus yielding (4.11). We now find the infimum of (4.11) which we first write as

$$P_{\underset{\sim}{\omega}}\left\{Y_{ij} < \frac{\mu_i - \mu_j}{\sqrt{\sigma_i^2/N_i + \sigma_j^2/N_j}}\ \forall\ i \in A,\ j \notin A;\ Z_k < (\sigma_{\ell}^2/\sigma_k^2)Z_{\ell}\ \forall\ k \in B,\ \ell \notin B\right\}. \tag{4.12}$$

Here the Y's and Z's are as defined in (4.6). Arguing as in Theorem 4.1, a lower bound on (4.12) is

$$P\left\{Y_{ij} < \frac{h}{\sqrt{\nu/Z_i + \nu/Z_j}},\ Y_{ik} < \frac{h}{\sqrt{\nu/Z_i + \nu/Z_k}};\ Z_k < \theta Z_i,\ Z_k < \theta Z_j\ \ \forall\ i \in A,\ j \notin A \cup B,\ k \in B\right\} \tag{4.13}$$

where, by using the Slepian inequality, the Y's are uncorrelated $N(0,1)$ rv's conditional on the Z's. To write an integral expression for (4.10) we relabel the Z's as follows: the $Z_{k^*} = \max\{Z_k: k \in B\}$ as W, the Z_j for $j \notin A \cup B$ as $(V_1,\ldots,V_d) = \underset{\sim}{V}$, the Z_k for $k \in B$, $k \neq k^*$

as $(U_1,\ldots,U_{b-1}) = \underset{\sim}{U}$ and the Z_i for $i \in A$ as $(T_1,\ldots,T_a) = \underset{\sim}{T}$. Also let T denote a generic χ^2_ν rv. Then by the usual conditioning arguments we obtain

$$b \int_0^\infty \left[\int_{w/\theta}^\infty \cdots \int_{w/\theta}^\infty \left\{ \int_0^w \cdots \int_0^w \left[\prod_{i=1}^a \int_{w/\theta}^\infty \Phi\left(\frac{h}{\sqrt{\nu/t_i + \nu/w}} \right) \prod_{k=1}^{b-1} \Phi\left(\frac{h}{\sqrt{\nu/t_i + \nu/u_k}} \right) \right.\right.\right.$$

$$\left.\left.\left. \times \prod_{j=1}^d \Phi\left(\frac{h}{\sqrt{\nu/t_i + \nu/v_j}} \right) dF_\nu(t_i) \right] \prod_{k=1}^{b-1} dF_\nu(u_k) \right\} \prod_{j=1}^d dF_\nu(v_j)\, dF_\nu(w) \right]$$

$$= b\, E_W \left[E_{\underset{\sim}{V}} \left\{ E_{\underset{\sim}{U}} \left[E_T \left\{ I(T > W/\theta)\, \Phi\left(\frac{h}{\sqrt{\nu/T + \nu/W}} \right) \right.\right.\right.\right.$$

$$\times \left[\prod_{k=1}^{b-1} I(U_k < W)\, \Phi\left(\frac{h}{\sqrt{\nu/T + \nu/U_k}} \right) \right]$$

$$\left.\left.\left.\left. \times \left[\prod_{j=1}^d I(V_j > W/\theta)\, \Phi\left(\frac{h}{\sqrt{\nu/T + \nu/V_j}} \right) \right] \right\} \right]^a \right\} \right] \tag{4.14}$$

where $I(E)$ denotes the indicator function of event E and E_X denotes the expectation wrt X. Now (4.14) is a $(p-a+1)$ - fold integral which is difficult to evaluate even on a computer. So we seek a computationally feasible lower bound on it. To this end we note in (4.14) that for fixed $\underset{\sim}{U}, \underset{\sim}{V}$ and W, the expectation with respect to T is $E_T\{\Pi_{i=1}^3 f_i(T)\}$ where

$$f_1(T) = I(T > W/\theta)\, \Phi\left(\frac{h}{\sqrt{\nu/T + \nu/W}} \right),$$

$$f_2(T) = \prod_{k=1}^b \left\{ I(U_k < W)\, \Phi\left(\frac{h}{\sqrt{\nu/T + \nu/U_k}} \right) \right\},$$

$$f_3(T) = \prod_{j=1}^d \left\{ I(V_j > W/\theta)\, \Phi\left(\frac{h}{\sqrt{\nu/T + \nu/V_j}} \right) \right\}.$$

Each $f_i(T)$ is increasing in T and therefore the $f_i(T)$ are associated rv's (conditioned on $\underset{\sim}{U}, \underset{\sim}{V}$ and W) as a consequence of properties P_3 and P_4 stated in Esary, Proschan and Walkup (1967). Therefore

$$E_T\{\prod_{i=1}^{3} f_i(T)\} \geq \prod_{i=1}^{3} E_T\{f_i(T)\}. \tag{4.15}$$

Using this fact and the mutual independence of the U_k and V_j we obtain the following lower bound on (4.14):

$$b\, E_W\left[\left\{E_T\left[I(T > W/\theta)\,\Phi\left(\frac{h}{\sqrt{\nu/T + \nu/W}}\right)\right]\right\}^a \right.$$

$$\times \prod_{k=1}^{b-1}\left[E_{U_k}\left\{E_T\, I(U_k < W)\,\Phi\left(\frac{h}{\sqrt{\nu/T + \nu/U_k}}\right)\right]\right\}^a\Bigg]$$

$$\left.\times \prod_{j=1}^{d}\left[E_{V_j}\left\{E_T\left[I(V_j > W/\theta)\,\Phi\left(\frac{h}{\sqrt{\nu/T + \nu/V_j}}\right)\right]\right\}^a\right]\right]$$

$$= b\int_0^\infty \left\{\int_{w/\theta}^\infty \Phi\left(\frac{h}{\sqrt{\nu/t + \nu/w}}\right) dF_\nu(t)\right\}^a \left\{\int_0^w\left[\int_0^\infty \Phi\left(\frac{h}{\sqrt{\nu/t + \nu/u}}\right) dF_\nu(t)\right]^a dF_\nu(u)\right\}^{b-1}$$

$$\times \left\{\int_{w/\theta}^\infty \left[\int_0^\infty \Phi\left(\frac{h}{\sqrt{\nu/t + \nu/v}}\right) dF_\nu(t)\right]^a dF_\nu(v)\right\}^d dF_\nu(w)$$

which upon substituting (4.8) gives (4.9). □

We next turn to the study of the infimum of $P_{\underset{\sim}{\omega}}(d_p | P_{II})$ over $\underset{\sim}{\omega} \in \Omega_p^*$.

<u>Theorem 4.3.</u> For given p and specified $\{a,b,\delta,\theta\}$ with $a + b \leq p$,

$$\inf_{\Omega_p^*} P_{\underset{\sim}{\omega}}(d_p | P_{II}) \geq \int_0^\infty \left[\int_0^{\theta w} \{H(v,\theta w)\}^{p-M-1}\, \{G(v,w)\}^{b-1}\right.$$

$$\left.\times \left\{(p-M)\,\Phi\left(\frac{h}{\sqrt{\nu/v + \nu/w}}\right) \{1 - F_\nu(w)\} + (a-m)\, H(v,\theta w)\right\} dF_\nu(v)\right]$$

$$\times \{1 - F_\nu(w)\}^{a-m-1}\, dF_\nu(w) \tag{4.16}$$

where $m = \min(a,b)$, $M = \max(a,b)$ and $G(\cdot,\cdot)$ and $H(\cdot,\cdot)$ are defined in (4.1) and (4.8), respectively.

Proof. Fix $\underset{\sim}{\omega} \in \Omega_p^*$ and thus $A = A(\underset{\sim}{\mu};a)$, $B = B(\underset{\sim}{\sigma}^2;b)$ and $C = A \cap B$ with $p \in C$. We obtain

$$P_{\underset{\sim}{\omega}}(d_p | P_{II}) = P_{\underset{\sim}{\omega}}\{p \in \hat{A} \cap \hat{B};\ \overline{X}_p > \overline{X}_j\ \forall\ j \in \hat{C},\ j \neq p\}$$

$$\geq P_{\underset{\sim}{\omega}}\{\overline{X}_p > \overline{X}_i\ \forall\ i \notin A;\ S_k^2 < S_\ell^2\ \forall\ k \in B,\ \ell \notin B;$$

$$\overline{X}_p > \overline{X}_j\ \forall j \in \hat{A} \cap B,\ j \neq p\}$$

$$\geq P_{\underset{\sim}{\omega}}\{\overline{X}_p > \overline{X}_i\ \forall\ i \notin A;\ \overline{X}_p > \overline{X}_j\ \forall\ j \in C,\ j \neq p;$$

$$S_k^2 < S_\ell^2\ \forall k \in B,\ \ell \notin B\}$$

$$\geq P\{Y_{pi} < \frac{h}{\sqrt{\nu/Z_p + \nu/Z_i}}\ \forall\ i \notin A;\ Y_{pj} < \frac{h}{\sqrt{\nu/Z_p + \nu/Z_j}}\ \forall\ j \in C,$$

$$j \neq p;\ Z_k < \theta Z_\ell\ \forall\ k \in B,\ \ell \notin B\} \tag{4.17}$$

where, by using the S pian inequality, the Y's are uncorrelated N(0,1) rv's conditional on the Z's. Now (4.17) is clearly minimized when c is maximized, i.e., $c = m = \min(a,b)$ and $a + b - c = M = \max(a,b)$. To obtain an integral expression for this probability, condition on $Z_{\ell^*} = \min_{\ell \notin B} Z_\ell = w$, $Z_p = v$ and $Z_i = u_i\ \forall\ i \notin A,\ i \neq \ell^*$ and $Z_j = u_j\ \forall j \in C,\ j \neq p$. Then considering the two cases: $\ell^* \in \overline{A} \cap \overline{B}$ and $\ell^* \in A \cap \overline{B}$ separately, and combining the two resulting integrals we obtain (4.16). □

Remark 4.2. For fixed ν as $h \to \infty$ we see that $G(v,w) \to F_\nu(\theta w)$, $H(v,w) \to 1 - F_\nu(w/\theta)\ \forall\ v,\ w$, and in particular $H(v,0) \to 1$. By applying the dominated convergence theorem we then see that as $h \to \infty$ (4.2), (4.9) and (4.16) all approach the expression

$$(p-b) \int_0^\infty \{F_\nu(\theta w)\}^b \{1 - F_\nu(w)\}^{p-b-1}\, dF_\nu(w)$$

$$= b \int_0^\infty \{F_\nu(w)\}^{b-1} \{1 - F_\nu(w/\theta)\}^{p-b}\, dF_\nu(w)$$

$$= P\{\hat{B} = B\} \tag{4.18}$$

where this probability is calculated under the slippage configuration $\sigma_{[1]}^2 = \cdots = \sigma_{[b]}^2 = \sigma_{[b+1]}^2/\theta = \cdots = \sigma_{[p]}^2/\theta$. Thus the procedure is limited by the accuracy with which it can estimate B in the first stage.

As $\nu \to \infty$, (4.18) converges to unity so that any preassigned γ can be attained by choosing the first stage sample size sufficiently large.

Furthermore it is easy to see that (4.2), (4.9) and (4.16) all approach unity as $h, \nu \to \infty$ so that $h/\sqrt{\nu} \to \infty$. Thus there exist many (h,ν) values which solve the equation obtained by setting each one of the above expressions equal to the specified value $\gamma \in (1/p,1)$. If ν_0 is the smallest integer which makes (4.18) $\geq \gamma$ then clearly one must have $\nu > \nu_0$ (i.e., $n > \nu_0 + 1 = n_0$, say) to assure a finite solution in h.

Remark 4.3. For given p and specified $\{a,b,\delta,\theta,\gamma\}$ the design constant h to be used in P_I is obtained by first choosing $\nu > \nu_0$ and then solving the equation obtained by setting (4.2) equal to γ.

Similarly the design constant h to be used in P_{II} is obtained by first choosing $\nu > \nu_0$, then determining h_1 (resp., h_2) as the solution to the equation obtained by setting (4.9) (resp., (4.16)) equal to γ, and finally letting $h = \max(h_1,h_2)$.

Remark 4.4. The question of determining an "optimal" choice of (h,ν) is not addressed in the present article. One could choose (h,ν) to minimize the expected total sample size $E_{\underline{\omega}}(N) = \sum_{i=1}^{p} E_{\underline{\omega}}(N_i)$. We note that it is possible to derive exact expressions for $E_{\underline{\omega}}(N)$ for P_I and P_{II}, although we have not given them here.

5. COMPUTATION OF CONSTANTS AND SIMULATION RESULTS FOR P_I AND P_{II}

5.1 Computation of Constants for P_I and P_{II}

The various multiple integrals required in the computation of the h values for P_I and P_{II} were evaluated using Gauss-Laguerre quadrature for the outer integral and Gauss-Legendre quadrature for the inner integrals (except, for the $H(v,0)$ function appearing in (4.9) for which the former method was used). In applying the Gauss-Legendre quadrature method if the upper limit of any inner integral was $+\infty$, it was truncated at a value U chosen so that the truncation error is less than 10^{-8}. Note that this error is on the conservative side, i.e., it makes the value of the integral smaller and hence the value of h larger. Thirty-two point quadrature formulae were used for both the methods, the weights and zeroes of the corresponding polynomials being obtained from Stroud and Secrest (1966).

A bisection method was used to solve the equations resulting from setting (4.2), (4.9) and (4.16) equal to γ, respectively. Programs for evaluating the integrals and solving the equations were written in APL and were run on an IBM PC-1 machine in single precision (8 byte reals). The programs are included in Santner and Tamhane (1983).

We studied the performance of P_I and P_{II} using a Monte Carlo simulation. For this purpose we first computed the h values (correct to second decimal place) for two cases which are given in Table 1.

We note that in Case II, $h = h_1 = \max(h_1,h_2)$ where h_1 (resp., h_2) is from Theorem 4.2 (resp., Theorem 4.3). It is not generally true, however, that $h_1 \geq h_2$.

5.2 Simulation Results

For simulating P_I (resp., P_{II}) the constants listed under Case I (resp., Case II) in Table 1 were used; δ was set equal to 0.2 in both cases. The performance of P_I was studied for six different $\underset{\sim}{\omega}$ while that of P_{II} was studied for twelve different $\underset{\sim}{\omega}$. These parameter configurations are listed in Table 2 for P_I and in Table 3A (the $C = \phi$ case) and Table 3B (the $C \neq \phi$ case) for P_{II}.

Both simulation programs were written in FORTRAN and are included in Santner and Tamhane (1983). The IMSL subroutines GGNML and GGCHS were used to generate a set of p independent standard normal and p independent chi-square random variables, respectively, in each run of the simulation experiment. The $\overline{X}_i$ and S_i^2 values for different parameter configurations were obtained from the same basic set of these random variables by making the usual transformations. A total of 5000 runs was used in each simulation experiment. All simulations were performed on Cornell University's IBM 3081 computer.

Table 1

The Cases Studied and the Corresponding Values of h

Case	Table(s)	p	a	b	θ	n_0	γ	h
I	2	5	3	3	2.0	45	0.90	2.89
II	3A,3B	5	2	2	2.0	45	0.90	3.59

Table 2

Monte Carlo Estimates for P_I[a]

$\underset{\sim}{\sigma}^2$	$\underset{\sim}{\mu}$	$P_{\underset{\sim}{\omega}}(CS)$	$E_{\underset{\sim}{\omega}}(N_1)$	$E_{\underset{\sim}{\omega}}(N_2)$	$E_{\underset{\sim}{\omega}}(N_3)$	$E_{\underset{\sim}{\omega}}(N_4)$	$E_{\underset{\sim}{\omega}}(N_5)$
	$\underset{\sim}{\mu}_1 = (5,5,0,0,0.2)$	0.8992 (0.0043)					
$\underset{\sim}{\sigma}_1^2 = (2,2,1,1,1)$	$\underset{\sim}{\mu}_2 = (0,0,0,0,0.2)$	0.9360 (0.0035)	50.94 (36.80)	51.78 (38.31)	205.15 (48.84)	204.31 (47.87)	203.05 (47.46)
	$\underset{\sim}{\mu}_3 = (-1,-1,-1,-1,0.2)$	0.9806 (0.0020)					
	$\underset{\sim}{\mu}_1 = (5,5,0,0,0.2)$	0.9544 (0.0030)					
$\underset{\sim}{\sigma}_2^2 = (3,3,1,1,1)$	$\underset{\sim}{\mu}_2 = (0,0,0,0,0.2)$	0.9554 (0.0029)	45.00 (0.00)	45.35 (9.48)	210.03 (45.19)	208.79 (44.75)	207.82 (44.21)
	$\underset{\sim}{\mu}_3 = (-1,-1,-1,-1,0.2)$	0.9996 (0.0003)					

[a] The estimated standard error of each estimate is given in parentheses below it. All estimates are based on 5000 runs.

The results for P_I are displayed in Table 2. We first note that the smallest P(CS) is attained at the configuration $(\underset{\sim}{\mu}_1, \underset{\sim}{\sigma}_1^2)$ which has $A = \{1,2,5\}$, $B = \{3,4,5\}$ and $C = \{5\}$. From the proof of Theorem 4.1 we see that this configuration is close to the least favorable configuration which theoretically minimizes the $P_{\underset{\sim}{\omega}}(CS)$ of P_I for $\underset{\sim}{\omega}$ in the preference zone when $\bar{A} \cap \bar{B} = \phi$, viz., $\mu_i = +\infty\ \forall\ i \in A \cap \bar{B}$, $\mu_p - \mu_i = \delta$ $\forall\ i \in B\ (i \neq p)$ and $\sigma_j^2/\sigma_i^2 = \theta\ \forall\ i \in B,\ j \notin B$. We further see that this smallest P(CS) nearly equals the nominal value of $\gamma = 0.90$ indicating that very little probability is lost by the use of the Slepian inequality and the inequality used at the second step of (4.7). The latter inequality is known to be quite sharp. The sharpness of the Slepian inequality here is due to the fact that conditional on the S_i^2 for $i \in B = \{3,4,5\}$ there is only a single correlation, that between $Y_{5,3}$ and $Y_{5,4}$, which is set equal to zero; it can be shown that the value of this correlation coefficient is approximately 1/2. In general, there are $\binom{b-1}{2}$ correlations between the Y_{pi} for $i \in B$, $i \neq p$, each approximately 1/2, which are set equal to zero by the Slepian inequality. Thus for larger values of b, use of the Slepian inequality will result in a greater loss in probability.

For P_I the estimates in Table 2 show that $\underset{\sim}{\mu}_1 < \underset{\sim}{\mu}_2 < \underset{\sim}{\mu}_3$ in order of increasing favorability of a CS for fixed $\underset{\sim}{\sigma}_i^2$, $i = 1,2$ whereas $\underset{\sim}{\sigma}_1^2 < \underset{\sim}{\sigma}_2^2$ in order of increasing favorability of a CS for fixed $\underset{\sim}{\mu}_j$, $j = 1,2,3$. These results can be explained intuitively as follows: For fixed $\underset{\sim}{\sigma}^2$, the distribution of $\hat{B}$ is identical for all $\underset{\sim}{\mu}$. However, if $\underset{\sim}{\mu}_2$ is viewed as the limit of the vectors $(\varepsilon,\varepsilon,0,0,0\cdot 2)$ as $\varepsilon \downarrow 0$ then $A = \{1,2,5\}$ for both $\underset{\sim}{\mu}_1$ and $\underset{\sim}{\mu}_2$ but if either Π_1 or Π_2 is in $\hat{B}$ then a CS is much more likely under $\underset{\sim}{\mu}_2$ than $\underset{\sim}{\mu}_1$. Similar heuristics apply to the comparison of $\underset{\sim}{\mu}_3$ with $\underset{\sim}{\mu}_2$. For fixed $\underset{\sim}{\mu}_j$, $j = 1,2,3$, the procedure has a higher probability of correctly identifying the true B set under $\underset{\sim}{\sigma}_2^2$ then under $\underset{\sim}{\sigma}_1^2$ and hence the value of the P(CS) is larger under $\underset{\sim}{\sigma}_2^2$.

Finally we note that for $i \in B$ (resp., $j \notin B$), $E(N_i)$ (resp., $E(N_j)$) roughly equals $\sigma_i^2(h/\delta)^2$ (resp., n_0) if σ_j^2/σ_i^2 is sufficiently large $\forall\ i \in B,\ j \notin B$.

We now turn to the simulation results for P_{II}. The configurations considered in Table 3A correspond to $A = \{4,5\}$, $B = \{1,2\}$ and $C = \phi$. The estimate of the P(CS) given for each $\underset{\sim}{\omega}$ is the proportion of times no population was selected. Here the configuration $(\underset{\sim}{\mu}_1, \underset{\sim}{\sigma}_1^2)$ is nearly least favorable and the achieved P(CS) in this configuration is

Table 3A

Monte Carlo Estimates for P_{II} when C is Empty[a]

$\underset{\sim}{\sigma}^2$	$\underset{\sim}{\mu}$	$P_{\underset{\sim}{\omega}}(CS)$	$E_{\underset{\sim}{\omega}}(N_1)$	$E_{\underset{\sim}{\omega}}(N_2)$	$E_{\underset{\sim}{\omega}}(N_3)$	$E_{\underset{\sim}{\omega}}(N_4)$	$E_{\underset{\sim}{\omega}}(N_5)$
$\underset{\sim}{\sigma}_1^2 = (1,1,2,2,2)$	$\underset{\sim}{\mu}_1 = (0,0,0,0.2,0.2)$	0.9388 (0.0034)	322.68 (68.05)	320.02 (68.45)	647.83 (139.10)	643.48 (137.98)	640.73 (136.30)
	$\underset{\sim}{\mu}_2 = (-1,-1,-1,0.2,0.2)$	0.9594 (0.0028)					
$\underset{\sim}{\sigma}_2^2 = (1,1,3,3,3)$	$\underset{\sim}{\mu}_1 = (0,0,0,0.2,0.2)$	0.9760 (0.0022)	322.68 (68.05)	320.02 (68.45)	971.49 (208.65)	964.98 (206.97)	960.83 (204.45)
	$\underset{\sim}{\mu}_2 = (-1,-1,-1,0.2,0.2)$	0.9988 (0.0005)					

[a] The estimated standard error of each estimate is given in parentheses below it. All estimates are based on 5000 runs.

Table 3B

Monte Carlo Estimates for P_{II} when C is Nonempty[a]

$\underset{\sim}{\sigma}^2$	$\underset{\sim}{\mu}$	$P_{\underset{\sim}{\omega}}(CS)$	$E_{\underset{\sim}{\omega}}(N_1)$	$E_{\underset{\sim}{\omega}}(N_2)$	$E_{\underset{\sim}{\omega}}(N_3)$	$E_{\underset{\sim}{\omega}}(N_4)$	$E_{\underset{\sim}{\omega}}(N_5)$
$\underset{\sim}{\sigma}_1^2 = (2,2,2,1,1)$	$\underset{\sim}{\mu}_1 = (5,0,0,0,0.2)$	0.9396 (0.0034)	644.86 (136.10)	639.54 (136.91)	647.83 (139.10)	321.99 (68.99)	320.61 (68.15)
	$\underset{\sim}{\mu}_2 = (1,0,0,0,0.2)$	0.9396 (0.0034)					
	$\underset{\sim}{\mu}_3 = (0,0,0,0,0.2)$	0.9580 (0.0028)					
	$\underset{\sim}{\mu}_4 = (-1,-1,-1,-1,0.2)$	0.9690 (0.0025)					
$\underset{\sim}{\sigma}_2^2 = (3,3,3,1,1)$	$\underset{\sim}{\mu}_1 = (5,0,0,0,0.2)$	0.9782 (0.0021)	967.05 (204.15)	959.07 (205.35)	971.49 (208.65)	321.99 (68.99)	320.61 (68.15)
	$\underset{\sim}{\mu}_2 = (1,0,0,0,0.2)$	0.9782 (0.0021)					
	$\underset{\sim}{\mu}_3 = (0,0,0,0,0.2)$	0.9892 (0.0015)					
	$\underset{\sim}{\mu}_4 = (-1,-1,-1,-1,0.2)$	0.9996 (0.0003)					

[a] The estimated standard error of each estimate is given in parentheses below it. All estimates are based on 5000 runs.

significantly higher than the nominal value of $\gamma = 0.90$ indicating that (4.9) gives a quite conservative lower bound on the actual P(CS). The explanation for this again lies in the fact that the lower bound (4.11) is sharp iff $\mu_i - \mu_j \to +\infty \; \forall \; i \in A, j \notin A$ and $\sigma_j^2/\sigma_i^2 \to +\infty \; \forall \; i \in B, j \notin B$; for any other configuration such as $(\underset{\sim}{\mu}_1, \underset{\sim}{\sigma}_1^2)$, the actual P(CS) is greater by an amount equal to the terms ignored in arriving at the lower bound (4.11). Some extra loss in P(CS) results by the use of the Slepian inequality and the inequality (4.15). Also note that $\underset{\sim}{\mu}_1 < \underset{\sim}{\mu}_2$ (resp., $\underset{\sim}{\sigma}_1^2 < \underset{\sim}{\sigma}_2^2$) in the order of favorability for a CS because for fixed $\underset{\sim}{\sigma}_i^2$, i = 1,2 (resp., $\underset{\sim}{\mu}_j$, j = 1,2) the probability of correctly identifying the true A (resp., B) set is higher under $\underset{\sim}{\mu}_2$ (resp., $\underset{\sim}{\sigma}_2^2$) than under $\underset{\sim}{\mu}_1$ (resp., $\underset{\sim}{\sigma}_1^2$).

The configurations considered in Table 3B correspond to $A = \{1,5\}$, $B = \{4,5\}$ and $C = \{5\}$. For configurations $\underset{\sim}{\mu}_3$ and $\underset{\sim}{\mu}_4$ one can also take $A = \{4,5\}$ and $C = \{4,5\}$. Once again, the configuration $(\underset{\sim}{\mu}_1, \underset{\sim}{\sigma}_1^2)$ is nearly least favorable and the explanation for the excess P(CS) in this configuration over the nominal value $\gamma = 0.90$ is similar to that given in the preceding case. For fixed $\underset{\sim}{\sigma}_1^2$ or $\underset{\sim}{\sigma}_2^2$, $\underset{\sim}{\mu}_1$ through $\underset{\sim}{\mu}_4$ are successively more favorable configurations. Similarly for fixed $\underset{\sim}{\mu}_j (1 \le j \le 4)$, $\underset{\sim}{\sigma}_2^2$ is more favorable than $\underset{\sim}{\sigma}_1^2$.

Finally we turn to the discussion of the estimates of the $E(N_i)$ in Tables 3A and 3B. Since P_{II} samples from all populations in both the stages, each $E(N_i)$ depends on σ_i^2 but not on σ_j^2 for $j \neq i$. This is in contrast to the case of P_I where each $E(N_i)$ depends on all the σ_j^2 because whether a population is sampled in the second stage or not in P_I depends on whether it enters set $\hat{B}$. Since in Tables 3A and 3B, n_0 is much smaller than $\sigma_i^2(h/\delta)^2 \; \forall \; i$, all the $E(N_i)$ roughly equal $\sigma_i^2(h/\delta)^2$.

6. CONCLUDING REMARKS

In this article we have provided one of many possible formulations of the general problem of selecting the normal population with a large mean and a small variance. Here a large mean (resp., a small variance) is defined relative to the set of means (resp., variances) of all populations at hand. Alternatively, one may wish to define these quantities relative to some known standards. Thus one may specify the values μ_0 and σ_0^2 and regard any $\mu_i \ge \mu_0$ (resp., $\sigma_i^2 \le \sigma_0^2$) as a large mean (resp., a small variance). In some situations one may have a control population which is normal with mean μ_0 and variance σ_0^2 both of which are unknown. In such cases, in addition to sampling from the p contending populations, one can also sample from the control population in two stages to obtain estimates of μ_0 and σ_0^2 with which the estimates of the μ_i and $\sigma_i^2 (1 \leq i \leq p)$ can then be compared. The authors are studying such a procedure.

SUMMARY

Consider $p \geq 2$ normal populations with the ith one having mean μ_i and variance σ_i^2 $(1 \leq i \leq p)$. All the parameters are unknown. Let $\mu_{[1]} \leq \cdots \leq \mu_{[p]}$ and $\sigma_{[1]}^2 \leq \cdots \leq \sigma_{[p]}^2$ be the ordered means and variances, respectively. For specified integers a and b $(1 \leq a,b \leq p)$ consider sets $A = \{i: \mu_i \geq \mu_{[p-a+1]}\}$, $B = \{i: \sigma_i^2 \leq \sigma_{[b]}^2\}$, and $C = A \cap B$. The experimenter's goal is to select the population associated with $\max\{\mu_i: i \in C\}$ if C is nonempty and select no population if C is empty. It is desired to design the experiment to guarantee a prespecified level of the probability of a correct selection (P(CS)) for $\underset{\sim}{\mu} = (\mu_1, \mu_2, \ldots, \mu_p)$ and $\underset{\sim}{\sigma}^2 = (\sigma_1^2, \sigma_2^2, \ldots, \sigma_p^2)$ lying in a suitably defined "preference-zone." We propose different two-stage procedures for the cases: (I) $a + b > p$ when C cannot be empty, and (II): $a + b \leq p$. The infimums of the P(CS) of the two procedures are studied. Computer programs are described for determining the design constants for the procedures based on these infimums. Simulation results are given to assess the conservatism of certain bounds employed in deriving the infimums.

REFERENCES

Bechhofer, R.E. (1954). A single-sample multiple decision procedure for ranking means of normal populations with known variances. Ann. Math. Statist., 25, 16-39.

Bechhofer, R.E. and Sobel, M. (1954). A single-sample multiple decision procedure for ranking variances of normal populations. Ann. Math. Statist., 25, 273-289.

Dantzig, G.B. (1940). On the non-existence of tests of "Student's" hypothesis having power functions independent of σ. Ann. Math. Statist., 11, 186-192.

Dudewicz, E.J. (1971). Non-existence of a single-sample selection procedure whose P(CS) is independent of the variances. S. Afr. Statist. J., 5, 37-39.

Dudewicz, E.J. and Dalal, S.R. (1975). Allocation of observations in ranking and selection with unequal variances. Sankhyā Ser. B., 37, 28-78.

Esary, J.D., Proschan, F., and Walkup, D.W. (1967). Association of random variables with applications. Ann. Math. Statist., 38, 1466-1474.

Handa, B.R. and Jain, V.K. (1982). New goal for ranking and selection of normal populations. Unpublished manuscript.

Rinott, Y. (1978). On two-stage procedures and related probability inequalities. Commun. Statist.-Theor. Meth. Ser. A, 8, 799-811.

Santner, T.J. and Tamhane, A.C. (1983). Designing experiments for selecting a normal population with a large mean and a small variance. School of OR/IE Tech. Report No. 606, Cornell University.

Slepian, D. (1982). On the one-sided barrier problem for Gaussian noise. Bell System Tech. J., 41, 463-501.

Stein, C. (1945). A two-sample test for a linear hypothesis whose power is independent of the variance. Ann. Math. Statist., 24, 243-258.

Stroud, A.H. and Secrest, D. (1966). Gaussian Quadrature Formulas. Englewood Cliffs, N.J.: Prentice Hall, Inc.

SEQUENTIAL LIKELIHOOD PROCEDURES FOR SELECTING A SUBSET AND FIXED SAMPLE SIZE ANALOGUES

Milton Sobel*

Department of Mathematics
University of California
at Santa Barbara
Santa Barbara, California

Hsiu-Fen Wu

Department of Mathematics
University of California
at Santa Barbara
Santa Barbara, California

1. SEQUENTIAL LIKELIHOOD PROCEDURES AND FIXED SAMPLE SIZE ANALOGUES

Given k populations we wish to consider a likelihood procedure for selecting a subset of k populations that contains the "best" one, i.e., the one with the largest population mean. The procedure is quite flexible and has both a sequential form $R_1 = R_{SL}$ and a fixed sample size form $R_2 = R_{FSS,L}$ considered separately below. In both cases we consider only independent sampling of one k-vector-at-a-time where the ith component is from population Π_i (i = 1,2,...,k) and the components in the same vector (as well as in different vectors) are all independent and identically distributed except possibly for the value θ_i (i = 1,2,...,k) of a scalar parameter θ. We use the normal distribution to illustrate the form of the procedure, noting that the same technique can be applied to any member of the exponential family in (1) below. Let S denote the size of the selected subset for all of the procedures.

Since we introduce an indifference zone, a straightforward comparison with the procedure of Gupta (1965) for the same problem is not possible. However, if we modify the procedure in Gupta (1965) by inserting an indifference zone and requiring that the probability of a correct selection P(CS) be at least P* at the so-called least favorable (LF) configura-

---

* This paper was supported by National Science Foundation Grant MCS82-02247.

tion (cf. (2) below), then a fair comparison can be made. Since this modification improves the performance of the procedure in Gupta (1965), it is difficult to argue that the resulting comparison is biased against the fixed sample size procedure in Gupta (1965). We refer to this procedure as $R_3 = R_{FSS,MG}$.

A fourth procedure $R_4 = R_{FSS,2R}$ included in our discussion is another fixed sample size procedure which uses two requirements: an upper bound on the expected subset size E(S) over the whole parameter space and a lower bound on the P(CS) in the LF configuration. This also provides a fair comparison with R_1.

The sequential form of the likelihood procedure R_1 has the advantage that we can avoid selecting the entire set of all k populations. In fact we can arbitrarily fix the size s $(1 \leq s < k)$ of the selected subset. In a special case of our general rule we accomplish this by taking observations a vector-at-a-time until our statistic indicates that the probability of a correct selection (in the LF configuration) is at least P*, i.e., that the s largest sample means contain (among them) the best population. With probability 1 this rule terminates in a finite number of steps if the largest population is different from the others. The probabilistic basis for the likelihood procedure R_1 described below is based on results in Bechhofer, Kiefer and Sobel (1968); we refer below to this book by BKS.

In this paper, for several, reasons, we have not done much with R_2 except to define it. First, it requires a great deal of Monte Carlo sampling to match up the probability of a correct selection P(CS) and the expected subset size E(S) with those of other procedures. Secondly, the procedure R_2 does not avoid the possibility of putting all the k populations into the selected subset.

2. THE LIKELIHOOD PROCEDURE R_1 (SEQUENTIAL FORM)

Consider the exponential family

$$f(x;\theta) = \exp\{h(x)Q(\theta) + r(x) + S(\theta)\} \tag{1}$$

where $Q(\theta)$ is a monotone function of the scalar quantity θ; in the normal case with common variance $\sigma^2 = 1$ we take $Q(\theta) = \theta$ (we usually denote it by μ) and $h(x) = x$. Let $\bar{x}_{im}$ denote the sample mean from Π_i $(i = 1,2,\ldots,k)$ based on the first m vector observations, and let $\bar{x}_{[1]} \leq \bar{x}_{[2]} \leq \cdots \leq \bar{x}_{[k]}$ denote the ordered sample means (suppressing m). Let $\mu_{[1]} \leq \mu_{[2]} \leq \cdots \leq \mu_{[k]}$

denote the ordered population means. For a preassigned $\delta^* > 0$, the LF configuration of interest is defined by

$$\mu_{[k]} = \mu + \delta^*, \; \mu_{[1]} = \mu_{[2]} = \cdots = \mu_{[k-1]} = \mu, \tag{2}$$

i.e., by letting one population slip to the right by the positive amount δ^*. If we associate the ordered $\bar{x}$-values and the ordered μ-values to form a likelihood, it was shown in BKS (1968) that the resulting likelihoods are ordered (from smaller to larger) according as we associate with $\mu + \delta^*$ the sample mean, $\bar{x}_{[1]}, \bar{x}_{[2]}, \ldots, \bar{x}_{[k]}$. (In the general case (1), these sample means would be replaced by the ordered values of the sufficient statistic $\sum_{i=1}^{m} h(x_i)$, assuming that the parameter $\tau = Q(\theta)$ is increasing in θ, so that we can replace θ by τ as the basis for ranking.) Note that the use of the LF configuration (2) has reduced the total number of likelihoods possible (by associating the μ's and the $\bar{x}$'s) from k! to k.

The likelihood for a subset of size s ($1 \leq s < k$) to contain the best population (in the LF configuration) attains its s largest values by associating $\mu + \delta^*$ with $\bar{x}_{[k-s+1]}, \bar{x}_{[k-s+2]}, \ldots, \bar{x}_{[k]}$. We form our likelihood statistic $L_s(\delta^*)$ by taking the ratio of the sum of these s largest likelihoods to the sum of all k likelihoods. After cancelling common factors we obtain a result that does not depend on μ, namely

$$L_s(\delta^*) = \sum_{i=k-s+1}^{k} e^{m\delta^*\bar{x}_{[i]}} \Big/ \sum_{j=1}^{k} e^{m\delta^*\bar{x}_{[j]}}, \tag{3}$$

where m is the number of vectors of observations taken (or the stage of experimentation). The sequential form of the likelihood procedure for selecting a subset of size s is to continue sampling a vector-at-a-time until (for the first time)

$$L_s(\delta^*) \geq P^* , \tag{4}$$

where P^* and δ^* are preassigned such that $s/k < P^* < 1$ and $\delta^* > 0$. It follows from BKS (1968) that this procedure has in the LF configuration a probability of a correct selection, P(CS), equal to at least P^*. In BKS (1968) this procedure was developed and studied for the special case $s = 1$, i.e., for selecting the best population, but essentially the same proof as given there holds for selecting a subset of any size s containing the best population.

In this sequential form, if $\mu_{[k]} > \mu_{[k-1]}$ then the procedure terminates with probability one and we end up with a subset of exactly size s. Below we describe a variation which allows us to terminate with the same P* requirement and a subset size of at most s*.

Suppose we alter the P*-requirement for the LF configuration by specifying s* pairs (s, P_s^*), $s = 1,2,\ldots,s^*$, with $1 \leq s^* < k$, which indicate that we are willing to accept a subset of size s if the $P\{CS|LF\} \geq P_s^*$, i.e., the value of P* changes (generally increases) with the subset size being considered. This is quite reasonable since the change (say) from $s = 1$ to $s = 2$ drastically reduces the final information obtained from the procedure. For example, we could set $P_s^* = (10s - 1)/10s$ $(s = 1,2,\ldots,s^*)$ for some $s^* < k$ and then s* becomes the upper bound on the size of the selected subset. This formulation has the property that we can specify an upper bound s* on the subset size and still take advantage of the possibility of having a subset size smaller than s* at stopping time.

The fixed sample size form of the likelihood procedure R_2 is to consider $L_1(\delta^*)$, $L_2(\delta^*)$, ... in that order (which must be increasing) and find the smallest s such that $L_s(\delta^*) \geq P^*$. Clearly such an s with $1 \leq s \leq k$ must exist since $L_k(\delta^*) \equiv 1$.

3. ASYMPTOTIC ANALYSES

Two types of asymptotic analyses are of interest, and these are (i) $P^* \to 1$, and (ii) $\delta^* \to 0$. We consider only the sequential likelihood procedure based on (4) that results in a fixed subset size s $(1 \leq s < k)$. Similar results apply for the other variation in Section 2, but this has not yet been worked out.

As $P^* \to 1$ for fixed δ^* the excess over the boundary tends to zero, and we use the "mean path" approximation (cf. Remark 6.4.2 in BKS (1968)) to estimate the expected number E(N) of observations needed from each population for termination, i.e., we get a first term asymptotic evaluation of E(N) for any configuration with $\mu_{[k]} - \mu_{[k-1]} = \delta$ and $\mu_{[k-1]} = \ldots = \mu_{[1]}$ by simply setting our statistic $L_s(\delta^*)$ equal to P* in (4) with $\bar{x}_{[i]}$ replaced by $\mu_{[i]}$ for this configuration, and solving for the resulting value of $m \approx E(N)$. This yields the equation

$$\frac{s - 1 + e^{m\delta\delta^*/\sigma^2}}{k - 1 + e^{m\delta\delta^*/\sigma^2}} = P^* \qquad (5)$$

or equivalently for $P^* \to 1$, solving for $m \approx E(N)$,

$$E(N) \approx [\frac{\sigma^2}{\delta\delta^*}] \ln[\frac{(k-1)P^* - s + 1}{1 - P^*}] \approx [\frac{\sigma^2}{\delta\delta^*}] \ln[\frac{k-s}{1-P^*}] . \tag{6}$$

This first term result is in agreement with (6.4.1) and (6.4.3) of BKS (1968), although we are dealing here with a different procedure for $s > 1$.

The corresponding E(N)-result for $\delta^* \to 0$ is of the form $(\sigma/\delta^*)^2 h_{k,s}^{(w)}(\delta/\delta^*,P^*)$ as in (6.3.5) of BKS (1968) (or, for the general case of the exponential family, as in (6.3.1c)); here $h_{k,s}^{(w)}(\delta/\delta^*,P^*)$ is the E(N)-result for the analogous problem of k Wiener Processes with $\delta^* = 1$ (cf. Section 6.3 of BKS (1968)).

4. ASYMPTOTIC RESULTS FOR THE MODIFIED GUPTA PROCEDURE R_3

Our modification is to set the P*-requirement at the LF configuration defined by (2). This yields the exact equation

$$P\{CS|LF\} = \int_{-\infty}^{\infty} \Phi^{k-1}[x + \frac{(d + \delta^*)}{\sigma} \sqrt{n}]\phi(x)dx = P^*, \tag{7}$$

where Φ (resp., ϕ) denote the standard normal cdf (resp., density). The selected populations are those with sample means in the closed interval $[\bar{x}_{[k]} - d, \bar{x}_{[k]}]$. If we replace x by -x in (7) and make the approximation

$$[1 - \Phi(x - \frac{(d + \delta^*)}{\sigma} \sqrt{n})]^{k-1} \approx 1 - (k-1)\Phi[x - \frac{(d + \delta^*)}{\sigma} \sqrt{n}] , \tag{8}$$

then we obtain our first equation for determining d and n, namely

$$[\frac{d + \delta^*}{\sigma}] \sqrt{n/2} \approx - \Phi^{-1}[\frac{1-P^*}{k-1}] . \tag{9}$$

Our plan is to set the expected subset size for this procedure equal to s at the LF configuration and solve the resulting two equations for n. Then we wish to compare this with E(N) needed for the sequential likelihood procedure R_1. Presently we look only at the leading term for both procedures.

The expected subset size E{S|LF} at the LF configuration for R_3 is given exactly by

$$E\{S|LF\} = P^* + (k-1) \int_{-\infty}^{\infty} \Phi^{k-2}(x + \frac{d}{\sigma} \sqrt{n})\Phi[x + (\frac{d-\delta^*}{\sigma}) \sqrt{n}]\phi(x)dx \tag{10}$$

and the same method as used in (8) yields the approximation

$$s \approx P^* + (k-1)\ \{1 - \Phi[-\ (\frac{d-\delta^*}{\sigma})\sqrt{n/2}] - (k-2)\Phi[\ -\frac{d}{\sigma}\ \sqrt{n/2}]\}. \tag{11}$$

Since we are interested in small δ^*, we consider $d > \delta^*$ and hence for large values of n the last term on the right in (11) can be neglected. Moreover we must have $d \to \delta^*$ since P^* is approaching 1 and s is strictly less than k. Then (11) yields are second equation for determining d and n, namely

$$[\frac{d-\delta^*}{\sigma}]\sqrt{n/2} \approx \Phi^{-1}[\frac{s-P^*}{k-1}] \approx \Phi^{-1}\ [\frac{s-1}{k-1}]\ . \tag{12}$$

From (9) and (12) we obtain

$$n \approx \frac{\sigma^2}{2(\delta^*)^2}\ [-\ \Phi^{-1}(\frac{1-P^*}{k-1}) - \Phi^{-1}(\frac{s-1}{k-1})]^2\ . \tag{13}$$

Using the Feller-Laplace expansion for the tail area of the standard normal distribution, $\Phi(-x) \approx \phi(x)/x$ for $x \to \infty$, the leading term of the approximation (13) becomes

$$n \approx [\frac{\sigma}{\delta^*}]^2\ \ln[\frac{k-1}{1-P^*}] \tag{14}$$

which is essentially the same as the result above in (6) for any $s > 1$ for the sequential likelihood procedure R_1, except that we have $\delta\delta^*$ in the denominator there. One way to get a fair comparison is to concentrate on the LF configuration, i.e., at $\delta = \delta^*$, where we equated the fixed and expected subset sizes. Hence for the same σ, k, P^* in the LF (or slippage) configuration with $\delta = \delta^*$ we can expect to find that the fixed n for procedure R_3 and E(N) for procedure R_1 are approximately equal.

5. FIXED SAMPLE SIZE PROCEDURE R_4 BASED ON TWO REQUIREMENTS

For this procedure R_4 we again pick up all populations with sample means in a closed interval $[\bar{x}_{[k]} - c, \bar{x}_{[k]}]$, but we use two requirements to determine the c-value and the common number n of observations required. The two requirements are:

$$\min_{\delta \geq \delta^*} P(CS) \geq P^*\ , \tag{15}$$

$$\max_{\Omega} E(S) \leq s\ , \tag{16}$$

where Ω denotes the total parameter space. In (15) the minimum occurs in the LF configuration where $\delta = \delta^*$ and in (16) the maximum occurs in the equal parameter (EP) configuration. Since the sequential likelihood procedure R_1 has $S = s$ and hence $E(S) \leq s$, another fair comparison is obtained by using the same s-value in (16).

The requirement (15) leads to

$$\int_{-\infty}^{\infty} \Phi^{k-1}\left(y + (c+\delta^*) \frac{\sqrt{n}}{\sigma}\right)\phi(y)dy = P^* , \tag{17}$$

and (16) leads to

$$k \int_{-\infty}^{\infty} \Phi^{k-1}\left(y + \frac{c\sqrt{n}}{\sigma}\right)\phi(y)dy = s . \tag{18}$$

Using the same asymptotic method as in Section 4 we solve for n the two equations

$$\left(\frac{c+\delta^*}{\sigma}\right)\sqrt{n/2} \approx - \Phi^{-1}\left(\frac{1-P^*}{k-1}\right) , \tag{19}$$

$$\frac{c}{\sigma}\sqrt{n/2} \approx - \Phi^{-1}\left(\frac{1-\frac{s}{k}}{k-1}\right) , \tag{20}$$

and obtain

$$n \approx 2\left[\frac{\Phi^{-1}\left(\frac{1-P^*}{k-1}\right) - \Phi^{-1}\left(\frac{1-sk^{-1}}{k-1}\right)}{\delta^*/\sigma}\right]^2 \approx 4\left(\frac{\sigma}{\delta^*}\right)^2 \ln\left(\frac{k-1}{1-P^*}\right) . \tag{21}$$

Although this asymptotic result is larger by a factor of 4 than the result in (14), it will be seen below that for moderate values of P^* and s the correct (small sample) answer is closer to a factor of 2.

6. MONTE CARLO RESULTS

Monte Carlo studies have confirmed the results in Section 4. In the first 5 cases with selected combinations of k, δ^* and P^*, each based on more than 1000 repetitions of the sequential experiment, the sequential result was smaller in 3 of the 5 cases, but in all 5 cases the difference was statistically not significant. There was some evidence that the sequential procedure improves relative to the fixed sample size procedure R_3 as $\delta^* \to 0$. Hence it would be desirable to see if a finer comparison based on second order terms substantiates this result.

To describe and give a table for our Monte Carlo results we have to explain how we obtain comparable results. For the fixed sample size procedure R_3 we used Table 1 in Deely and Gupta (1968), where for the nominal value $P^* = .75$ (in the equal means configuration), the expected proportion of populations selected and the P{CS} are both given for selected slippage configurations. All these cases have the proportion $s/k > 1/2$ and hence the case $s/k \leq 1/2$ has not yet been investigated. Most of the P{CS} values were too close to one (say, .999 or 1.000) and we had to avoid such cases, where the table accuracy is not sufficient and also the costs of sequential sampling become prohibitive to make good comparisons.

For example, with $P^* = .75$ and $k = 3$, we find by interpolation in Table 1 of Deely and Gupta (1968) for $k = 3$ and $s = 2$ (so that $s/k = .6667$) that $\delta\sqrt{n} = 1.1875$ and the corresponding P{CS} = .9430. Hence for $\delta = \delta^* = .3$, .2 and .1 the fixed sample size n-value needed per population is 15.745, 35.427 and 141.708, respectively, to attain a P{CS} of at least .9430 in the slippage configuration with $\mu_{[3]} - \mu_{[2]} = \mu_{[3]} - \mu_{[1]} = \delta^*$.

In order to attain .9430 in the sequential procedure (4) for the same k,s and δ^* we want to set P^* on the right side of (4) a bit below .9430, so that the average value of the left side of (4) at stopping time is closer to .9430. Four values of P^*, each involving 1000 repetitions of the sequential experiment, were used to get our results; thus for $\delta^* = .2$ we used $P^* = .932$, .933, .934 and .935 with interpolation to attain .9430. Interpolation in the four $\bar{n}$-values for each δ^* in the sequential procedure gave the results 15.213, 34.293 and 135.329 for $\delta^* = .3$, .2 and .1, respectively. Although all three are better (i.e., smaller) than the fixed sample size results given above, none of these is significantly better than the results for R_3. All the results in Table 1 above are significantly better than those of R_4.

For the equal means case the expected number of observations needed per population is of course larger than for the slippage configuration. To illustrate the type of increase that occurs we consider the first case of the above table where $k = 3$, $s = 2$ and $\delta^* = .3$. Monte Carlo experimentation (like that above) gives the result 18.79 in the EP configuration for the expected number of observations required per population and the 2 sigma confidence interval for this is (17.97, 19.61). This also represents the maximum of E{N} over all configurations. Thus we note that the maximum E{N} need not be much larger than that required by the slippage configuration. This result is somewhat analogous (but not comparable) to the fact that under equal means the expected subset size E{S} for equal

Table 1

Fair Comparisons of the Amount of Sampling Required (with Sequential Results[a] for Procedure R_1 Based on Monte Carlo Trials)

Cases					Sequential result for E(N) with fixed subset size s under R_1	Fixed sample size result for R_3 with E(S)=s in slippage configuration	Fixed sample size result for R_4 with Max E(S) $\leq$ s over all configurations
	k	s	δ^*	P^{*b}			
1	3	2	.3	.9430	15.21	15.75	25.28
2	3	2	.2	.9430	34.29	35.43	56.89
3	3	2	.1	.9430	135.33	141.71	227.46
4	6	4	.2	.9550	39.24	39.06	60.09
5	6	4	.1	.9950	156.53	156.25	240.37
6	3	1	.2	.9500	93.66	-	183.69
7	3	1	.2	.9750	131.39	-	244.85
8	3	1	.2	.9888	151.39	-	316.69
9	10	1	.2	.9925	234.15	-	479.20
10	25	1	.2	.9474	215.99	-	357.57

[a] Some of these P*-values were obtained from Deely, Gupta (1968) to make the results comparable; others were taken from BKS (1968) to avoid some interpolation.

[b] Quadratic interpolation was used throughout.

parameters is 3(.943) = 2.83 under the fixed sample size procedure R_3; the corresponding result for R_1 is of course E{S} = s = 2.

SUMMARY

A general method for sequential selection and/or identification problems based on updating at each stage the likelihoods associated with various decisions (and stopping when any one of them reaches a preassigned P*-value) was developed in Bechhofer, Kiefer and Sobel (1968). This sequential method also applies to the problem of subset selection, although this application was never developed. This paper develops the application to subset selection and preliminary results based on both asymptotic theory and Monte Carlo studies are used to make fair comparisons between fixed-sample-size results and sequential results obtained by the above procedure. Some of the numerical differences are significant, and others are not significant. The fact that the sequential procedure represents an application of a general technique with even more possible applications is important. A proof of the least favorable property of the sequential procedure is given in the Appendix.

APPENDIX

Least Favorable Property of the Procedure R_1

In the following proof we use a similar notation and level of generality that was used in Chapter 6 of BKS (1968); we refer the reader to the latter reference for the background of our methodology and introduce as few new definitions as possible. We consider here only the case of selecting the best one; the corresponding proof for the more general problem of selecting the best t will be considered separately.

The likelihood statistic for selecting a subset (the populations having the s largest sample means) which contains the best population is

$$Q(\underset{\sim}{\tau}, \underset{\sim}{y}_m) = \frac{\sum_{j=k-s+1}^{k} \sum_{\alpha\in S_{k-1,1}} \exp\{\underset{\sim}{\tau}\cdot\alpha\gamma_j \underset{\sim}{y}_m\}}{\sum_{\alpha\in S_k} \exp\{\underset{\sim}{\tau}\cdot\alpha \underset{\sim}{y}_m\}}$$

$$= \frac{\sum_{j=k-s+1}^{k} \sum_{c=1}^{C} \exp\{\underset{\sim}{\tau}\cdot\alpha_c\gamma_j \underset{\sim}{y}_m\}}{\sum_{i=1}^{k} \sum_{c=1}^{C} \exp\{\underset{\sim}{\tau}\cdot\alpha_c\beta_i \underset{\sim}{y}_m\}}, \qquad (22)$$

where $\underset{\sim}{\tau} = (\tau_1, \tau_2, \ldots, \tau_k); \tau_1 \le \tau_2 \le \cdots \le \tau_k$; $\underset{\sim}{y}_m = (y_{1m}, y_{2m}, \ldots, y_{km})$; $y_{1m} \le y_{2m} \le \cdots \le y_{km}$; γ_j and β_i are transpositions defined by $\gamma_j = (j,k)(j = k\text{-}s\text{+}1, \ldots, k)$ and $\beta_i = (i,k)(i = 1, \ldots, k)$, $C = (k-1)!$, S_k is the full symmetric group on the k letters 1, 2, ..., k, $S_{k-1,1}$ is the subset of S_k which permutes only the first k-1 letters (which is identical with S_{k-1} for t = 1) and $\alpha_c \in S_{k-1,1}$.

We first note that putting $\underset{\sim}{\tau}_{k-1,1} = (\tau_{k-1}, \ldots, \tau_{k-1}, \tau_k)$ in (22) gives after cancellation

$$Q(\underset{\sim}{\tau}_{k-1,1}, \underset{\sim}{y}_m) = \frac{\sum_{j=k-s+1}^{k} \exp\{\underset{\sim}{\tau}_{k-1,1} \cdot \gamma_j \underset{\sim}{y}_m\}}{\sum_{i=1}^{k} \exp\{\underset{\sim}{\tau}_{k-1,1} \cdot \beta_i \underset{\sim}{y}_m\}} . \tag{23}$$

We want to show that for any $\underset{\sim}{\tau}$ and $\underset{\sim}{\tau}_{k-1,1}$ (defined as above in terms of $\underset{\sim}{\tau}$), any vector of sufficient statistics $\underset{\sim}{y}_m$ (and any m)

$$Q(\underset{\sim}{\tau}, \underset{\sim}{y}_m) \geq Q(\underset{\sim}{\tau}_{k-1,1}, \underset{\sim}{y}_m). \tag{24}$$

This is the so-called least favorable property. Writing out (24) gives

$$\frac{\sum_{j=k-s+1}^{k} \sum_{c=1}^{C} \exp\{\underset{\sim}{\tau} \cdot \alpha_c \gamma_j \underset{\sim}{y}_m\}}{\sum_{i=1}^{k} \sum_{c=1}^{C} \exp\{\underset{\sim}{\tau} \cdot \alpha_c \beta_i \underset{\sim}{y}_m\}} \geq \frac{\sum_{j=k-s+1}^{k} \exp\{\underset{\sim}{\tau}_{k-1,1} \cdot \gamma_j \underset{\sim}{y}_m\}}{\sum_{i=1}^{k} \exp\{\underset{\sim}{\tau}_{k-1,1} \cdot \beta_i \underset{\sim}{y}_m\}} . \tag{25}$$

Cross multiplication in (25) gives yet another form

$$\sum_{j=k-s+1}^{k} \sum_{i=1}^{k-s} A(j,i) \geq \sum_{j=k-s+1}^{k} \sum_{i=1}^{k-s} A(i,j) \tag{26}$$

where

$$A(j,i) = \sum_{c=1}^{C} \exp\{\underset{\sim}{\tau} \cdot \alpha_c \gamma_j \underset{\sim}{y}_m + \underset{\sim}{\tau}_{k-1,1} \cdot \beta_i \underset{\sim}{y}_m\} \quad ; \tag{27}$$

the terms for $i > k-s$ on both sides of (26) were identical and cancelled each other. Note that we now have $1 \leq i \leq k-s < j \leq k$ and in particular that $i < j$, with strict inequality. Our plan is to show the inequality in (26) on a term-by-term basis, i.e. we show that $A(j,i) \geq A(i,j)$ for the above limits on i and j or equivalently

$$\sum_{c=1}^{C} \exp\{\underset{\sim}{\tau} \cdot \alpha_c \gamma_j \underset{\sim}{y}_m + \underset{\sim}{\tau}_{k-1,1} \cdot \beta_i \underset{\sim}{y}_m\} \geq \sum_{c=1}^{C} \exp\{\underset{\sim}{\tau} \cdot \alpha'_c \beta_i \underset{\sim}{y}_m + \underset{\sim}{\tau}_{k-1,1} \cdot \gamma_j \underset{\sim}{y}_m\} , \tag{28}$$

where we have written α'_c (instead of α_c) to indicate that we are going to reorder the $C = (k-1)!$ permutations on the right side of (28). Again we make a term-by-term comparison within each exponential or, equivalently, we wish to show that

$$\underset{\sim}{\tau}\cdot\alpha_c\gamma_j\underset{\sim}{y}_m - \underset{\sim}{\tau}\cdot\alpha_c'\beta_i\underset{\sim}{y}_m \geq \underset{\sim}{\tau}_{k-1,1}\cdot\gamma_j\underset{\sim}{y}_m - \underset{\sim}{\tau}_{k-1,1}\cdot\beta_i\underset{\sim}{y}_m \quad . \tag{29}$$

Treating the 2 sides (RHS and LHS) of (29) separately, we obtain

$$\begin{aligned} \text{RHS} &= \underset{\sim}{\tau}_{k-1,1}\cdot[(j,k) - (i,k)]\,\underset{\sim}{y}_m \\ &= \sum_{w=1}^{k} \tau_{k-1}y_w + \delta y_j - \Big(\sum_{w=1}^{k} \tau_{k-1}y_w + \delta y_i\Big) \\ &= \delta(y_j - y_i)\ , \end{aligned} \tag{30}$$

where $\delta = \tau_k - \tau_{k-1}$,

$$\begin{aligned} \text{LHS} &= \underset{\sim}{\tau}\cdot\alpha_c(j,k)\underset{\sim}{y}_m - \underset{\sim}{\tau}\cdot\alpha_c'(i,k)\underset{\sim}{y}_m \\ &= \sum_{\substack{w=1\\ \neq j,k}}^{k}\tau_{\alpha_c(w)}y_w + \tau_{\alpha_c(j)}y_k + \tau_{\alpha_c(k)}y_j \\ &\quad - \Big(\sum_{\substack{w=1\\ \neq i,k}}^{k}\tau_{\alpha_c'(w)}y_w + \tau_{\alpha_c'(i)}y_k + \tau_{\alpha_c'(k)}y_i\Big) \\ &= (\tau_{\alpha_c(i)} - \tau_{\alpha_c'(k)})y_i + (\tau_{\alpha_c(k)} - \tau_{\alpha_c'(j)})y_j \\ &\quad + (\tau_{\alpha_c(j)} - \tau_{\alpha_c'(i)})y_k. \end{aligned} \tag{31}$$

We now consider two cases according as $j = k$ or $k\text{-}s\text{+}1 \leq j < k$.

<u>Case 1</u>: $j = k$ or $\gamma_j = (k,k) =$ identity transposition.
Then we set

$$\alpha_c'(w) = \alpha_c(w) \qquad w = 1,2,\ldots,k\ , \tag{32}$$

look at the LHS of (29) in the last line of (31) and obtain

$$\begin{aligned} \text{LHS} &= (\tau_{\alpha_c(i)} - \tau_{\alpha_c(k)})y_i + (\tau_{\alpha_c(k)} - \tau_{\alpha_c(i)})y_k \\ &= (\tau_{\alpha_c(k)} - \tau_{\alpha_c(i)})(y_j - y_i) \geq \delta(y_j - y_i) = \text{RHS}, \end{aligned} \tag{33}$$

since $\alpha_c(k) = k$, $\alpha_c(i) < k$ and $\tau_k - \tau_{\alpha_c(i)} \geq \tau_k - \tau_{k-1}$.

Case 2: $k-s+1 \leq j < k$.
Then we set

$$\alpha_c'(w) = \alpha_c(w) \quad \text{for all } w \text{ except } i \text{ and } j$$

$$\alpha_c'(i) = \alpha_c(j),\ \alpha_c'(j) = \alpha_c(i), \tag{34}$$

look again at the LHS of (29) in the last line of (31) and obtain

$$\begin{aligned} \text{LHS} &= (\tau_{\alpha_c(i)} - \tau_{\alpha_c(k)})y_i + (\tau_{\alpha_c(k)} - \tau_{\alpha_c(i)})y_j \\ &= (\tau_{\alpha_c(k)} - \tau_{\alpha_c(i)})(y_j - y_i) = (\tau_k - \tau_{\alpha_c(i)})(y_j - y_i) \\ &\geq \delta(y_j - y_i) = \text{RHS} \end{aligned} \tag{35}$$

since $\alpha_c(i) < k$ and $\tau_k - \tau_{\alpha_c(i)} \geq \delta$.

Since the LHS $\geq$ RHS of (29) in both cases the likelihood inequality needed for the least favorable property is proved.

ACKNOWLEDGMENT

This paper was delivered at a Statistical Research Conference dedicated to the memory of Jack Kiefer and Jacob Wolfowitz. The Conference was held at Cornell University, July 6-9, 1983, and was a special meeting of the Institute of Mathematical Statistics, co-sponsored by the American Statistical Association. Financial support was provided by the National Science Foundation, the Office of Naval Research, the Army Research Office-Durham, and units of Cornell University.

REFERENCES

Bechhofer, R. E., Kiefer, J. and Sobel., M. (1968). Sequential Identification and Ranking Procedures. Statistical Research Monographs, Vol. III, IMS and University of Chicago.

Deely, J. J. and Gupta, S. S. (1968). On the properties of subset selection procedures. Sankhya, Series A, Vol. 30A, 37-50.

Gupta, S. S. (1965). On some multiple decision (selection and ranking) rules. Technometrics, Vol. 7, 225-245.

ALLOCATION OF OBSERVATIONS FOR SELECTING THE BEST NORMAL POPULATION*

Y.L. Tong

Department of Mathematics
and Statistics
University of Nebraska
Lincoln, Nebraska

David E. Wetzell[†]

Department of Mathematics
and Statistics
University of Nebraska
Lincoln, Nebraska

1. INTRODUCTION

Let there be k normal populations with unknown means μ_i and (known or unknown) variances $\sigma_i^2 (i=1,\dots,k)$, and consider the problem of selecting the best population as studied in the pioneering paper by Bechhofer (1954). If n_i observations are to be taken from the i-th population, then a legitimate question is how the n_i values should be determined such that, subject to a fixed $N = \sum_{i=1}^{k} n_i$, the probability of a correct selection (P(CS)) is maximized or nearly maximized. This problem, originally due to Bechhofer (1954), has generated some interesting discussions and rather unexpected results in the literature. One such result is that, when $k > 2$ and when the population means are close to each other in a certain fashion, then an increase in the sample size from the best population alone results in a decrease in P(CS) (Tong and Wetzell (1979)). In fact, this result demonstrates the importance of the allocation problem; for an inefficient allocation not only will waste resources, but also may have a negative effect on P(CS). On the other hand, the proof of this result actually illustrates the difficulty of obtaining a

*This work was partially supported by NSF Grant No. MCS81-00775, A01.

[†]Now at Texas Instruments, Inc.

suitable solution to the allocation problem. Namely, a change in the sample size from one of the populations always changes the correlation structure of the underlying multivariate normal distribution which makes it difficult to assess the true effect on P(CS). Thus, to our knowledge, an exact solution to such an allocation problem has not yet been obtained except for $k = 2$.

In this paper we provide a partial solution to this problem for the large sample case. This partial solution is based on the solution to a certain maximization problem which involves functions of the standard normal cdf; this latter problem is solved in the Appendix using certain majorization inequalities. Since the (asymptotically) optimal allocation depends on the means and the variances which are unknown, an adaptive sequential allocation procedure is then considered. This procedure is developed on the basis of estimating the unknown parameters and identifying the leading population sequentially.

We note in passing that recently Bechhofer and Tamhane (1982) also studied such an allocation problem, and gave some nice results. They basically treated this problem from a max-min viewpoint (where the min is taken over the preference zone and the max is taken over all allocations); hence their approach is different from ours. As a result, as they pointed out in their technical report, the asymptotic results they obtained do not agree with ours in all cases.

2. ALLOCATION OF OBSERVATIONS

Let $\bar{X}_i$ denote the sample mean from the ith population with mean μ_i, variance σ_i^2, and sample size $n_i (i=1,\ldots,k)$. As usual, let $\mu_{[1]} \le \cdots \le \mu_{[k]}$ be the ordered population means and $\delta_i = \mu_{[k]} - \mu_{[i]} (i=1,\ldots,k-1)$. Let $((1),\ldots,(k))$ denote a (unknown) permutation of $(1,\ldots,k)$ such that $\mu_{[i]} = \mu_{(i)}$. Then it is well-known that, under the Bechhofer (1954) selection rule, the probability of a correct selection function is

$$\begin{aligned} P(CS) &= P\left[\bar{X}_{(i)} \le \bar{X}_{(k)},\ i=1,\ldots,k-1\right] \\ &= \int_{-\infty}^{\infty} \prod_{i=1}^{k-1} \Phi\left[(\sqrt{n_{(i)}}/\sigma_{(i)})(\sigma_{(k)}z/\sqrt{n_{(k)}} + \delta_i)\right] d\Phi(z) \\ &= \beta(\underset{\sim}{n}, n_{(k)}) \quad \text{(say)} \end{aligned} \tag{2.1}$$

where $\Phi(\cdot)$ is the standard normal cdf. The r.h.s. of (2.1) is exactly of the form in (A.12) with $p = k-1$, $a_i = \sigma_{(i)}^{-1}$, and $b_M = \sqrt{n_{(k)}}/\sigma_{(k)}$. Under the least-favorable configuration when $\delta_i = \delta^* > 0$, (2.1) reduces to

$$\beta(\underset{\sim}{n}, n_{(k)}) = \int_{-\infty}^{\infty} \prod_{i=1}^{k-1} \Phi\left[(\sqrt{n_{(i)}}/\sigma_{(i)})(\sigma_{(k)} z/\sqrt{n_{(k)}} + \delta^*)\right] d\Phi(z). \tag{2.2}$$

The integrals (2.1) and (2.2) can be given in the form of multivariate normal probabilities. After examining the correlation behavior of the underlying distribution, which depends on $(\sqrt{n_{(1)}},\ldots,\sqrt{n_{(k)}})$, Tong and Wetzell (1979) proved that the r.h.s. of (2.1) is increasing in $n_{(i)}$ for $i \le k-1$ for each fixed $n_{(k)}$, and is decreasing in $n_{(k)}$ when the δ_i values are sufficiently small.

In this section we first consider the allocation problem, for large $N = \sum_{i=1}^{k} n_i$, when $n_{(k)}$ is large but preassigned. The solution that we obtain is nonimplementable since it depends on the unknown permutation ((1), (2),...,(k)) in addition to the unknown parameters δ_i and $\sigma_i^2 (1 \le i \le k)$. However, the solution to this artificial problem suggests a method for constructing an asymptotically optimal adaptive procedure (discussed in the next section) in which the unknown quantities are sequentially estimated.

For every (large) fixed c let us define

$$\beta_c(\underset{\sim}{n}, n_{(k)}) = \int_{-c}^{c} \prod_{i=1}^{k-1} \Phi\left[(\sqrt{n_{(i)}}/\sigma_{(i)})(\sigma_{(k)} z/\sqrt{n_{(k)}} + \delta_i)\right] d\Phi(z).$$

Let $\underset{\sim}{n}_c^* = (n_{(1)}^*,\ldots,n_{(k-1)}^*) \in \Lambda_M$ (where $M = N - n_{(k)}$) and $\underset{\sim}{n}' = (n'_{(1)},\ldots,n'_{(k-1)}) \in \Lambda_M$ denote two allocations such that

$$\sup_{\underset{\sim}{n} \in \Lambda_M} \beta_c(\underset{\sim}{n}, n_{(k)}) = \beta_c(\underset{\sim}{n}_c^*, n_{(k)}).$$

We shall say that $\underset{\sim}{n}'$ is an asymptotically c-optimal allocation, for given $n_{(k)}$, if for every fixed small $\varepsilon > 0$ and every fixed c there exists an $M_{\varepsilon,c}$ such that $M > M_{\varepsilon,c}$ implies that

$$\max_{i \ne j} \left| \frac{n_{(i)}^*}{n_{(j)}^*} - \frac{n'_{(i)}}{n'_{(j)}} \right| < \varepsilon, \quad 1 \le i \ne j < k.$$

Applying Theorem A.1, we immediately obtain the following:

<u>Theorem 2.1</u>. Let $\underset{\sim}{n}'$ satisfy

$$\frac{n'_{(i)}}{n'_{(j)}} = \left[\left(\frac{\sigma_{(i)}}{\delta_{(i)}}\right)/\left(\frac{\sigma_{(j)}}{\delta_{(j)}}\right)\right]^2 \quad \text{for all } i \ne j. \tag{2.3}$$

Then $\underset{\sim}{n}'$ is an asymptotically c-optimal allocation.

The allocation $\underset{\sim}{n}'$ is, of course, intuitively appealing; because when the mean of a given population is close to that of the best population (i.e., δ_i is small), and/or when its variance is large, then a relatively larger portion of observations should be taken from that population.

The second allocation problem concerns the optimal ratio $n^*_{(k)}/N$, or equivalently, the ratio M/N. We do not have any useful analytical results on this problem, but feel that a previous result of Bechhofer (1969) can be applied to yield an approximate solution. Bechhofer's (1969) result, which concerns the allocation problem when comparing several treatments with a control, implies that when $\delta_i = \delta > 0$ $(i=1,\ldots,k-1)$ then the asymptotically optimal allocation must satisfy

$$\frac{n^*_{(k)}}{N} \to [1 + (\sum_{i=1}^{k-1} \sigma^2_{(i)}/\sigma^2_{(k)})^{\frac{1}{2}}]^{-1} \equiv \gamma^*_k \text{ as } N \to \infty.$$

Thus, when the δ_i's are not necessarily all equal, the following two allocations seem to be reasonable alternatives:

$$\frac{n_{(k)}}{N} = \gamma^*_k , \tag{2.4a}$$

$$\frac{n_{(k)}}{N} = \gamma^{**}_k \equiv [1 + \bar{\delta}(\sum_{i=1}^{k-1} \sigma^2_{(i)}/\sigma^2_{(k)}\delta^2_{(i)})^{\frac{1}{2}}]^{-1}. \tag{2.4b}$$

Probably (2.4b) should be used rather than (2.4a) if there are reasons to believe that the δ_i's are likely to be unequal. In the special case when all the δ_i's are equal then, of course, (2.4a) and (2.4b) agree with each other.

It should be noted that we do not claim that the above solution is optimal. The exact optimal solution appears to be very complicated; therefore we have obtained an approximation to the optimal solution. Our approximate optimal solution can be briefly described as follows: For large N, the proportion of observations taken from the best population should be approximately γ^*_k or γ^{**}_k (depending on which one is used); then the rest of the $M = N - n_{(k)}$ observations should be allocated to the other populations according to (2.3).

3. AN ADAPTIVE PROCEDURE FOR SEQUENTIAL ALLOCATION OF OBSERVATIONS

The solution to the allocation problem treated in Section 2 involves the unknown parameters $\delta_{(i)}$ and $\sigma^2_{(i)}$. Since in reality the true values of these parameters are unknown, a direct implementation of such an allocation procedure is not possible except for the special case of k=2 with known variances. Consequently, in the following we consider an adaptive approach to this problem.

The basic idea behind the (sequential) adaptive approach is this: At each stage, the means and variances of all of the populations are estimated from the data already collected. On the basis of the estimated values of the parameters, a decision is made on which population the next observation should be taken from, so that (a fixed number of) N observations are allocated sequentially, one at a time. Such a procedure is described below:

Procedure R. (a) First take n_0 observations from each population. (It is recommended that $n_0 \geq 4$ in order to have sufficiently precise initial estimates.) (b) After observing $\{X_{ij}\}$ for $j=1,\ldots,n_i$ and $i=1,\ldots,k$, compute

$$\bar{X}_i = \frac{1}{n_i}\sum_{j=1}^{n_i} X_{ij}, \quad S_i^2 = \frac{1}{n_i-1}\sum_{j=1}^{n_i} (X_{ij} - \bar{X}_i)^2 \tag{3.1}$$

for $i=1,\ldots,k$. Denote the order statistics by $(\bar{X}_{[1]},\ldots,\bar{X}_{[k]})$, and compute

$$\begin{aligned} \hat{\delta}_i &= \bar{X}_{[k]} - \bar{X}_{[i]} \quad \text{for } i=1,\ldots,k-1, \\ \bar{\hat{\delta}} &= \frac{1}{k-1}\sum_{i=1}^{k-1} \hat{\delta}_i. \end{aligned} \tag{3.2}$$

Then call the population which yields $\bar{X}_{[k]}$ the leading population, and let K denote the integer such that $\bar{X}_K = \bar{X}_{[k]}$. (c) Compute

$$\hat{\gamma}_k^* = [1 + (\sum_{i \neq K} S_i^2 / S_K^2)^{\frac{1}{2}}]^{-1} \tag{3.3a}$$

or

$$\hat{\gamma}_k^{**} = [1 + \bar{\hat{\delta}}(\sum_{i \neq K} S_i^2 / S_K^2 \hat{\delta}_i^2)^{\frac{1}{2}}]^{-1} \tag{3.3b}$$

depending on whether (2.4a) or (2.4b) is used, respectively. Take the next observation from the leading population if $n_K / \sum_{i=1}^{k} n_i < \hat{\gamma}_k^*$ (or $\hat{\gamma}_k^{**}$); otherwise, take the next observation from the i-th population where $i \neq K$ satisfies

$$n_i \hat{\delta}_i / S_i = \min_{j \neq K} (n_j \hat{\delta}_j / S_j).$$

(d) Stop sampling when $\sum_{i=1}^{k} n_i = N$, recompute $\bar{X}_i$ as in (3.1), and select the population which yields the largest sample mean $\bar{X}_{[k]}$.

As for the asymptotic behavior of this procedure, we have obtained a proof of the following result: If $\delta_i > 0$ and $\sigma_i^2 > 0$ for each i then, as $N \to \infty$, $P(CS) \to 1$ and the (random) sample sizes $n_1, \ldots, n_k$ satisfy $n_{(k)}/N \to \gamma_k^*$ a.s. (or $\hat{\gamma}_k^{**}$ a.s., depending on which one is used), and

$$\frac{n_{(i)}}{n_{(j)}} \to \left(\frac{\sigma_{(i)}}{\delta_{(i)}}\right)^2 / \left(\frac{\sigma_{(j)}}{\delta_{(j)}}\right)^2 \text{ a.s.}$$

for $i \neq K$, $j \neq K$. Thus, at least asymptotically, the $n_{(i)}$ do behave as desired. The proof of this result is quite involved and, in order not to overload this paper, it is not given here.

It should be noted that the statement of the above procedure is given for the general case when all the parameters are unknown. If some of the parameters (such as the variances) are known, then changes in the procedure can be made through obvious modifications. Also, we have treated the problem under the assumption that $k > 2$; because for $k = 2$ with known variances the probability of a correct selection can be expressed as a univariate normal probability; in that case the solution to the allocation problem is well-known and is not discussed here.

The procedure R can also be modified to allow N to be random. That is, the total sample size is not preassigned and we now have a complete sequential setting, and the sampling terminates according to a specified stopping rule. Such a sequential procedure, developed along the direction of the adaptive ranking procedures considered by Tong (1978) and studied by Wetzell (1979), is not treated here.

Some Monte Carlo studies were carried out to investigate the behavior of the Procedure R. One of such results is given here: For $k = 4$, $\mu_i = i$ $(i=1,\ldots,4)$, $\sigma_1 = \sigma_2 = \sigma_3 = 10$ and $\sigma_4 = 1$, and for $N = 400$, we carried out 200 simulation runs. The observed probability of a correct selection was 0.875 (using $\hat{\gamma}_k^*$) and the observed ratios of n_i/n_j were quite close to the theoretical expectations. In this situation if one ignores the sequential allocation idea, and takes an equal number of 100 observations from each of the four populations, then the probability of a correct selection is

$$P(CS) = P[Z_1 \le 30/\sqrt{101},\ Z_2 \le 20/\sqrt{101},\ Z_3 \le 10/\sqrt{101}], \tag{3.4}$$

where (Z_1, Z_2, Z_3) is a multivariate normal variable with the Z_i having zero means, unit variances, and a common correlation $\rho = 0.01$. Since ρ is very close to zero, one can approximate the r.h.s. of (3.4) by

$$\Phi(30/\sqrt{101})\Phi(20/\sqrt{101})\Phi(10/\sqrt{101}) \doteq 0.82.$$

In this case, the improvement in the probability of a correct selection over the equal-sample-size allocation is about 0.055. It is believed that under certain configurations the improvement can be much more significant. However, no additional Monte Carlo results to support this claim are available at this time.

APPENDIX

Maximization of Certain Functions Involving the Standard Normal Distribution

In this appendix we study the problems of maximization of certain functions which involve the standard normal distribution.

First consider the case of two variables. Denote by $\Phi(\cdot)$ the standard normal cdf, and define

$$f(n_1,n_2) = \Phi(a_1\sqrt{n_1})\Phi(a_2\sqrt{n_2}). \tag{A.1}$$

For arbitrary but fixed M let

$$\Lambda_M = \{(n_1,n_2) | n_i \geq 0,\ i=1,2,\ \sum_{i=1}^{2} n_i = M\},$$

and let (n_1',n_2'), (n_1'',n_2'') be in Λ_M such that

$$(n_1'/n_2') = (a_2/a_1)^2, \quad (n_1''/n_2'') = (a_1/a_2)^2.$$

Let $(n_1^*,n_2^*) \in \Lambda_M$ be an optimal solution which satisfies

$$\sup_{(n_1,n_2) \in \Lambda_M} f(n_1,n_2) = f(n_1^*,n_2^*).$$

In the following lemma we show that (n_1^*,n_2^*) is a convex combination of (n_1',n_2') and (n_1'',n_2'') and that, as $M \to \infty$, $(n_1^*/n_2^*) \to (a_2/a_1)^2$.

Lemma A.1. For preassigned $a_1 \geq a_2 > 0$ and for every fixed M, there exists an $\alpha \in [0,1]$ such that

$$(n_1^*,n_2^*) = (1-\alpha)(n_1',n_2') + \alpha(n_1'',n_2''). \tag{A.2}$$

Further, $\alpha \to 0$ as $M \to \infty$.

Proof. It is obvious that there always exists an α satisfying (A.2). We need to show that this α value is between 0 and 1. Consider the maximization of

$$\log f(n,M-n) = \log \Phi(a_1\sqrt{n}) + \log \Phi(a_2\sqrt{M-n})$$

(which is equivalent to the maximization of $f(n,M-n)$), and define

$$g(n) = a_1\sqrt{n} + a_2\sqrt{M-n}, \quad n \geq 0$$

as a function of n. It is easy to verify that, for each fixed M, g(n) is a concave function of n and attains a unique maximum at $n = M/[1+(a_2/a_1)^2]$, at which point one has $n/(M-n) = (a_1/a_2)^2$. Thus for (n_1,n_2) satisfying

$$(n_1,n_2) = (1-\alpha)(n_1',n_2') + \alpha(n_1'',n_2''), \tag{A.3}$$

one must have

$$(a_1\sqrt{n_1},a_2\sqrt{n_2}) \overset{\omega}{>} (a_1\sqrt{n_1'},a_2\sqrt{n_2'}) \quad \text{for all } \alpha < 0,$$

$$(a_1\sqrt{n_1},a_2\sqrt{n_2}) \overset{\omega}{>} (a_1\sqrt{n_1''},a_2\sqrt{n_2''}) \quad \text{for all } \alpha > 1,$$

where the symbol "$\overset{\omega}{>}$" stands for weak supermajorization (Marshall and Olkin (1979, p.10)). Since $\Phi(x)$ is a log-concave function of x (Tong (1978)), and is continuous and strictly increasing, it follows from a result in Marshall and Olkin (1979, p.10) that for (n_1,n_2) satisfying (A.3),

$$\log f(n_1,n_2) < \log f(n_1',n_2') \quad \text{for all } \alpha < 0,$$

$$\log f(n_1,n_2) < \log f(n_1'',n_2'') \quad \text{for all } \alpha > 1.$$

This implies that in (A.2), α must be bounded by 0 and 1.

If $a_1 = a_2$, then the above result immediately yields $n_1^* = n_2^*$ for all M. Thus in the rest of the proof it will be assumed that $a_1 > a_2$. To show that $\alpha \to 0$ as $M \to \infty$, let α_r (which is also in [0,1] and depends on M) satisfy

$$\begin{aligned}(n_1^*/n_2^*) &= (1-\alpha_r)(n_1'/n_2') + \alpha_r(n_1''/n_2'') \\ &= (1-\alpha_r)(a_2/a_1)^2 + \alpha_r(a_1/a_2)^2;\end{aligned} \tag{A.4}$$

we show instead that $\alpha_r \to 0$ as $M \to \infty$. Consider the maximization of $\log f(n,M-n)$ by imposing the condition $\frac{d}{dn} \log f(n,M-n) = 0$. It follows that $\frac{d}{dn} \log f(n,M-n) = 0$ if and only if

$$\sqrt{n}\, a_2/(\sqrt{M-n}\, a_1) = \frac{\phi(a_1\sqrt{n})\Phi(a_2\sqrt{M-n})}{\phi(a_2\sqrt{M-n})\Phi(a_1\sqrt{n})}$$

where $\phi(\cdot)$ is the density function corresponding to $\Phi(\cdot)$. Taking logs on both sides and simplifying, we have

$$\begin{aligned}&\log(na_2^2/(M-n)a_1^2) \\ &= -na_1^2 + (M-n)a_2^2 + 2\log(\Phi(\sqrt{M-n}\, a_2)/\Phi(\sqrt{n}\, a_1)).\end{aligned} \tag{A.5}$$

We now use

$$(n/(M-n)) = (1-\alpha_r)(a_2/a_1)^2 + \alpha_r(a_1/a_2)^2$$

to obtain

$$n = n(\alpha_r, M) = M[a_2^4 + (a_1^4 - a_2^4)\alpha_r]/[a_1^2 a_2^2 + (a_1^4 - a_2^4)\alpha_r + a_2^4].$$

Next substituting in (A.5) for n and M-n, we obtain

$$\begin{aligned} &- \log[((a_1^4 - a_2^4)\alpha_r + a_2^4)/a_1^4] \\ &= M\alpha_r[a_1^2(a_1^4 - a_2^4)/(a_1^2 a_2^2 + (a_1^4 - a_2^4)\alpha_r + a_2^4)] \\ &-2\log[\Phi(a_2(M - n(\alpha_r, M))^{\frac{1}{2}})/\Phi(a_1(n(\alpha_r, M))^{\frac{1}{2}})] \\ &= M\alpha_r b_1(\alpha_r) + b_2(\alpha_r) \text{ (say).} \end{aligned} \tag{A.6}$$

It is easy to see that for all $\alpha_r \in [0,1]$ one has

$$0 < a_1^2 - a_2^2 \leq b_1(\alpha_r) \leq a_1^2(a_1^2 - a_2^2)/a_2^2,$$

$$0 \leq -\log[((a_1^4 - a_2^4)\alpha_r + a_2^4)/a_1^4] \leq 4\log(a_1/a_2),$$

and that, by $(n/(M-n)) \geq (a_2/a_1)^2$, $b_2(\alpha_r) \geq 0$. Consequently one has, for α_r satisfying (A.6),

$$\begin{aligned} 0 \leq \alpha_r &< [(Mb_1(\alpha_r)]^{-1} 4\log(a_1/a_2) \\ &\leq \frac{4}{M} \cdot \log(a_1/a_2)/(a_1^2 - a_2^2), \end{aligned} \tag{A.7}$$

and the right-hand side in (A.7) goes to 0 as $M \to \infty$. □

<u>Lemma A.2</u>. For i=1,...,p let $a_i > 0$ be arbitrary but preassigned and denote

$$\Lambda_M = \{\underset{\sim}{n} = (n_1, \ldots, n_p) | n_i \geq 0, \quad i=1,\ldots,p, \quad \sum_{i=1}^{p} n_i = M\}, \tag{A.8}$$

$$f(\underset{\sim}{n}) = \prod_{i=1}^{p} \Phi(a_i \sqrt{n_i}). \tag{A.9}$$

Let $\underset{\sim}{n}^* = (n_1^*, \ldots, n_p^*)$ be such that

$$\sup_{\underset{\sim}{n} \in \Lambda_M} f(\underset{\sim}{n}) = f(\underset{\sim}{n}^*).$$

Then one must have

$$\left(\frac{a_j}{a_i}\right)^2 \leq \left(\frac{n_i^*}{n_j^*}\right) \leq \left(\frac{a_i}{a_j}\right)^2 \quad \text{for all } a_i \geq a_j, \tag{A.10}$$

and

$$\lim_{M \to \infty} \left(\frac{n_i^*}{n_j^*}\right) = \left(\frac{a_j}{a_i}\right)^2. \tag{A.11}$$

Proof. The proof follows from Lemma A.1 by treating two n_i variables at a time while holding the others fixed. □

Next, we turn our attention to integrals of the form

$$H(\underset{\sim}{n}) = \int_{-\infty}^{\infty} h(z,\underset{\sim}{n})d\Phi(z), \tag{A.12}$$

where

$$h(z,\underset{\sim}{n}) = \prod_{i=1}^{p} \Phi(a_i\sqrt{n_i}(z/b_M + \delta_i)),\ z\in(-\infty,\infty),\ \underset{\sim}{n}\in\Lambda_M, \tag{A.13}$$

$a_i > 0$, $\delta_i > 0$ are preassigned, $b_M \to \infty$, and $n_i/b_M \to \lambda_i > 0$ as $M \to \infty$. Such integrals are, of course, directly related to the selection and ranking problems.

Theorem A.1. Let $\underset{\sim}{n}_z^* = (n_1^*,\dots,n_p^*) \in \Lambda_M$ satisfy

$$\sup_{\underset{\sim}{n}\in\Lambda_M} h(z,\underset{\sim}{n}) = h(z,\underset{\sim}{n}_z^*).$$

Then, for every fixed $\varepsilon > 0$ and $c > 0$, there exists an $M_{\varepsilon,c}$ such that

$$\max_{i\neq j}\left|\frac{n_i^*}{n_j^*} - \frac{(a_j\delta_j)^2}{(a_i\delta_i)^2}\right| < \varepsilon \tag{A.14}$$

holds for all $M > M_{\varepsilon,c}$ and all $z \in [-c,c]$.

Proof. Let M_1 be such that $M > M_1$ implies $c < b_M \cdot \min_{1\le i\le p}\delta_i$. Then $a_i' = a_i(z/b_M + \delta_i) > 0$ for all $z \in [-c,c]$ for each i. Thus, by Lemma A.2, there exists an $M_2 \geq M_1$ such that $M > M_2$ implies that

$$\max_{i\neq j}\left|\frac{n_i^*}{n_j^*} - \frac{[a_j(z/b_M + \delta_j)]^2}{[a_i(z/b_M + \delta_i)]^2}\right| < \frac{\varepsilon}{2}\ .$$

Let M_3 be large enough so that

$$\max_{i\neq j}\left|\frac{[a_j(z/b_M + \delta_j)]^2}{[a_i(z/b_M + \delta_i)]^2} - \frac{(a_j\delta_j)^2}{(a_i\delta_i)^2}\right| < \frac{\varepsilon}{2}$$

for all $z \in [-c,c]$ (this is possible because $b_M \to \infty$ as $M \to \infty$). Let $M_{\varepsilon,c} = \max(M_2,M_3)$. Then one has

$$\max_{i\neq j}\left|\frac{n_i^*}{n_j^*} - \frac{(a_j\delta_j)^2}{(a_i\delta_i)^2}\right|$$

$$\leq \max_{i \neq j} \left| \frac{n_i^*}{n_j^*} - \frac{[a_j(z/b_M + \delta_j)]^2}{[a_i(z/b_M + \delta_i)]^2} \right|$$

$$+ \max_{i \neq j} \left| \frac{[a_j(z/b_M + \delta_j)]^2}{[a_i(z/b_M + \delta_i)]^2} - \frac{(a_j\delta_j)^2}{(a_i\delta_i)^2} \right|$$

$$\leq \varepsilon$$

for all $M > M_{\varepsilon,c}$ and all $z \in [-c,c]$. □

An application of Theorem A.1 yields the following result.

Theorem A.2. For preassigned $a_i > 0$ and $\delta_i > 0$ let $\underset{\sim}{n}^* \in \Lambda_M$ satisfy

$$H(\underset{\sim}{n}^*) = \sup_{\underset{\sim}{n} \in \Lambda_M} H(\underset{\sim}{n}),$$

and let $\underset{\sim}{n}' = (n_1', \ldots, n_p') \in \Lambda_M$ satisfy

$$\frac{n_i'}{n_j'} = \left(\frac{a_j\delta_j}{a_i\delta_i}\right)^2 \quad \text{for all } i \neq j.$$

If $b_M \to \infty$ and $n_i/b_M \to \lambda_i$ for some $\lambda_i > 0$ for all i, then for every given $\varepsilon > 0$ there exists an M_ε such that $M > M_\varepsilon$ implies that

$$|H(\underset{\sim}{n}^*) - H(\underset{\sim}{n}')| < \varepsilon . \tag{A.15}$$

Proof. It is easy to see that the following inequality holds for all $c > 0$:

$$\begin{aligned} &|H(\underset{\sim}{n}^*) - H(\underset{\sim}{n}')| \\ &\leq \left|\int_{-\infty}^{\infty} h(z,\underset{\sim}{n}^*)d\Phi(z) - \int_{-c}^{c} h(z,\underset{\sim}{n}^*)d\Phi(z)\right| \\ &+ \left|\int_{-c}^{c} h(z,\underset{\sim}{n}^*)d\Phi(z) - \int_{-c}^{c} h(z,\underset{\sim}{n}')d\Phi(z)\right| \\ &+ \left|\int_{-c}^{c} h(z,\underset{\sim}{n}')d\Phi(z) - \int_{-\infty}^{\infty} h(z,\underset{\sim}{n}')d\Phi(z)\right| \\ &\equiv \Delta_1 + \Delta_2 + \Delta_3 \quad \text{(say)}. \end{aligned}$$

Choosing $c = \Phi^{-1}(1-\varepsilon/8)$ and using the fact that $0 < h(z,\underset{\sim}{n}^*) < 1$, $0 < h(z,\underset{\sim}{n}') < 1$ one has

$$\Delta_1 < \varepsilon/4, \quad \Delta_3 < \varepsilon/4.$$

Then it follows from Theorem A.1 that, for sufficiently large M_ε, $M > M_\varepsilon$ implies that

$$h(z,\underset{\sim}{n}^*) \geq 1 - \varepsilon/2, \; h(z,\underset{\sim}{n}') \geq 1 - \varepsilon/2$$

for all $z \in [-c,c]$. Thus one has

$$\Delta_2 \leq \int_{-c}^{c} |h(z,\underset{\sim}{n}^*) - h(z,\underset{\sim}{n}')| d\Phi(z)$$

$$\leq \int_{-c}^{c} (\varepsilon/2) d\Phi(z) < \varepsilon/2,$$

and the proof of the theorem is complete. □

SUMMARY

This paper concerns the allocation of observations for selecting the best of k normal populations when the total sample size is considered large but fixed. Since the asymptotic solution depends on the population means and variances which are unknown, a sequential allocation scheme is then developed by estimating these parameters and identifying the leading population sequentially.

ACKNOWLEDGMENT

We wish to thank Professor A. C. Tamhane for his useful comments on an earlier draft of this paper.

REFERENCES

Bechhofer, R. E. (1954). A single-sample multiple decision procedure for ranking means of normal populations with known variances. Ann. Math. Statist., 25, 16-39.

Bechhofer, R. E. (1969). Optimal allocation of observations when comparing several treatments with a control. In Multivariate Analysis II, P. R. Krishnaiah, ed., Academic Press, Inc.: New York, 463-473.

Bechhofer, R. E. and Tamhane, A. C. (1982). Optimal sample size allocation for selecting the best of several normal populations with known, unequal variances. Technical Report No. 567, School of Operations Research and Industrial Engineering, Cornell University, Ithaca, NY.

Marshall, A. W. and Olkin, I. (1979). Inequalities: Theory of Majorization and its Applications. Academic Press Inc.: New York.

Tong, Y. L. (1978). An adaptive solution to ranking and selection problems. Ann. Statist., 6, 658-672.

Tong, Y. L. and Wetzell, D. E. (1979). On the behaviour of the probability function for selecting the best normal population. Biometrika, 66, 174-176.

Wetzell, D. E. (1979). Allocation of observations in ranking and selection problems via majorization and other related inequalities. Ph.D. thesis, Department of Mathematics and Statistics, University of Nebraska-Lincoln, Lincoln, NE.

A CLASS OF SEQUENTIAL PROCEDURES FOR K-SAMPLE PROBLEMS CONCERNING NORMAL MEANS WITH UNKNOWN UNEQUAL VARIANCES

Bruce W. Turnbull*

School of Operations Research
and Industrial Engineering
Cornell University
Ithaca, New York

Lionel I. Weiss

School of Operations Research
and Industrial Engineering
Cornell University
Ithaca, New York

1. INTRODUCTION

Let $\Pi_1, \Pi_2, \ldots, \Pi_k$ be $k \geq 2$ normal populations with unknown means $\mu_1, \mu_2, \ldots, \mu_k$ and unknown variances $\sigma_1^2, \sigma_2^2, \ldots, \sigma_k^2$. Among typical statistical problems concerning the means $\{\mu_i\}$ are the following:

(1) Obtaining a fixed width confidence interval for a linear combination of the $\{\mu_i\}$;

(2) Testing homogeneity of means in a one-way or multi-way layout with specified size α and power β at a fixed alternative.

(3) Selecting the best population (largest mean) with preassigned probability P*. (Bechhofer 1954).

(4) Selecting the best population provided it is better than a specified standard (Bechhofer and Turnbull 1978, Wilcox 1983).

(5) Selecting a subset of populations containing the best population with preassigned probability (Gupta 1965).

(6) Selecting a subset of populations better than a control population or a specified standard (Gupta and Sobel 1958, Turnbull 1976).

(7) Partitioning a set of normal populations by their locations with respect to a control (Tong 1969, Wilcox 1979).

When the variances $\{\sigma_i^2\}$ are all known and equal, fairly simple single-stage fixed sample size procedures are available. In general, when

*This work was supported in part by NIH Grant GM 26384.

the $\{\sigma_i^2\}$ are unknown, two-stage procedures can be used analogous to those first proposed by Stein (1945) and extended by Chapman (1950) and Ruben (1953). (For problem 5, a single-stage procedure can still be used if the unknown variances are assumed equal.) However, two-stage procedures are generally inefficient with regard to expected sample size because only the first stage observations are utilized to estimate the variances. In addition there can be difficulties in the choice of the first stage sample sizes especially if a common first stage sample size is required from all populations and yet the variances are thought a priori to be widely differing--see Section 5. A final drawback of two-stage procedures is that fairly extensive tables are needed to implement the procedures; however few exist in the literature and these assume a common initial sample size. (Tables are more widely available for two-stage procedures for the case when the unknown variances are assumed equal.) Some specific references will be given in Sections 3 and 4 where Problems (2) and (3), respectively, are discussed.

In this paper we study a class of simple sequential sampling rules where the sample size on Π_i is determined by when the sample variance for that population crosses a certain boundary. Once sampling has terminated, one proceeds as if the populations all have a common known variance; thus tables of constants for the corresponding single-stage procedures can be used--all of which are easily available. However, the procedure is not exact and error probability requirements are only guaranteed asymptotically.

This type of sequential procedure was first suggested by Anscombe (1952, 1953) and later by Robbins (1959). Since then, over forty papers have appeared examining the performance of the procedure in various problems; most treat the case of estimation or testing with $k = 1$ or 2; some look at the multivariate extension; some change or relax the normality assumption. A survey for the normal case $k = 1$ has been given by Ghosh and Mukhopadhyay (1976). In this paper we attempt to supply a unified approach to statistical problems involving several populations. Although the procedures can be applied in many fields, we envisage that they would be particularly useful in the area of simulation where such problems are important (Tocher 1963 p. 114, Kleijnen 1974, Law and Kelton 1982, Chap. 9) and where sequential sampling rules are easy to implement.

In the next section, we examine the sampling rules in detail; in Sections 3 and 4 we give specific applications to the problems (2) and (3) listed earlier, namely testing for homogeneity, and selection of the best population. In Section 5 we compare the sequential procedures with the corresponding two-stage procedures.

2. SAMPLE SIZE DETERMINATION

We assume that observations $\{X_{ij};\ j \geq 1\}$ are available sequentially from population Π_i $(i \leq i \leq k)$. We let n^* denote that common sample size required by the appropriate single-stage procedure to satisfy the error probability requirements in the case when all the variances are known and equal to unity. When the variances are unknown and unequal, we proceed as follows. For each population Π_i $(1 \leq i \leq k)$

(A) Select an initial sample size n_i^0; thereafter take observations sequentially one at a time.

(B) Terminate sampling from Π_i at a sample size $N_i = n_i$ $(\geq n_i^0)$, when the computed within population sample variance $S_{i,n_i}^2 = \sum_{j=1}^{n_i}(X_{ij}-\overline{X}_{i.})^2/(n_i-1)$ first satisfies

$$S_{i,n_i}^2/n_i \leq 1/n^* \tag{1}$$

Here $\overline{X}_{i.} = \sum_{j=1}^{n_i} X_{ij}/n_i$.

(C) When sampling has terminated on all k populations, proceed as if $\overline{X}_1,\overline{X}_2,\ldots,\overline{X}_k$ were independent, normal with common variance $1/n^*$.

We denote this procedure by P. In (A) the choice of n_i^0 (≥ 2) for $1 \leq i \leq k$ is arbitrary but typically it will be small e.g. $n_i^0 = 3$ or 5. In contrast to two-stage procedures, since all the data are used to estimate the $\{\sigma_i^2\}$, it is not necessary to have a large initial sample size, nor is it necessary that the n_i^0 $(1 \leq i \leq k)$ be equal. In fact the choice of $\{n_i^0\}$ will not affect the asymptotic properties of our procedure (to be described below), nor will such variants as sampling two observations at a time or taking five, say, additional observations after the stopping criterion (B) is satisfied (cf. Starr 1966). Also in (B), the r.h.s. of (1) can be replaced by $f(n_i)$ where $(f(n);\ n = 1,2,\ldots)$ is any non-decreasing sequence lending to the limit $1/n^*$. Note also that sampling terminates almost surely since $S_{i,n}^2/n \to 0$ a.s. as $n \to \infty$ for $1 \leq i \leq k$.

We define $n_i^* = n^*\,\sigma_i^2$ $(1 \leq i \leq k)$. These would be the required sample sizes in a single-stage procedure if the true values of the $\{\sigma_i^2\}$ were known. For the first theorem we generalize (B) to allow sampling r_i observations at a time from Π_i as mentioned above.

<u>Theorem 1</u>. Define a stopping rule for sampling in population Π_i by

$$N_i = \min\{n \mid n \geq n_i^0,\ n = n_i^0 + r_i\ell\ (\ell = 0,1,2,\ldots) \text{ and } S_{in}^2 \leq n/n^*\}. \tag{2}$$

Suppose $r_i \leq n_i^0$, then

$$E(N_i) - n_i^* \leq n_i^0 \quad \text{and} \tag{3a}$$

$$E(N_i^2) - (n_i^*)^2 \leq (n_i^0)^2 + n_i^* (1 + r_i + n_i^0). \tag{3b}$$

Hence the "regret" $E(N_i) - n_i^*$ is uniformly bounded for all σ_i^2 unlike the two-stage procedures. A similar result holds for $r_i > n_i^0$--see below.

Proof. Let $c = 1/n^*$ and define a stopping time M_i by $M_i = \min\{n | n \geq n_i^0,\ n = n_i^0 + r_i \ell \ (\ell = 0,1,2,\ldots) \text{ and } \sum_{j=1}^n (X_{ij} - \mu_i)^2 \leq cn(n-1)\}$. Since $(n-1) S_{in}^2 = \sum_{j=1}^n (X_{ij} - X_{i.})^2 \leq \sum_{j=1}^n (X_{ij} - \mu_i)^2$, we have $N_i \leq M_i$, $EN_i \leq EM_i$ and $EN_i^2 \leq EM_i^2$.

Now

$$c(M_i - r_i)(M_i - r_i - 1) \leq \sum_{j=1}^{M_i - r_i} (X_{ij} - \mu_i)^2 + c(n_i^0 - r_i)(n_i^0 - r_i - 1)$$

where if $M_i = r_i$, the empty sum is defined to be zero. Taking expectations and using Wald's Lemma, we obtain:

$$\begin{aligned}(EM_i - r_i)\,\sigma_i^2 + c(n_i^0 - r_i)(n_i^0 - r_i - 1) &\geq cE[(M_i - r_i)^2] - cE(M_i - r_i) \\ &\geq c[E(M_i - r_i)]^2 - cE(M_i - r_i)\end{aligned} \tag{4}$$

Since $M_i \geq n_i^0 \geq r_i$ we have

$$(\sigma_i^2/c) + [(n_i^0 - r_i)(n_i^0 - r_i - 1)/(EM_i - r_i)] \geq E(M_i - r_i) - 1$$

and thus $$EM_i \leq \sigma_i^2/c + n_i^0 = n_i^* + n_i^0.$$

Substituting this result in (4), we find that the r.h.s. of (3b) is an upper bound for $E(M_i^2) - (n_i^*)^2$ and hence also for $E(N_i^2) - (n_i^*)^2$. Strictly, for Wald's Lemma, we first need $EM_i < \infty$ but as in Robbins et al. (1968), we can show the same results hold for $M_t = \min(M_i, t)$ for all t and note that $E(M_t) \uparrow EM_i$ at $t \to \infty$). This proof was suggested by methods in an unpublished paper by Srivastava and Bhargava (1969) (which followed earlier work by Simons (1968)). We remark that the bound cannot be improved as is apparent by considering $\sigma_i^2 \to 0$.

In the less usual case of $r_i > n_i^0$, it can similarly be shown that $EN_i - n_i^*$ is uniformly bounded by $2r_i + 1$. One simply notes that

$$\sum_{j=1}^{M} (X_{ij} - \mu_i)^2 \geq c(M_i - r_i)(M_i - r_i - 1) \geq cM_i(M_i - 2r_i - 1)$$

and uses Wald's Lemma and $E(M_i^2) \geq (EM_i)^2$ as before. This also leads to the upper bound for $E(N_i^2) - n_i^{*2}$ of $(2r_i + 1)(2r_i + 1 + n_i^*)$.

It is possible, using the technique of Robbins (1959), to obtain the exact distribution of the stopping variables $\{N_i\}$. For completeness we describe this technique here. First note that the $\{N_i\}$ are independent.

For computational convenience, we will consider a modified procedure P' that is a special case of P in which n_i^0 is odd and thereafter we sample two observations at a time $(r_i \equiv 2)$ from each population. This leads to consideration of χ_2^2 random variables. This special form of an exponential distribution enables the following recursions to be developed (cf. Robbins et al., 1968).

For $1 \leq i \leq k$ and $m = 1,2,3,\ldots$, we define $a_{im} = 0$ $(m \leq (n_i^0 - 1)/2)$ and $a_{im} = (m-1)(2m-1)/n_i^*$ for $m \geq (n_i^0 + 1)/2$. Now let $h_{i1}(\cdot) \equiv 1$ and compute successively

$$h_{im}(a_{in}) = \sum_{j=1}^{m-1} (a_{in} - a_{im})^j \, h_{i,m-j}(a_{im})/j!$$

for $m = 2,3,\ldots$ and $n = m+1, m+2,\ldots$. Then for $1 \leq i \leq k$ and $m = 1,2,\ldots$

$$P(N_i = 2m+1) = P_{im} = c_{im} - c_{i,m+1} \tag{5}$$

where $c_{i1} = 1$ and

$$c_{im} = \exp(-a_{im}) \sum_{j=1}^{m-1} h_{i,m-j}(a_{im}) \qquad m \geq 2.$$

We can use these results to obtain exact expressions for the means (ASN's) and variances of the sample sizes (see Table 1). These results also make possible exact calculations of error rates which can be compared with those nominal rates predicted by asymptotic theory (see e.g. Section 4 below).

In the remainder of this section we consider the asymptotic properties of the sequential procedure P and show that the appropriate error probability requirements are met in an asymptotic sense. We consider a sequence of problems indexed by a parameter δ^*, say, where the error probability requirements become more stringent as $\delta^* \to 0$ causing $n^* \to \infty$ for the corresponding unit variance single-stage procedure. Here n_i^0 and r_i remain fixed. We then have the following results:

Table 1

Properties of Procedure P' for k = 3. Populations are Selected Parameter Values

$\delta^*/\bar{\sigma}$	.50	.40	.30	.25	.20
Sample size statistics					
Pop. 1 n_1^*	18.55	28.99	51.54	74.22	115.97
$E(N_1)$	16.88	26.80	49.58	72.68	114.85
$Var(N_1)$	50.78	95.42	168.29	217.56	297.57
Pop. 2 n_2^*	27.83	43.49	77.31	111.33	173.95
$E(N_2)$	25.66	41.36	75.82	110.18	173.05
$Var(N_2)$	90.68	146.24	223.56	288.63	411.42
Pop. 3 n_3^*	41.75	65.23	115.97	166.99	260.93
$E(N_3)$	39.59	63.54	114.85	166.08	260.02
$Var(N_3)$	140.94	199.43	297.46	397.64	568.91
Test of homogeneity of means					
10% level test: actual size	.194(.013)	.156(.011)	.138(.011)	.130(.011)	.116(.010)
actual power (nominal = .620)	.647(.015)	.617(.015)	.617(.015)	.612(.015)	.647(.015)
5% level test: actual size	.128(.011)	.093(.009)	.083(.009)	.069(.008)	.054(.007)
actual power (nominal = .495)	.533(.016)	.511(.016)	.504(.016)	.485(.016)	.516(.016)
1% level test: actual size	.060(.008)	.040(.006)	.032(.006)	.019(.004)	.014(.004)
actual power (nominal = .266)	.308(.015)	.292(.014)	.285(.014)	.267(.014)	.271(.014)
Selection problem					
PCS (nominal = 0.950)	0.914	0.923	0.937	0.942	0.946

Parameter used: $k = 3$, $\sigma_1^2 : \sigma_2^2 : \sigma_3^2 = 1 : 1.5 : (1.5)^2$, $n_i^0 \equiv 5$, $r_i^0 \equiv 2$, $\lambda = 2.7101$.
All results are exact except for error rates for test of homogeneity of means which are estimated from 1000 Monte Carlo simulations. (Standard errors shown in parentheses)
Power is for δ^* slippage alternative when $\eta^2 = 2(\delta^*)^2/3$.

Theorem 2. For $1 \leq i \leq k$, as $n_i^* \to \infty$

(a) $\lim(N_i/n_i^*) = 1$ a.s.

(b) $E(N_i) \sim n_i^*$, $Var(N_i) \sim 2n_i^*$, $\gamma_1(N_i) \sim -(2/n_i^*)^{1/2}$

where $\gamma_1(N) = E[(N - EN)^3]/[Var\ N]^{3/2}$ is the skewness.

(c) $(2n_i^*)^{-1/2}(N_i - n_i^*) \to N(0,1)$ in distribution.

Property (a) is proved by Chow and Robbins (1965), while (b) and (c) are given by Anscombe (1952, 1953).

The final step needed is to show the procedure P does indeed asymptotically guarantee the appropriate error probability requirements.

This depends on the particular problem at hand (e.g. problems (1)-(7) of Section 1) but in each case the method is the same. One first notes that the $\{N_i\}$ depend on the sample variances and are thus independent of the sample means. Then one uses Theorem 2(a) and the dominated convergence theorem. In the sections that follow, we consider two examples in detail, namely testing for homogeneity of means in a one-way layout and selecting the best of k populations.

3. TESTING HOMOGENEITY OF MEANS IN A ONE-WAY LAYOUT

Consider the problem of testing $H_0: \mu_1 = \mu_2 = \ldots = \mu_k$ with size and power requirements:

$$P[\text{Reject } H_0 | H_0] = \alpha \tag{6a}$$

$$P[\text{Reject } H_0 | \sum(\mu_i - \bar{\mu})^2 = \eta^2] = 1 - \beta \tag{6b}$$

where α,β,η are specified and $\bar{\mu} = \sum\mu_i/k$. Note that for the slippage configuration in which $\mu_1 - \delta^* = \mu_2 = \ldots = \mu_k$ we have $\eta^2 = (k-1)(\delta^*)^2/k$. An exact two-stage procedure has been given by Bishop and Dudewicz (1978), but as mentioned in Section 1, such a procedure is inefficient.

Following procedure P, at step C we behave at termination as if all the sample means $\{\bar{X}_{i.}\}$ have common variance $1/n^*$ and thus H_0 is rejected if

$$n^* \sum_{i=1}^{k} (\bar{X}_{i.} - \bar{X}_{..})^2 > \kappa \tag{7}$$

where $\bar{X}_{..} = \sum_{i=1}^{k} \bar{X}_{i.}/k$ and κ is the $(1-\alpha)$'th percentage point of the central chi-square distribution with $k-1$ degrees of freedom (i.e. $P[\chi^2_{k-1} > \kappa] = \alpha$). The value of n^*, the common sample size if all the $\{\sigma_i^2\}$ were known equal to one, is given by $n^* = (\lambda/\eta)^2$ where λ satisfies

$$P[\chi^2_{k-1}(\lambda^2) > \kappa] = 1 - \beta.$$

Here $\chi^2_{k-1}(\lambda^2)$ denotes the non-central χ^2 distribution with $k - 1$ degrees of freedom and non-centrality parameter λ^2. Convenient tables for obtaining λ^2 can be found in Haynam, Govindarajulu and Leone (1970).

Now, using the fact that the $\{N_i\}$ are independent of the $\{\bar{X}_{i.}\}$, we have

$$\begin{aligned} P[\text{Reject } H_0] &= E\{P[n^* \sum(\bar{X}_{i.} - \bar{X}_{..})^2 > \kappa]\} \\ &= E\{P[\sum(Y_i - \bar{Y}_.)^2 > \kappa]\} \end{aligned}$$

where $Y_i \sim N(\mu_i, n^* \sigma_i^2/N_i)$

Applying Theorem (2a) and dominated convergence now give the required result that requirements (6a,b) are met asymptotically as $n^* \to \infty$ ($\eta \to 0$ or $\beta \to 1$).

Table 1 gives some simulation results for sample size and achieved error probabilities for moderate values of n^*. Let $\bar{\sigma}^2 = \sum_{i=1}^k \sigma_i^2/k$. Then the performance characteristics depend only on the k quantities $\sigma_2^2/\bar{\sigma}^2, \sigma_3^2/\bar{\sigma}^2, \ldots, \sigma_k^2/\bar{\sigma}^2$ and $\eta/\bar{\sigma}$ (also k,α,β of course). For $k = 3$, $\sigma_1^2 : \sigma_2^2 : \sigma_3^2 = 1 : (1.5)^2 : (1.5)^2$, Table 1 gives summary information on sample size distribution of N_1, N_2, N_3 and error probabilities. These are for $n_i^0 \equiv 5$ $(1 \leq i \leq 3)$, two a time sampling $(r \equiv 2)$, and $\lambda = 2.7101$ and selected values of $\eta = \delta^* \sqrt{2/3}$. (Simulation results, not shown, indicate that sample sizes and error rates for one at a time sampling $(r_i \equiv 1)$ are very similar. Results of a more extensive investigation of Monte Carlo simulation for a wider set of parameter values are available from the authors upon request.) The Monte Carlo results are based on 1000 replications--standard errors are given in parentheses. Note that even though we sample two-at-a-time and can obtain an exact calculation for the distribution of the $\{N_i\}$, simulation to obtain the error probabilities is still necessary because the exact distribution of $\sum(\bar{X}_{i\cdot} - \bar{X}_{\cdot\cdot})^2$, given $N_1, N_2, \ldots, N_k$, is intractable. This is in contrast to the PCS calculation in the selection problem to be described in the next section.

4. SELECTING THE BEST OF k NORMAL POPULATIONS WITH UNKNOWN, UNEQUAL VARIANCES

Let $\mu_{[1]} \leq \mu_{[2]} \leq \cdots \leq \mu_{[k]}$ denote the ranked values of the means. The pairing of the Π_i with the $\mu_{[j]}$ is assumed completely unknown. It is desired to select that population associated with $\mu_{[k]}$. The procedure is to provide the following error probability guarantee

$$P\{CS\} \geq P^* \quad \text{whenever} \quad \mu_{[k]} - \mu_{[k-1]} \geq \delta^* \tag{8}$$

Here CS denotes the event of correctly selecting the population associated with $\mu_{[k]}$; $\{\delta^*, P^*\}$ with $0 < \delta^* < \infty$, $1/k < P^* < 1$ are preassigned constants specified by the experimenter before the start of experimentation.

For the case when the σ_i^2 $(1 \leq i \leq \kappa)$ are <u>known</u>, the single-stage procedure of Bechhofer (1954, p. 24) can be employed. If the variances are <u>equal</u> and <u>known</u>, then the two-stage procedure of Tamhane and Bechhofer (1977, 1979), the closed sequential procedure of Paulson (1964), or the open sequential procedure of Bechhofer, Kiefer and Sobel (1968) are available. If the variances are equal but their common value is unknown, then the two-stage procedure of Bechhofer, Dunnett and Sobel, the three-stage procedure of Tamhane (1976) or the sequential procedure of Paulson (1964, p. 178) can be used. Finally, if the values of the variances are completely unknown, then either the two-stage procedure of Dudewicz and Dalal (1975) or the two-stage procedure of Rinott (1978) can be used. All of the procedures referred to above have been proved analytically to guarantee (8).

In addition to those procedures, Srivastava (1966), Srivastava and Ogilvie (1968), and Robbins, Sobel, Starr (1968) have proposed procedures of a type similar to P which asymptotically $(\delta^* \to 0)$ guarantee (8) for the common unknown variance case.

Following the procedure P given in Section 2, the decision at step (C) is to select as best that population which produced the largest sample mean. For given $(n_1, n_2, \ldots, n_k)$ let $\psi_i(n_i, n_2, \ldots, n_k)$ denote the probability of selecting Π_i. Then

$$\psi_i(n_1,n_2,\ldots,n_k) = \int \prod_{\substack{j=i \\ j\neq i}}^{k} \Phi\left[\frac{n_j^{1/2}}{\sigma_j}\left(\mu_i - \mu_j + y\sigma_i/n_i^{1/2}\right)\right] d\Phi(y) \qquad (9)$$

$$= P[Z_i > Z_j \ (j \neq i)]$$

where $Z_\ell \sim N(\mu_\ell, \sigma_\ell^2/n_\ell)$ $1 \leq \ell \leq k$. Here Φ denotes the standard normal cdf.

From Bechhofer (1954) we have that $n^* = (\omega/\delta^*)^2$ where ω is the solution of the equation

$$\int_{-\infty}^{\infty} \Phi^{k-1}(\omega + y)\, d\Phi(y) = P^* \qquad (10)$$

Values of $\omega = \omega(k, P^*)$ for selected (k, P^*) are given in Table I of Bechhofer (1954). From (9) and the fact that the $\{N_i\}$ are independent of the $\overline{X}_{i\cdot}$, we have that the P[CS] is no less than

$$E \int_{-\infty}^{\infty} \prod_{i=1}^{k-1} \Phi\left[\frac{N_j^{1/2}}{\sigma_j}\left(\delta^* + y\sigma_k/N_k^{1/2}\right)\right] d\Phi(y) \qquad (11)$$

whenever $\mu_k - \mu_i \geq \delta^*$. (Here we have assumed w.l.o.g. that Π_k is best i.e. $\mu_{[k]} = \mu_k$). By Theorem 2(a), as $\delta^* \to 0$ $(n^* \to \infty)$

$$\frac{N_j^{1/2}}{\sigma_j}\left(\delta^* + y\sigma_k/N_k^{1/2}\right) \to \omega + y \quad \text{a.s.}$$

and using dominated convergence and (10) we have that the probability requirement (8) is met asymptotically. Mukhopadhyay (1979) obtained this result in the setting of $k = 2$.

For the two-at-a time sampling rule $(r_i = 2)$ an exact expression for the probability of selecting Π_i as best is given by

$$\sum_{m_1=1}^{\infty} \sum_{m_2=2}^{\infty} \cdots \sum_{m_k=1}^{\infty} \phi_i(2m_1+1,\ 2m_2+1,\ldots,2m_k+1)\ P_{1m_1} P_{2m_2} \cdots P_{km_k} \tag{12}$$

where ϕ_i is given by (9) and P_{im_i} by (5). Denote by $P_i(CS|LF)$ the value of (12) when $\mu_i - \mu_j = \delta^*$ for $j \neq i$ $(1 \leq i \leq k)$.

For $k = 3$, $P^* = 0.95$ and selected parameter values, Table I gives summary information on the exact sample size distributions of N_1, N_2, N_3 and values of $P_2(CS|LF)$--labelled PCS. (Values of $P_1(CS|LF)$ were within .001 less than that stated and $P_3(CS|LF)$ within .001 more than that stated for the parameter combinations shown. More extensive tabulations are available from the authors upon request.) The integral $\phi_i(n_1,n_2,n_3)$ was evaluated by re-expressing it as a bivariate normal probability and using the IMSL routine MDBNOR. Simulation results, not shown, indicate that the performance of one-at-a-time sampling rules $(r_i \equiv 1)$ is very similar.

5. COMPARISON WITH TWO-STAGE PROCEDURE OF DUDEWICZ AND DALAL

Dudewicz and Dalal (1975) (hereafter referred to as DD) proposed a two-stage procedure that guarantees (8) exactly. An initial sample of size n_0 is taken from each population Π_i. Then $N_i - n_0$ additional observations are taken from Π_i where

$$N_i = \max\left\{n_0 + 1,\ \left[\left(\frac{S_{i0}h}{\delta^*}\right)^2\right]\right\} \tag{13}$$

where $h = h(n_0,k,P^*)$ is a constant tabulated in Section 8 of DD, S_{i0}^2 is the unbiased estimate of σ_i^2 based on the initial sample from Π_i. Next weighted means

$$\tilde{X}_i = \sum_{j=1}^{N_i} a_{ij} X_{ij}$$

are calculated, where the $\{a_{ij}\}$ satisfy

$$\sum_{j=1}^{N_i} a_{ij} = 1,\ a_{i1} = \ldots = a_{in_0} \quad \text{and} \quad S_{i0}^2 \sum_{j=1}^{N_i} a_{ij}^2 = (\delta^*/h)^2 .$$

See DD for a particular choice of $\{a_{ij}\}$. Finally select as best that population associated with the largest observed weighted mean $\tilde{X}_i$. Following Seelbinder (1953), $E(N_i)$ can be found as follows. (Seelbinder actually uses n_0 not $n_0 + 1$ in the definition of N_i). Let $c_i = n_0(n_0 - 1)(\delta^*)^2/\sigma_i^2 h^2$ and let F denote the c.d.f. of a χ^2 random variable with $n_0 - 1$ degrees of freedom. Also let

$$\kappa_i = \frac{2\cdot(c_i/2)^{(n_0-1)/2}}{(n_0-1)\Gamma((n_0-1)/2)} \cdot \exp(-\frac{c_i}{2})$$

Then

$$E(N_i^{DD}) = (n_0+1)\ F(c_i) + (h\sigma_i/\delta^*)^2(1 + \kappa_i - F(c_i)) \qquad (14)$$

Higher moments of N_i can be found analogously as in Moshman (1958). Seelbinder (1953) shows that

$$[E_{\delta^*}(N_i^{DD}) - (h\sigma_i/\delta^*)^2] \to 0 \quad \text{as} \quad \delta^* \to 0$$

This implies that, as $\delta^* \to 0$,

$$E_{\delta^*}(N_i^{DD})/n_i^* \to (h/\omega)^2 > 1,$$

since $h = h(n_0,k,P^*) \downarrow \omega$ as $n_0 \to \infty$. Comparing this with the result of Section 2, which stated that for procedure P $E(N_i)/n_i^* \to 1$ as $\delta^* \to 0$, we see that the DD procedure is asymptotically inefficient in the sense of Chow and Robbins (1965).

We now investigate the efficiency of the two procedures for small and moderate sample sizes. We first address the question of initial sample size. For procedure P, as mentioned previously, the n_i^0 should be chosen fairly small e.g. 3,5,11, but for the DD procedure the choice of n_0 can be crucial. Here choices of n_0 that are too large or too small will result in large expected total sample sizes. We considered two ways of choosing n_0, namely

(I) Arbitrarily set $n_0 = 30$. This is the largest value of n_0 for which DD tabulate the constant h, and it is a value recommended in the NBS Handbook (Natrella 1963 Section 2-3.3), also Anscombe (1953 p. 3).

(II) If we have prior "guesses" $\{\tilde{\sigma}_i^2\}$ for each of the $\{\sigma_i^2\}$, choose $n_0 = n_0(\tilde{\sigma}_1,\ldots,\tilde{\sigma}_k)$ to minimize $\text{ASN} = \sum_{i=1}^k E(N_i)$ where $E(N_i)$ is given by (14) with $\tilde{\sigma}_i^2$ replacing σ_i^2 $(1 \leq i \leq k)$.

To study the performance of the latter method (II), a numerical study was undertaken. Complete details are available from the authors. For example, if the true values are $\delta^*/\bar{\sigma} = .5$, $\sigma_1^2\colon \sigma_2^2\colon \sigma_3^2 = 1\colon (1.5)\colon (1.5)^2$ and $P^* = .95$ is specified then a perfect initial guess $(\tilde{\sigma}_i = \sigma_i,\ 1 \leq i \leq 3)$ leads to an optimal choice of $n_0 = 20$ and an ASN of 102.41. If $\delta^*/\bar{\sigma}$ is incorrectly guessed as 1.0, while the ratio $\sigma_1^2\colon \sigma_2^2\colon \sigma_3^2$ remains correct then the choice of n_0 will be 8 and ASN = 124.09. If $\delta^*/\bar{\sigma}$ is guessed correctly but $\tilde{\sigma}_1^2\colon \tilde{\sigma}_2^2\colon \tilde{\sigma}_3^2$ is incorrectly assumed to be 1:3:9 then $n_0 = 15$ and ASN = 104.65. From Table 1 these ASN's can be compared with the ASN of 82.13 for procedure P. However because sample sizes are small, the guaranteed PCS is .914--less than the nominal .95. Thus it is of interest to consider the DD procedure with nominal PCS $P^* = .914$. In this case the optimal choices of n_0 and resulting ASN's for the three sets of $\tilde{\sigma}^2$'s mentioned above are (15, 78.82), (7, 93.68) and (12, 80.37) respectively. The known variance procedure would require a total of 65.82 observations. In general, the choice of optimal n_0 appeared somewhat insensitive to poor guesses for the ratio $\sigma_1^2\colon \sigma_2^2\colon \sigma_3^2$. This was perhaps somewhat surprising because the DD procedure does require a common initial sample size n_0, and a good choice of n_0 for Π_1 might be expected to a poor one for Π_2 and vice versa. Choice of n_0 is sensitive to the initial guess of the overall magnitude $\bar{\sigma}$. In general a poor guess can cost dearly (cf. Starr 1966, p. 46).

It is conjectured that the two-stage procedure of Rinott (1978) would yield results similar to that of DD. In fact ASN's would probably be higher since the method is conservative (a PCS higher than the nominal P^*).

Mukhopadhyay (1979) shows that Rinott's procedure is also asymptotically inefficient for any fixed n_0 but by allowing the initial sample size n_0 to depend on δ^* such that $n_0 \to \infty$ as $\delta^* \to 0$, asymptotic efficiency can be achieved $(EN_i/n_i^* \to 1)$, at least for $k = 2$. However it has <u>not</u> been proved that the procedure has uniformly bounded regret ($EN_i - n^*$ uniformly bounded). This is true for P (Theorem 1). Rinott's procedure involves a constant h^* which is the solution of a complicated double integral equation (see his Equation (13)) and he does not provide any tables for h^*.

6. CONCLUDING REMARKS

The asymptotic allocation of population sample sizes in proportion to their variances is not optimal for $k = 2$ where sample sizes should be proportional to the standard deviations. Mukhopadhyay (1979) and Hayre and Gittins (1981) have examined this problem further. For $k \geq 3$ some unpublished work of Bechhofer and Tamhane suggests that allocation proportional to variances _is_ minimax optimal in the selection problem provided the $\{n_i^*\}$ are sufficiently large.

SUMMARY

This paper presents a survey of results on a class of sequential procedures and provides a unified approach to a variety of statistical problems that concern inference about means of several normal populations when their variances are unknown and unequal. As examples, the problem of testing for homogeneity of means in a one-way layout and the problem of selection of the largest mean are given special attention. Some comparisons are made with two-stage procedures.

ACKNOWLEDGMENT

It is most appropriate that our paper should appear in this volume in honor of Robert Bechhofer, because it was with his interest and inspiration that the research leading to this paper was undertaken. The authors are very grateful to Robert Bechhofer not only for the ideas that he contributed to this paper but for his continuing assistance and encouragement that he has given us over many years as a close friend and colleague.

This research was supported in part by a grant from the National Institutes of Health.

REFERENCES

Anscombe, F.J. (1952). Large sample theory of sequential estimation. _Proc. Camb. Phil. Soc._ _48_, 600-607.

Anscombe, F.J. (1953). Sequential estimation. _J. Roy. Statist. Soc. B_ _15_, 1-21.

Bechhofer, R.E. (1954). A single-sample multiple decision procedure for ranking means of normal populations with known variances. Ann. Math. Statist. 25, 16-19.

Bechhofer, R.E. and Turnbull, B.W. (1978). Two (k+1)-decision selection procedures for comparing k normal means with a control. J. Am. Statist. Assoc. 73, 385-392.

Bishop, T.A. and Dudewicz, E.J. (1978). Exact analysis of variance with unequal variances: Test procedures and tables. Technometrics 20, 419-425.

Chapman, D.G. (1950). Some two sample tests. Ann. Math. Statist. 21, 601-606.

Chow, Y.S. and Robbins, H. (1965). On the asymptotic theory of fixed-width sequential confidence intervals for the mean. Ann. Math. Statist. 36, 457-462.

Dudewicz, E.J. and Dalal, S.R. (1975). Allocation of observations in ranking and selection with unequal variances. Sankhyā, Ser. B, 37, 28-78.

Ghosh, M. and Mukhopadhyay, N. (1979). Sequential point estimation of the mean when the distribution is unspecified. Commun. Statist.-Theor. Meth. A8(7), 637-652.

Gupta, S.S. (1965). On some multiple decision (selection and ranking) rules. Technometrics, 7, 225-245.

Gupta, S.S. and Sobel, M. (1958). On selecting a subset which contains all populations better than a standard. Ann. Math. Statist. 29, 235-244.

Haynam, G.E., Govindarajulu, Z. and Leone, F.C. (1970). Selected tables in Mathematical Statistics, Vol. 1, Ed. by H.L. Harter and D.B. Owen, Chicago: Markham Publ. Co.

Hayre, L.S. and Gittins, J.C. (1981). Sequential selection of the larger of two normal means. J. Am. Statist. Assoc. 76, 696-700.

Kleijnen, J.P.C. (1974). Statistical Techniques in Simulation. Dekker: New York.

Law, A.M. and Kelton, W.D. (1982). Simulation modelling and analysis. McGraw-Hill, New York.

Moshman, J. (1958). A method for selecting the size of the initial sample in Stein's two-sample procedure. Ann. Math. Statist. 29, 1271-1275.

Mukhopadhyay, N. (1979). Some comments on two-stage selection procedures. Commun. Statist-Theor. Meth. A8(7), 671-683.

Natrella, M.G. Experimental Statistics. Handbook 91, National Bureau of Standards, U.S. Dept. of Commerce, Washington D.C.

Paulson, E. (1964). A sequential procedure for selecting the population with the largest mean from k normal populations. Ann. Math. Statist. 35, 174-180.

Rinott, Y. (1978). On two-stage selection procedures and related probability inequalities. Commun. Statist.-Theory and Methods, A7, 799-811.

Robbins, H. (1959). Sequential estimation of the mean of a normal population. In Probability and Statistics - The Harald Cramér Volume 235-245. Almquist and Wiksell, Uppsala, Sweden.

Robbins, H., Sobel, M. and Starr, N. (1968). A sequential procedure for selecting the largest of k means. Ann. Math. Statist. 39, No. 1 88-92.

Ruben, H. (1953). Discussion of paper by F. Anscombe. J. Roy. Statist. Soc. B15, 23-25.

Seelbinder, B.M. (1953). On Stein's two-stage sampling scheme. Ann. Math. Statist. 24, 640-649.

Simons, G. (1968). On the cost of not knowing the variance when making a fixed width confidence interval for the mean. Ann. Math. Statist. 39, 1946-1952.

Srivastava, M.S. (1966). Some asymptotically efficient sequential procedures for ranking and slippage problems. J. Roy. Statist. Soc. B28, 370-380.

Srivastava, M.S. and Bhargava, R.P. (1969). On fixed-width confidence region for the mean. Unpublished.

Srivastava, M.S. and Ogilvie, J. (1968). The performance of some sequential procedures for a ranking problem. Ann. Math. Statist. 39, No. 3, 1040-7.

Starr, N. (1966). The performance of a sequential procedure for the fixed-width interval estimation of the mean. Ann. Math. Statist. 37, 36-50.

Stein, C. (1945). A two-sample test for a linear hypothesis whose power is independent of the variance. Ann. Math. Statist. 16, 243-258.

Tamhane, A.C. (1976). A three-stage elimination type procedure for selecting the largest normal mean (common unknown variance). Sankhyā, Ser. B, 38, 339-349.

Tamhane, A.C. and Bechhofer, R.E. (1977). A two-stage minimax procedure with screening for selecting the largest normal mean. Commun. Statist. - Theor. Meth. A6, 1003-1033.

Tamhane, A.C. and Bechhofer, R.E. (1979). A two-stage minimax procedure with screening for selecting the largest mean (II): an improved PCS lower bound and associated tables. Commun. Statist. - Theor. Meth. A8, 337-358.

Tocher, K.D. (1963). The Art of Simulation. Princeton, NJ, Van Nostrand.

Tong, Y.L. (1969). On partitioning a set of normal populations by their locations with respect to a control. Ann. Math. Statist. 40, 1300-1324.

Turnbull, B.W. (1976). Multiple decision rules for comparing several populations with a fixed known standard. Commun. Statist. - Theor. Meth. A5, 1225-1244, Amendment (1980), A9, 673.

Wilcox, R.R. (1979). Comparing examinees to a control. Psychometrika, 44, 55-68.

Wilcox, R.R. (1983). Selecting the best population provided it is better than a standard: The unequal variance case. Unpublished.

TESTING FOR THE SAME ORDERING IN SEVERAL GROUPS OF MEANS*

Roger L. Berger

Department of Statistics
North Carolina State University
Raleigh, North Carolina

1. INTRODUCTION

In many situations an experimenter is interested in questions regarding the ordering of some parameters. Interest may focus on the largest or smallest parameter. This interest is due, at least in part, to the fact that orderings often determine which populations are "best" or "worst" in some sense.

Over the past thirty years, many statistical procedures have been developed to answer questions regarding the ordering of parameters. Some of the early work of Bechhofer [2] dealt with completely ranking parameters and finding the largest parameter. These and other methods of selecting the largest or smallest parameters and ranking parameters are discussed in Gupta and Panchapakesan [7] which also includes a fairly complete bibliography.

Much of the literature regarding testing hypotheses about ordered parameters falls into the "isotonic regression" literature. An early book involving testing hypotheses about ordered parameters is that of Barlow, Bartholomew, Bremner, and Brunk [1]. Later papers regarding hypothesis testing include Robertson and Wegman [9] and Dykstra and Robertson [5].

---

*This research was supported in part by the U.S. Army Research Office Grant DAAG 29-82-K-0168 at Florida State University. The United States Government is authorized to reproduce and distribute reprints for governmental purposes.

In this paper we consider a hypothesis testing problem regarding ordered parameters which does not fall in the isotonic regression area. We consider a problem in which there are two groups of normal means, $\mu_{11},\ldots,\mu_{1K}$, and $\mu_{21},\ldots,\mu_{2K}$. The question we consider is, "Are these two groups in the same order?" Specifically we consider testing

$$H_0: (\mu_{1i} - \mu_{1j})(\mu_{2i} - \mu_{2j}) \leq 0 \quad \text{for some} \quad 1 \leq i < j \leq K$$

versus

$$H_1: (\mu_{1i} - \mu_{1j})(\mu_{2i} - \mu_{2j}) > 0 \quad \text{for every} \quad 1 \leq i < j \leq K.$$

The alternative states that the two groups of means are in exactly the same order. The multiplicative way of writing H_0 and H_1 is convenient notationally. But note that it might be more explanatory to write H_1 as this: There exists a permutation $(\pi(1),\ldots,\pi(K))$ of $(1,\ldots,K)$ such that $\mu_{1\pi(1)} > \cdots > \mu_{1\pi(K)}$ and $\mu_{2\pi(1)} > \cdots > \mu_{2\pi(K)}$. This choice of null and alternative hypotheses is appropriate if one wishes to demonstrate that the two groups are in the same order and is going to follow the usual practice of controlling the Type I error probability at a small value.

An example in which the experimenter might wish to test H_0 versus H_1 is the following. An agricultural researcher might be interested in K varieties of a crop which are to be grown in two dissimilar regions. The researcher might hypothesize that despite the dissimilarities, the rankings of the mean yields for the K varieties will be the same in the two regions. That is, the same variety will have the highest mean yield, the same variety will have the second highest mean yield, etc. To confirm his hypothesis, he might test H_0 versus H_1.

In Section 2, the assumptions of our model are presented. The size α likelihood ratio test (LRT) of H_0 versus H_1 is derived in Section 3. The LRT's for two related problems, testing whether several (more than two) groups of means are in the same order and testing whether a group of means is in a specified order, are presented in Section 4.

2. NOTATION AND ASSUMPTIONS

We will consider testing H_0 versus H_1 based on the following sample. Let X_{rst}; $r = 1,2$; $s = 1,\ldots,K$; $t = 1,\ldots,n_{rs}$; be independent normal observations. The mean and variance of X_{rst} are μ_{rs} and σ^2, respectively. Let $\overline{X}_{rs} = \sum_{t=1}^{n_{rs}} X_{rst}/n_{rs}$ be the sample mean from the (r,s) population. Let $SS_X = \sum_r\sum_s\sum_{t=1}^{n_{rs}}(X_{rst} - \overline{X}_{rs})^2$ (we will use the notation

$\sum_r$ for $\sum_{r=1}^{2}$ and $\sum_s$ for $\sum_{s=1}^{K}$ throughout). Let $N = \sum_r \sum_s n_{rs}$ denote the total sample size. Then $S^2 = SS_x/(N-2K)$ is the usual pooled unbiased estimate of σ^2 and S^2 is independent of the sample means.

For every $r = 1,2$ and $1 \leq i < j \leq K$, define the statistic

$$T_{ij}^r = \frac{\overline{X}_{ri} - \overline{X}_{rj}}{S(1/n_{ri} + 1/n_{rj})^{1/2}} . \tag{2.1}$$

This is the usual t-statistic for testing hypotheses like H_0^*: $\mu_{ri} \leq \mu_{rj}$. If T_{ij}^1 and T_{ij}^2 are both large positive numbers, this might be taken as evidence that $\mu_{1i} > \mu_{1j}$ and $\mu_{2i} > \mu_{2j}$. Similarly if T_{ij}^1 and T_{ij}^2 are both small negative numbers this might be taken as evidence that $\mu_{1i} < \mu_{1j}$ and $\mu_{2i} < \mu_{2j}$. It will be seen in Section 3 that the LRT of H_0 versus H_1 is based on this reasoning.

Consider the hypothesis H_0^{ij}: $(\mu_{1i} - \mu_{1j})(\mu_{2i} - \mu_{2j}) \leq 0$. This hypothesis states that the ith and jth means are not in the same order in the two sets. We shall use the convention that H_0, H_1, and H_0^{ij} refer to both the statements about the parameters and the subsets of the parameter space defined by the statements. Using this convention we can write that

$$H_0 = \bigcup_{1 \leq i < j \leq K} H_0^{ij}. \tag{2.2}$$

Hypotheses such as H_0 which can be conveniently expressed as unions have been considered in Gleser [6], Berger [3], and Berger and Sinclair [4].

The LRT statistic for H_0 is defined as

$$\lambda(\underset{\sim}{x}) = \frac{\sup_{H} L(\underset{\sim}{x};\underset{\sim}{\mu},\sigma)}{\sup_{H_0} L(\underset{\sim}{x};\underset{\sim}{\mu},\sigma)} \tag{2.3}$$

where $\underset{\sim}{x}$ is the sample of N observations, $\underset{\sim}{\mu}$ is the vector of $2K$ population means, L is the likelihood function from the normal model and $H = H_0 \cup H_1$. A LRT of H_0 versus H_1 is any test which rejects H_0 for large values of λ, that is, any test with a rejection region of the form $\{\underset{\sim}{x}: \lambda(\underset{\sim}{x}) \geq c\}$ where $c \geq 1$ is a specified constant.

The LRT statistic for H_0^{ij} is defined as

$$\lambda^{ij}(\underset{\sim}{x}) = \frac{\sup_{H} L(\underset{\sim}{x};\underset{\sim}{\mu},\sigma)}{\sup_{H_0^{ij}} L(\underset{\sim}{x};\underset{\sim}{\mu},\sigma)} . \tag{2.4}$$

Because of (2.2), it is clear that $\lambda(\underset{\sim}{x}) = \min_{1 \leq i < j \leq K} \lambda^{ij}(\underset{\sim}{x})$. Thus a LRT of H_0 versus H_1 is any test which rejects H_0 if

$$\min_{1 \leq i < j \leq K} \lambda^{ij}(\underset{\sim}{x}) \geq c \tag{2.5}$$

where $c \geq 1$ is a specified constant. In the next section, (2.5) will be used to derive a simple form for the LRT and determine critical values so that the LRT has a specified size.

3. DERIVATION OF THE LRT

In this section we prove the following two theorems which describe the LRT for H_0 versus H_1.

Theorem 3.1. The test which rejects H_0 in favor of H_1 iff, for every $1 \leq i < j \leq K$, either $\min(T^1_{ij}, T^2_{ij}) \geq t$ or $\max(T^1_{ij}, T^2_{ij}) \leq -t$ where t is a positive constant is a LRT of H_0 versus H_1.

Theorem 3.2. Let $0 \leq \alpha < .5$. If in Theorem 3.1, $t = t_{\alpha, N-2K}$, the upper αth percentile from a Student's t distribution with $N-2K$ degrees of freedom, then the resulting LRT is a size α test.

Lemma 3.1. For any $c > 1$ and $1 \leq i < j \leq K$, $\{\underset{\sim}{x}: \lambda^{ij}(\underset{\sim}{x}) \geq c\}$ is the same set as $\{\underset{\sim}{x}: \min(T^1_{ij}, T^2_{ij}) \geq t \text{ or } \max(T^1_{ij}, T^2_{ij}) \leq -t\}$ where $t = ((c^{2/N}-1)(N-2K))^{1/2} > 0$.

Proof. By standard analysis for LRT statistics for normal models, $\lambda^{ij}(\underset{\sim}{x}) = (\lambda^{ij}_*(\underset{\sim}{x}))^{N/2}$ where

$$\lambda^{ij}_*(\underset{\sim}{x}) = \frac{\inf_{H_0^{ij}} \sum_r \sum_s \sum_{t=1}^{n_{rs}} (x_{rst} - \mu_{rs})^2}{SS_x}$$

$$= 1 + \frac{\inf_{H_0^{ij}} \sum_r \sum_s n_{rs}(\bar{x}_{rs} - \mu_{rs})^2}{SS_x}$$

$$= 1 + \frac{\inf_{H_0^{ij}} \sum_r [n_{ri}(\bar{x}_{ri} - \mu_{ri})^2 + n_{rj}(\bar{x}_{rj} - \mu_{rj})^2]}{SS_x} . \tag{3.1}$$

Equality (3.1) is true since all the $n_{rs}(\bar{x}_{rs} - \mu_{rs})^2$ terms, except for

$s = i,j$, can be set to zero by choosing $\mu_{rs} = \bar{x}_{rs}$. This choice is permissible since the μ_{rs}, $s \neq i,j$, are unconstrained by H_0^{ij}.

If $(\bar{x}_{1i} - \bar{x}_{1j})(\bar{x}_{2i} - \bar{x}_{2j}) \leq 0$, $\lambda^{ij}(\underset{\sim}{x}) = \lambda_*^{ij}(\underset{\sim}{x}) = 1$ because the infimum in (3.1) is zero. This infimum is attained at the parameter point with $\mu_{rs} = \bar{x}_{rs}$ $r = 1,2$, and $s = i,j$, a parameter point in H_0^{ij}. So a sample point with $(\bar{x}_{1i} - \bar{x}_{1j})(\bar{x}_{2i} - \bar{x}_{2j}) \leq 0$ is in neither of the sets specified in the Lemma.

Consider calculation of the infimum in (3.1) if $(\bar{x}_{1i} - \bar{x}_{1j})(\bar{x}_{2i} - \bar{x}_{2j}) > 0$. First suppose $\bar{x}_{1i} < \bar{x}_{1j}$ and $\bar{x}_{2i} < \bar{x}_{2j}$. Consider calculating the infimum in (3.1) on the set of μ's such that $\mu_{1i} \leq \mu_{1j}$ and $\mu_{2i} \geq \mu_{2j}$. This infimum is clearly attained at a point with $\mu_{1i} = \bar{x}_{1i}$ and $\mu_{1j} = \bar{x}_{1j}$. Thus the infimum is just

$$\inf_{\mu_{2i} \geq \mu_{2j}} n_{2i}(\bar{x}_{2i} - \mu_{2i})^2 + n_{2j}(\bar{x}_{2j} - \mu_{2j})^2. \tag{3.2}$$

Using the fact that $\bar{x}_{2i} < \bar{x}_{2j}$, it is easily verified that (3.2) is attained when $\mu_{2i} = \mu_{2j} = (n_{2i}\bar{x}_{2i} + n_{2j}\bar{x}_{2j})/(n_{2i} + n_{2j})$. Thus (3.2) is equal to $(\bar{x}_{2i} - \bar{x}_{2j})^2/(1/n_{2i} + 1/n_{2j})$. Now consider calculating the infimum in (3.1) on the set of $\underset{\sim}{\mu}$'s such that $\mu_{1i} \geq \mu_{1j}$ and $\mu_{2i} \leq \mu_{2j}$. The same reasoning as above yields that this infimum is $(\bar{x}_{1i} - \bar{x}_{1j})^2/(1/n_{1i} + 1/n_{1j})$. Since H_0^{ij} is the union of these two sets of $\underset{\sim}{\mu}$'s, the infimum in (3.1) is the minimum of these two infima. Thus we have that if $\bar{x}_{1i} < \bar{x}_{1j}$ and $\bar{x}_{2i} < \bar{x}_{2j}$,

$$\lambda_*^{ij}(\underset{\sim}{x}) = 1 + \min_{r=1,2} \frac{(\bar{x}_{ri} - \bar{x}_{rj})^2}{SS_x(1/n_{ri} + 1/n_{rj})}. \tag{3.3}$$

The calculation of $\lambda_*^{ij}(\underset{\sim}{x})$ for a sample point with $\bar{x}_{1i} > \bar{x}_{1j}$ and $\bar{x}_{2i} > \bar{x}_{2j}$ yields the same value of $\lambda_*^{ij}(\underset{\sim}{x})$ as in (3.3).

The condition that $(\bar{x}_{1i} - \bar{x}_{1j})(\bar{x}_{2i} - \bar{x}_{2j}) > 0$ is equivalent to the condition that T_{ij}^1 and T_{ij}^2 have the same sign. Thus the following inequalities are equivalent:

$$\lambda^{ij}(\underset{\sim}{x}) \geq c$$

$$\lambda_*^{ij}(\underset{\sim}{x}) \geq c^{2/N}$$

$$\min((T_{ij}^1)^2, (T_{ij}^2)^2) \geq (c^{2/N}-1)(N-2K) \text{ and } (\bar{x}_{1i} - \bar{x}_{1j})(\bar{x}_{2i} - \bar{x}_{2j}) > 0$$

$$\min(T_{ij}^1, T_{ij}^2) \geq t \text{ or } \max(T_{ij}^1, T_{ij}^2) \leq -t.$$

This proves Lemma 3.1. □

Proof of Theorem 3.1. Theorem 3.1 follows from Lemma 3.1, representation (2.5), and noting that any $t > 0$ corresponds to a $c > 1$ in (2.5). □

Theorem 3.2 describes how the critical value t should be chosen to yield a size α LRT. Lemma 3.2 will be used in the proof of Theorem 3.2.

Lemma 3.2. Let $0 \le \alpha < .5$, $1 \le i < j \le K$, and $\nu = N - 2K$. Then

$$\sup_{H_0^{ij}} P_{\underset{\sim}{\mu},\sigma}(\min(T_{ij}^1, T_{ij}^2) \ge t_{\alpha,\nu} \quad \text{or} \quad \max(T_{ij}^1, T_{ij}^2) \le -t_{\alpha,\nu}) \le \alpha. \tag{3.4}$$

Proof. Let $F(s;\sigma)$ denote the distribution function of S. Let $Y_r = (\overline{X}_{ri} - \overline{X}_{rj})/(1/n_{ri} + 1/n_{rj})^{1/2}$, $r = 1,2$. Then $T_{ij}^r = Y_r/S$. Furthermore Y_1, Y_2, and S are independent and Y_r is normal with mean $\xi_r = (\mu_{ri} - \mu_{rj})/(1/n_{ri} + 1/n_{rj})^{1/2}$ and variance σ^2. The probability in (3.4) is equal to

$$\int_0^\infty P_{\underset{\sim}{\xi},\sigma}(\min(Y_1,Y_2) \ge st_{\alpha,\nu} \quad \text{or} \quad \max(Y_1,Y_2) \le -st_{\alpha,\nu})dF(s;\sigma). \tag{3.5}$$

If $(\underset{\sim}{\mu},\sigma) \in H_0^{ij}$ then either $\xi_1 \ge 0 \ge \xi_2$ or $\xi_1 \le 0 \le \xi_2$. In either case $\underset{\sim}{\xi}^* = (\xi_1 + \xi_2, 0)$ is majorized by $\underset{\sim}{\xi}$. Since the joint density of (Y_1,Y_2) is $g_\sigma(\underset{\sim}{y}-\underset{\sim}{\xi})$ where g_σ is a Schur concave function, and since the set $\{(y_1,y_2): \min(y_1,y_2) \ge st_{\alpha,\nu}$ or $\max(y_1,y_2) \le -st_{\alpha,\nu}\}$ is a Schur concave set, by Theorem 2.1 of Marshall and Olkin [8], the integrand in (3.5) is less than or equal to

$$P_{\underset{\sim}{\xi}^*,\sigma}(\min(Y_1,Y_2) \ge st_{\alpha,\nu} \quad \text{or} \quad \max(Y_1,Y_2) \le -st_{\alpha,\nu}) \tag{3.6}$$

if $(\underset{\sim}{\mu},\sigma) \in H_0^{ij}$. So the Lemma will be proved if we can show that (3.5), with $\underset{\sim}{\xi}$ replaced by $\underset{\sim}{\xi}^*$, is at most α.

$$\begin{aligned} &\int_0^\infty P_{\underset{\sim}{\xi}^*,\sigma}(\min(Y_1,Y_2) \ge st_{\alpha,\nu} \quad \text{or} \quad \max(Y_1,Y_2) \le -st_{\alpha,\nu})dF(s;\sigma) \\ &= \int_0^\infty P_{\underset{\sim}{\xi}^*,\sigma}(Y_1 \ge st_{\alpha,\nu}, Y_2 \ge st_{\alpha,\nu}) + P_{\underset{\sim}{\xi}^*,\sigma}(Y_1 \le -st_{\alpha,\nu}, Y_2 \le -st_{\alpha,\nu})dF(s;\sigma). \end{aligned} \tag{3.7}$$

The fact that $st_{\alpha,\nu} \ge 0$ was used to obtain the equality. Since Y_1 and Y_2 are independent normal random variables and the mean of Y_2 is zero, (3.7) equals

$$\begin{aligned} &\int_0^\infty P_{\underset{\sim}{\xi}^*,\sigma}(Y_1 \ge st_{\alpha,\nu}, Y_2 \ge st_{\alpha,\nu}) + P_{\underset{\sim}{\xi}^*,\sigma}(Y_1 \le -st_{\alpha,\nu}, Y_2 \ge st_{\alpha,\nu})dF(s;\sigma) \\ &\le \int_0^\infty P_{\underset{\sim}{\xi}^*,\sigma}(Y_2 \ge st_{\alpha,\nu})dF(s;\sigma) \\ &= P_{\underset{\sim}{\xi}^*,\sigma}(Y_2/S \ge t_{\alpha,\nu}) = \alpha. \end{aligned}$$

The fact that $st_{\alpha,\nu} \geq 0$ was used to obtain the inequality. At $\underset{\sim}{\xi}^*$, Y_2/S has a central t distribution since the mean of Y_2 is zero. This yields the last equality. Thus Lemma 3.2 is proved.

Proof of Theorem 3.2. First we will show that for any $(\underset{\sim}{\mu},\sigma) \in H_0$, the probability of rejecting H_0 is at most α. Fix $(\underset{\sim}{\mu},\sigma) \in H_0$. By (2.2), there is an i^* and j^* with $1 \leq i^* < j^* \leq K$, such that $(\underset{\sim}{\mu},\sigma) \in H_0^{i^*j^*}$. Then

$$P_{\underset{\sim}{\mu},\sigma}(\text{reject } H_0) = P_{\underset{\sim}{\mu},\sigma}(\min(T_{ij}^1,T_{ij}^2) \geq t_{\alpha,\nu} \text{ or } \max(T_{ij}^1,T_{ij}^2) \leq -t_{\alpha,\nu} \text{ for every } 1 \leq i < j \leq K)$$

$$\leq P_{\underset{\sim}{\mu},\sigma}(\min(T_{i^*j^*}^1,T_{i^*j^*}^2) \geq t_{\alpha,\nu} \text{ or } \max(T_{i^*j^*}^1,T_{i^*j^*}^2) \leq -t_{\alpha,\nu})$$

$$\leq \alpha,$$

the last inequality being true by Lemma 3.2 since $(\underset{\sim}{\mu},\sigma) \in H_0^{i^*j^*}$.

Now we will show that there exists a sequence of parameter points in H_0 such that the probabilities of rejecting H_0 approach a limit which is at least α. This will complete the proof of Theorem 3.1. Let $\underset{\sim}{\mu}_n$ be a sequence of mean vectors (with elements μ_{rsn}) such that $\mu_{11n} = \mu_{12n}$ for all n, $\mu_{1sn} - \mu_{1(s+1)n} \to \infty$ as $n \to \infty$ for $s = 2,\ldots,K-1$, and $\mu_{2sn} - \mu_{2(s+1)n} \to \infty$ as $n \to \infty$ for $s = 1,\ldots,K-1$. This implies that for $r = 1,2$, and $1 \leq i < j \leq K$ (except if $r = 1$, $i = 1$, and $j = 2$) $\mu_{rin} - \mu_{rjn} \to \infty$ as $n \to \infty$. All such $\underset{\sim}{\mu}_n$ are in H_0 since $\mu_{11n} - \mu_{12n} = 0$. Fix $\sigma > 0$ and let $\nu = N-2K$. Then (all limits are as $n \to \infty$)

$$\begin{aligned}
\lim P_{\underset{\sim}{\mu}_n,\sigma}(\text{reject } H_0) &\geq \lim P_{\underset{\sim}{\mu}_n,\sigma}(\min(T_{ij}^1,T_{ij}^2) \geq t_{\alpha,\nu},\ 1 \leq i < j \leq K) \\
&= \lim P_{\underset{\sim}{\mu}_n,\sigma}(T_{ij}^r \geq t_{\alpha,\nu},\ r = 1,2,\ 1 \leq i < j \leq K) \\
&= 1 - \lim P_{\underset{\sim}{\mu}_n,\sigma}\Big(\bigcup_{\substack{r=1,2\\1\leq i<j\leq K}} \{T_{ij}^r < t_{\alpha,\nu}\}\Big) \\
&\geq 1 - \lim \sum_{\substack{r=1,2\\1\leq i<j\leq K}} P_{\underset{\sim}{\mu}_n,\sigma}(T_{ij}^r < t_{\alpha,\nu}).
\end{aligned} \tag{3.8}$$

Since $\mu_{11n} = \mu_{12n}$, $P_{\underset{\sim}{\mu}_n,\sigma}(T_{12}^1 < t_{\alpha,\nu}) = 1-\alpha$ for every n. For any other choice of r, i, and j, since $\mu_{rin} - \mu_{rjn} \to \infty$, $\lim P_{\underset{\sim}{\mu}_n,\sigma}(T_{ij}^r < t_{\alpha,\nu}) = 0$. Thus the expression in (3.8) is α, as was to be shown. □

4. RELATED PROBLEMS

In this section we consider two problems which are similar to testing H_0 versus H_1. In each case we state a theorem which gives the LRT for the problem. The analyses are very similar to that in Section 3 and thus the proofs of the theorems are omitted.

4.1 Testing Whether More than Two Groups of Means Are in the Same Order

This problem is the same as that considered in the previous section except that rather than two groups of means there are $R > 2$ groups of means. We are interested in whether all R groups are in exactly the same order. The means are denoted by μ_{rs}; $r = 1,\ldots,R$; $s = 1,\ldots,K$. We are interested in testing

$$H_0^R\colon (\mu_{ui} - \mu_{uj})(\mu_{vi} - \mu_{vj}) \leq 0 \quad \text{for at least one choice of } 1 \leq u < v \leq R \text{ and } 1 \leq i < j \leq K$$

versus

$$H_1^R\colon (\mu_{ui} - \mu_{uj})(\mu_{vi} - \mu_{vj}) > 0 \quad \text{for every } 1 \leq u < v \leq R \text{ and } 1 \leq i < j \leq K.$$

If $T_{ij}^r = (\overline{X}_{ri} - \overline{X}_{rj})/S_R(1/n_{ri} + 1/n_{rj})^{1/2}$, where S_R is the usual, pooled estimate of σ based on all the observations, then we have the following result.

Theorem 4.1. Let $0 \leq \alpha < .5$. The test which rejects H_0^R iff, for every $1 \leq i < j \leq K$, either $\min(T_{ij}^1,\ldots,T_{ij}^R) \geq t_{\alpha,N-RK}$ or $\max(T_{ij}^1,\ldots,T_{ij}^R) \leq -t_{\alpha,N-RK}$ is the size α LRT of H_0^R versus H_1^R.

4.2 Testing Whether a Group of Means is in a Specified Order

We now consider the problem in which there is only one group of means, $\mu_1,\ldots,\mu_K$, of interest. We are interested in whether these means are in a specified order. Without loss of generality, we can assume the populations have been labeled so that the hypotheses of interest are these:

$$H_0^1\colon \mu_{i+1} \leq \mu_i \quad \text{for some} \quad i = 1,\ldots,K-1$$

$$H_1^1\colon \mu_{i+1} > \mu_i \quad \text{for every} \quad i = 1,\ldots,K-1.$$

If $T_i = (\overline{X}_{i+1} - \overline{X}_i)/S_1(1/n_i + 1/n_j)^{1/2}$, where S_1 is the usual, pooled estimate of σ based on the observations from the K populations, then we have the following result.

Theorem 4.2. Let $0 \leq \alpha < .5$. The test which rejects H_0^1 iff $\min(T_1,\ldots,T_{K-1}) \geq t_{\alpha,N-K}$ is the size α LRT of H_0^1 versus H_1^1.

Robertson and Wegman [9] consider a class of problems, one of which is testing the null hypothesis H_1^1 versus the alternative H_0^1. Thus they have the roles of the null and alternative reversed from our formulation. (They also have the equality signs in H_1^1 but this is unimportant. The same tests result in both formulations regardless of the location of the equality signs.) Our formulation is more appropriate if one wishes to control the probability of deciding the means are in the specified order when they are not so ordered.

ACKNOWLEDGMENT

The problem of testing H_0 versus H_1 was suggested in a conversation with Professor Sue Leurgans of the University of Wisconsin, Madison.

REFERENCES

Barlow, R.E., Bartholomew, D.J., Bremner, J.M. and Brunk, H.D. (1972). Statistical Inference under Order Restrictions. Wiley, New York.

Bechhofer, R.E. (1954). A single-sample multiple decision procedure for ranking means of normal populations with known variances. Ann. Math. Statist., 25, 16-39.

Berger, R.L. (1982). Multiparameter hypothesis testing and acceptance sampling. Technometrics, 24, 295-300.

Berger, R.L. and Sinclair, D.F. (1984). Testing hypotheses concerning unions of linear subspaces. J. Amer. Statist. Assoc., 79, 158-163.

Dykstra, R.L. and Robertson, T. (1982). Order restricted statistical tests on multinomial and Poisson parameters: the starshaped restriction. Ann. Statist., 10, 1246-1252.

Gleser, L.J. (1973). On a theory of intersection-union tests. IMS Bulletin, 2, 233. (Abstract).

Gupta, S.S. and Panchapakesan, S. (1979). Multiple Decision Procedures: Theory and Methodology of Selecting and Ranking Populations. Wiley, New York.

Marshall, A.W. and Olkin, I. (1974). Majorization in multivariate distributions. Ann. Statist., 2, 1189-1200.

Robertson, T. and Wegman, E.J. (1978). Likelihood ratio tests for order restrictions in exponential families. Ann. Statist., 6, 485-505.

SUBSET ESTIMATORS OF THE LARGEST MEAN

Saul Blumenthal*

Department of Statistics
Ohio State University
Columbus, Ohio

1. INTRODUCTION

Given independent observations $X_1,\ldots,X_k$ from k distinct location parameter distributions with parameters $\theta_1,\ldots,\theta_k$ respectively, it is well known that to estimate

$$\gamma = \max(\theta_1,\ldots,\theta_k) \tag{1.1}$$

the intuitively appealing estimator

$$X_{max} = \max(X_1,\ldots,X_k) \tag{1.2}$$

tends to have a large upward bias, especially if k is fairly large. For the normal distribution, various maximum likelihood and maximum probability estimators of γ have been developed by Blumenthal and Cohen [1,2] and by Dudewicz [4,5,6]. All involve the solution of a transcendental equation to evaluate the estimate for a given set of observations.

Generalized Bayes estimators using a uniform (improper) prior were examined for $k = 2$ in [1,2] and require the evaluation of the normal density and c.d.f. in the process of calculating the estimate. For $k > 2$, a multivariate normal orthant probability must be evaluated in order to compute the generalized Bayes estimate of γ. Dhariyal, Dudewicz and Blumenthal [3] looked at procedures which correct X_{max} (for $k = 2$) for

*Research of this author was supported in part by National Science Foundation Grant MCS 80-02340, Department of Mathematics, University of Illinois at Urbana-Champaign.

bias by subtracting estimates of the bias. These too, involve the normal c.d.f. in the estimate and tend to have poor mean squared error curves.

Present day computers notwithstanding, it would be convenient to have estimators of γ which improve on the excessive upward bias of X_{max} for large k yet which share with X_{max} the attribute of computational simplicity. In this paper, such an attempt is made by examining what we shall call subset estimators defined as follows. Let $X_{[1]} \leq \cdots \leq X_{[k]}$ be the ordered observations, and for a given δ, let m be such that

$$X_{[k-m]} < X_{[k]}-\delta \leq X_{[k+1-m]}. \tag{1.3}$$

Then we define the estimator X^* by

$$X^* = \frac{1}{m} \sum_{i=1}^{m} X_{[k+1-i]}. \tag{1.4}$$

For $k = 2$, X^* is the "hybrid estimator," δ_H, considered in [2]. In [2], numerical integrations showed that X^* with $\delta = (1/\sqrt{2})$ has mean squared error and bias properties similar to those of the maximum likelihood estimator, and X^* with $\delta = 1$ has very low bias with a slight increase in mean squared error. Monte Carlo results in [3] show that X^* with $\delta = 1$ has a bias which compares favorably with that of the maximum probability estimator (t(.5,.5)) and the bias corrected estimator and mean squared error comparable to t(.5,.5) and much better than that of the bias corrected estimator. On the basis of these comparisons for $k = 2$, there is good reason to hope that for larger k, X^* will compare favorably with its more complex competitors when δ is chosen appropriately.

In this paper, bias and mean squared error formulas are developed for X^* for general k. These are specialized to the case

$$\theta_{[1]} = \cdots = \theta_{[k-1]} = \theta_{[k]}-\Delta \tag{1.5}$$

(where $\theta_{[1]} \leq \cdots \leq \theta_{[k]}$ are the ordered parameters), which lends itself more readily to numerical computations. It might be hoped that (1.5) represents a "least favorable" configuration in the sense of maximizing the bias over all configurations of means such that $\theta_{[k]} - \theta_{[k-1]} \geq \Delta$. While this is the case for X_{max} (see Lemma 2.1), we shall see that it does not hold in general for X^*.

For $k = 3$, and normal distributions, we shall find a δ value which seems to be good for the (1.5) configuration in terms of reducing bias (Section 2) and one which seems good from a mean squared error viewpoint

(Section 3). In Section 4 we recommend a choice of δ. At least for $k = 3$, and we conjecture that for $k \geq 4$ the conclusion will be even stronger, we can choose δ so that X^* does achieve the goal of improving on X_{max} without a substantial increase in computational complexity.

2. BIAS OF THE SUBSET ESTIMATOR

Without loss of generality, we shall assume

$$\theta_1 \geq \theta_2 \geq \cdots \geq \theta_k \tag{2.1}$$

when deriving properties of the estimators. Let S be the set selected by the rule (1.3), and let M be its size. When X_i has density $f(x;\theta_i)$ and c.d.f. $F(x;\theta_i)$ with no particular form specified, we can obtain the expectation of X^* as given by (1.4) from

$$EX^* = \sum_{m=1}^{k} (1/m) \sum_{j=1}^{k} \Big(\sum_{C(j)} I^j(\theta_1,\ldots,\theta_k;\ i_1,\ldots,i_{m-1}) + \sum_{\substack{s=1\\ s\neq j}}^{k} \sum_{C(j,s)} I^{js}(\theta_1,\ldots,\theta_k;\ i_1,\ldots,i_{m-2})\Big) \tag{2.2}$$

where $\sum_{C(j)}$ extends over all $\binom{k-1}{m-1}$ subsets of indices $(i_1,\ldots,i_{m-1})$ which exclude the index j, $\sum_{C(j,s)}$ extends over the $\binom{k-2}{m-2}$ index sets excluding both j and s, and

$$I^j(\theta_1,\ldots,\theta_k;\ i_1,\ldots,i_{m-1}) = \int_{-\infty}^{\infty} xf(x;\theta_j) \prod_{t=1}^{m-1} [F(x;\theta_{i_t}) - F(x-\delta;\theta_{i_t})] \times \prod_{\substack{r\neq j\\ r\notin(i_1,\ldots,i_{m-1})}} F(x-\delta;\theta_r)dx \tag{2.3}$$

$$I^{js}(\theta_1,\ldots,\theta_k;\ i_1,\ldots,i_{m-2}) = \int_{x=-\infty}^{\infty} \int_{y=x-\delta}^{x} yf(y;\theta_s)f(x;\theta_j) \times \prod_{t=1}^{m-2} [F(x;\theta_{i_t}) - F(x-\delta;\theta_{i_t})] \prod_{\substack{r\neq j,s\\ r\notin(i_1,\ldots,i_{m-2})}} F(x-\delta;\theta_r)dxdy. \tag{2.4}$$

The special case

$$\theta_1 = \theta,\ \theta_2 = \theta_3 = \cdots = \theta_k = \theta_\Delta (< \theta) \tag{2.5}$$

will receive considerable attention, so we give here the specialization of (2.2) to this case, namely

$$EX^* = \sum_{m=1}^{k} (1/m)\{\binom{k-1}{m-1}I_1 + (k-1)\binom{k-2}{m-2}(I_2+I_3) + (k-1)\binom{k-2}{m-1}I_4 + (k-1)(k-2)\binom{k-3}{m-3}I_5\} \tag{2.6}$$

where

$$I_1 = \int_{-\infty}^{\infty} xf(x;\theta)U^{m-1}(x,\delta,\theta_\Delta)F^{k-m}(x-\delta;\theta_\Delta)dx,$$

$$I_2 = \int_{-\infty}^{\infty}\int_{x-\delta}^{x} y[f(y;\theta_s)f(x;\theta)+f(y;\theta)f(x;\theta_\Delta)]U^{m-2}(x,\delta,\theta_\Delta)F^{k-m}(x-\delta;\theta_\Delta)dydx,$$

$$I_3 = \int_{-\infty}^{\infty} xf(x;\theta_\Delta)U(x,\delta,\theta)U^{m-2}(x,\delta,\theta_\Delta)F^{k-m}(x-\delta;\theta_\Delta)dx,$$

$$I_4 = \int_{-\infty}^{\infty} xf(x;\theta_\Delta)U^{m-1}(x,\delta,\theta_\Delta)F(x-\delta;\theta)F^{k-m-1}(x-\delta;\theta_\Delta)dx,$$

$$I_5 = \int_{-\infty}^{\infty}\int_{x-\delta}^{x} yf(y;\theta_\Delta)f(x;\theta_\Delta)U^{m-3}(x;\delta;\theta_\Delta)F^{k-m-1}(x-\delta;\theta_\Delta) \cdot[U(x;\delta;\theta_\Delta)F(x-\delta;\theta_\Delta) + \frac{(k-m)}{(m-2)}U(x;\delta;\theta_\Delta)F(x-\delta;\theta)]dydx,$$

$$U(x;\delta;\theta) = F(x;\theta) - F(x-\delta;\theta),$$

and we take $\binom{a}{b}$ to be zero in (2.6) if $b > a$ or $b < 0$. Using the combinatorial identities

$$(k-i+1)\binom{k-i}{m-j} = (m-j+1)\binom{k-i+1}{m-j+1} \tag{2.7}$$

and

$$(1/m)\binom{k-2}{m-2} = (1/(k(k-1)))[k\binom{k-1}{m-1} - \binom{k}{m}] \tag{2.8}$$

gives

$$EX^* = \sum_{m=1}^{k} [(1/k)\binom{k}{m}(I_1-I_2-I_3+2I_5) + \binom{k-1}{m-1}(I_2+I_3-2I_5)] + \sum_{m=1}^{k-1} \binom{k-1}{m}I_4 + (k-1)\sum_{m=2}^{k} \binom{k-2}{m-2}I_5. \tag{2.9}$$

Next, moving the sums in (2.9) inside the integrals and applying the binomial expansion to obtain expressions such as

$$\sum_{m=1}^{k} \binom{k}{m}U^m(x,\delta,\theta_\Delta)F^{k-m}(x-\delta;\theta_\Delta) = F^k(x;\theta_\Delta) - F^k(x-\delta;\theta_\Delta) \equiv U_k(x;\delta;\theta_\Delta) \tag{2.10}$$

we obtain after some manipulation that

$$EX^* = \int (x/U(x;\delta;\theta_\Delta))\{f(x;\theta_\Delta)[F(x;\theta)F^{k-1}(x;\theta_\Delta)-F(x-\delta;\theta)F^{k-1}(x-\delta;\theta_\Delta)$$
$$- (U_k(x;\delta;\theta_\Delta)U(x;\delta;\theta)/kU(x;\delta;\theta_\Delta))] + (U_k(x;\delta;\theta_\Delta)f(x;\theta)/k)\}dx$$
$$+ \int \int_{x-\delta}^{x} (\frac{y}{U(x;\delta;\theta_\Delta)})\{f(y;\theta_\Delta)f(x;\theta_\Delta)[(k-1)F(x;\theta)F^{k-2}(x;\theta_\Delta) \tag{2.11}$$
$$- (\frac{1}{U(x;\delta;\theta_\Delta)})(U(x;\delta;\theta)F^{k-1}(x;\theta_\Delta)+F(x;\theta)F^{k-1}(x;\theta_\Delta)-F(x-\delta;\theta)F^{k-1}(x-\delta;\theta_\Delta))$$
$$+ 2(U_k(x;\delta;\theta_\Delta)U(x;\delta;\theta)/kU^2(x;\delta;\theta_\Delta))] + [f(y;\theta_\Delta)f(x;\theta)+f(y;\theta)f(x;\theta_\Delta)]$$
$$\times [F^{k-1}(x;\theta_\Delta)-(U_k(x;\delta;\theta_\Delta)/kU(x;\delta,\theta_\Delta))]]\}dydx$$

Now we specialize to the location parameter case where

$$F(x;\theta) = G(x-\theta);\ f(x;\theta) = g(x-\theta) \tag{2.12}$$

for a density $g(x)$ with

$$\int_{-\infty}^{\infty} xg(x)dx = 0. \tag{2.13}$$

Also, for $\Delta > 0$,

$$\theta_\Delta = \theta-\Delta. \tag{2.14}$$

Using (2.12) and (2.14) in (2.11), changing variables to $u = (x-\theta+\Delta)$, $v = (y-\theta+\Delta)$, collecting the terms involving $(\theta-\Delta)$, noting that

$$\int_{u-\delta}^{u} g(v-\Delta)dv = G(u-\Delta) - G(u-\Delta-\delta) \equiv U(u;\delta,\Delta) \tag{2.15}$$

and

$$\int_{-\infty}^{\infty} [g(u)G^{k-1}(u+\Delta)+(k-1)g(u+\Delta)G(u)G^{k-2}(u+\Delta)]du = G(u)G^{k-1}(u+\Delta)\Big|_{-\infty}^{\infty} = 1 \tag{2.16}$$

we find after considerable simplification that

$$EX^* = \theta + (B-\Delta)$$

where

$$B = \int_{-\infty}^{\infty} (u/U(u;\delta))\{g(u)[G(u-\Delta)G^{k-1}(u)-G(u-\Delta-\delta)G^{k-1}(u-\delta) - (\frac{U(u,\delta;\Delta)U_k(u;\delta)}{kU(u;\delta)})]$$
$$+ \frac{g(u-\Delta)}{k} U_k(u;\delta)\}du$$
$$+ \int_{-\infty}^{\infty} \int_{u-\delta}^{u} (v/U(u;\delta))\{g(v)g(u)[(k-1)G(u-\Delta)G^{k-2}(u) - (\frac{1}{U(u;\delta)})(U(u;\delta,\Delta)G^{k-1}(u)$$
$$+ G(u-\Delta)G^{k-1}(u) - G(u-\Delta-\delta)G^{k-1}(u-\delta)) + \frac{2U(u;\delta,\Delta)}{KU^2(u;\delta)} U_k(u;\delta)]$$
$$+ [g(v)g(u-\Delta)+g(v-\Delta)g(u)][G^{k-1}(u) - \frac{U_k(u;\delta)}{KU(u;\delta)}]\}dudv$$

and

$$U(u;\delta) = U(u;\delta;0)$$

$$U_k(u;\delta) = G^k(u) - G^k(u-\delta).$$

For the general location parameter distributions, the properties of B are not readily apparent from (2.17) and to study the bias further, we specialize to the normal distribution with known variance (taken to be unity for notational convenience), so that henceforth,

$$g(x) = \frac{1}{\sqrt{2\pi}} e^{-x^2/2}. \tag{2.18}$$

Noting for the normal that

$$\int_{u-\delta}^{u} vg(v)dv = g(u-\delta) - g(u) \tag{2.19}$$

to reduce the double integral to a single one and then using integration by parts with $dV = ug(u)$ to obtain expressions which do not involve the factor u, leads after considerable algebraic manipulation to

$$\begin{aligned} B = \int \{ & \Delta[\frac{g(x-\Delta)U_k(x;\delta)}{kU(x;\delta)} - \frac{g(x)U(x;\delta;\Delta)}{U(x;\delta)}[\frac{U_k(x;\delta)}{kU(x;\delta)} - G^{k-1}(x)]] \\ & + [g(x-\delta)g(x-\Delta)+g(x)g(x-\delta-\Delta)][U_{k-1}(x;\delta)/U(x;\delta)] \qquad (2.20) \\ & + g(x-\delta)g(x)U^{-2}(x;\delta)[G(x-\Delta)((k-2)G^{k-1}(x)-(k-1)G(x-\delta)G^{k-2}(x)+G^{k-1}(x-\delta)) \\ & + G(x-\delta-\Delta)(G^{k-1}(x)-(k-1)G(x)G^{k-2}(x-\delta)+(k-2)G^{k-1}(x-\delta))]\}dx. \end{aligned}$$

Although (2.20) has been simplified and condensed to a high degree, numerical integration is still necessary to determine just how the bias varies with δ and Δ. About the only analytic result obtainable is the trivial one that when $\Delta = 0$ (equal means), the bias decreases monotonically with increasing δ (i.e., it is best to average all k of the observations).

For numerical evaluation, (2.20) is not a good form to use. Expressions such as $\frac{U_k(x;\delta)}{U(x;\delta)}$ look like zero over zero for large x values and lead to numerical instability even though perfectly good analytic limits exist. For computing, it is better to use forms for B which involve sums but not ratios with small denominators. Applying to (2.20) the finite geometric series, so that we write

$$\frac{b^k-a^k}{b-a} = \frac{b^{k-1}(1-(a/b)^k)}{(1-(a/b))} = b^{k-1}\sum_{i=0}^{k-1}(a/b)^i, \quad (b > a)$$

we obtain the computationally stable formula

$$\begin{aligned} B = \int \{(\Delta/k)g(x-\Delta) \sum_{i=0}^{k-1} G^i(x-\delta)G^{k-1-i}(x) \\ + (g(x)g(x-\Delta-\delta) + g(x-\delta)g(x-\Delta)) \sum_{i=0}^{k-2} G^i(x-\delta)G^{k-2-i}(x) \\ + (k-1)g(x)g(x-\delta)G(x-\delta-\Delta) \sum_{i=0}^{k-3} G^i(x-\delta)G^{k-3-i}(x) \\ + g(x)g(x-\delta)U(x;\delta;\Delta) \sum_{i=2}^{k-1} (i-1)\binom{k-1}{i}G^{k-1-i}(x-\delta)U^{i-2}(x;\delta) \\ + (\Delta/k)g(x)U(x;\delta;\Delta) \sum_{i=2}^{k} (i-1)\binom{k}{i}G^{k-i}(x-\delta)U^{i-2}(x;\delta)\}dx. \end{aligned} \tag{2.21}$$

A great deal of attention has been devoted to developing bias formulae for the case (1.5), not only because the formulae simplify greatly in that situation, but also because of the intuitive feeling that it is the "least favorable" situation in the sense that the bias of X^* is maximized for (1.5) among all configurations in which $\theta_1 = \theta$, $\theta_2 \le \theta_\Delta$. We shall now demonstrate that for location parameter distributions and $\delta = 0$ ($X^* = X_{max}$), this intuition is correct.

Lemma 2.1. For the location parameter distributions (2.12), the bias of X_{max} is maximized by (1.5) among configurations with $\theta_1 - \theta_2 \ge \Delta$.

Proof: For location parameters, we use the invariance of the bias function to take $\theta = 0$, $\theta_i = \theta - \Delta_i$ with $\Delta_1 = 0$, $\Delta_i \ge 0$, $(2 \le i \le k)$. Write $\underset{\sim}{\Delta} = (0, \Delta_2, \ldots, \Delta_k)$, $\underset{\sim}{0} = (0, \ldots, 0)$ and note that for $\theta = 0$, the bias of X_{max} is just EX_{max} which in the location parameter case is

$$E_{\underset{\sim}{\Delta}}[\max(X_1, \ldots, X_k)] = E_{\underset{\sim}{0}}[\max(X_1, X_2-\Delta_2, \ldots, X_k-\Delta_k)]. \tag{2.22}$$

Since $\max(X_1, X_2-\Delta_2, \ldots, X_k-\Delta_k) \ge X_1$, we conclude $E_{\underset{\sim}{\Delta}}X_{max} \ge E_{\underset{\sim}{0}}X_1 = 0$, so that bias is always positive, and since $\max(X_1, X_2-\Delta_2, \ldots, X_k-\Delta_k)$ is non-increasing in each Δ_i $(2 \le i \le k)$, the result follows.

Also, from (2.22) with $\Delta_2 = \ldots = \Delta_k = \Delta$ (the least favorable configuration), it is evident that the bias decreases monotonically with Δ. Taking the limit as $\Delta \to \infty$ inside the expectation on the right side of (2.22), noting that the limit is X_1, we find that the bias approaches zero with increasing Δ.

For the normal distribution, in the Least Favorable Configuration ($\theta = 0$),

$$\text{Bias}(X_{max}) = EX_{max} = (k-1)\int g(x)G^{k-3}(x)[2g(x-\Delta)G(x) + (k-2)g(x)G(x-\Delta) - \Delta G(x)G(x-\Delta)]dx. \tag{2.23}$$

On the basis of Lemma 2.1, bias comparisons among X^* for various δ will be made under (1.5). Ideally, we would like Lemma 2.1 to hold for all X^* regardless of δ and it might seem intuitively that Lemma 2.1 should hold for all δ. However, this is not the case as can be seen with $k = 3$, in the normal case.

With $\theta_1 = 0$, $\theta_2 = -\Delta$, $\theta_3 = -(\Delta+\nu)$, the basic formula for EX^* can be written as

$$EX^* = \int [A(x)-\Delta B(x)-\nu C(x)]dx \tag{2.24}$$

where

$$\begin{aligned} A(x) = \frac{1}{2}\{&(G(x+\Delta+\nu) + G(x+\Delta+\nu-\delta))(g(x)g(x+\Delta-\delta) + g(x+\Delta)g(x-\delta)) \\ &+ (G(x+\Delta) + G(x+\Delta-\delta))(g(x)g(x+\Delta+\nu-\delta) + g(x-\delta)g(x+\Delta+\nu)) \\ &+ (G(x) + G(x-\delta))(g(x+\Delta)g(x+\Delta+\nu-\delta) + g(x+\Delta+\nu)g(x+\Delta-\delta)) \end{aligned}$$

$$\begin{aligned} B(x) = \frac{1}{6}\{&g(x)[4G(x+\Delta)G(x+\Delta+\nu) - G(x+\Delta)G(x+\Delta+\nu-\delta) - G(x+\Delta-\delta)G(x+\Delta+\nu) \\ &\quad - 2G(x+\Delta-\delta)G(x+\Delta+\nu-\delta)] \\ &+ g(x+\Delta)[4G(x)G(x+\Delta+\nu) - G(x)G(x+\Delta+\nu-\delta) + 2G(x-\delta)G(x+\Delta+\nu) \\ &\quad + G(x-\delta)G(x+\Delta+\nu-\delta)] \\ &+ g(x+\Delta+\nu)[4G(x)G(x+\Delta) - G(x)G(x+\Delta-\delta) + 2G(x-\delta)G(x+\Delta) \\ &\quad + G(x-\delta)G(x+\Delta-\delta)]\} \end{aligned}$$

$$\begin{aligned} C(x) = \frac{1}{6}\{&g(x+\Delta+\nu)[G(x)G(x+\Delta-\delta) + 2G(x)G(x+\Delta) + 2G(x-\delta)G(x+\Delta-\delta) \\ &\quad + G(x-\delta)G(x+\Delta)] \\ &+ (G(x+\Delta+\nu) - G(x+\Delta+\nu-\delta))[g(x)(2G(x+\Delta) + G(x+\Delta-\delta)) \\ &\quad + g(x+\Delta)(2G(x) + G(x-\delta))]\}. \end{aligned}$$

Numerical integration of (2.24) verified that Lemma 2.1 is not universally true. Some pertinent values are given in Table 1 and will be discussed in the sequel.

Extensive computations have been made with (2.21) for $k = 3$. These are summarized in Figure 18.1. Note that for small δ, the bias is positive and decreases monotonically as Δ increases. This phenomenon is readily explained by noting that for $\delta = 0$, $(\theta = 0)$, positive values of X_1 will tend to give $X_{max} = X_1$, but negative values of X_1 may lead to $X_{max} = (X_2 \text{ or } X_3) > X_1$, so that $X_{max} \geq X_1$ and X_1 is unbiased. The likelihood of $X_{max} > X_1$ will decrease as the separation Δ between population means increases. The same reasoning (fewer competitors to be

Table 1

Bias of X^* as ν Varies for Selected Values of δ and Δ, with $k = 3$

δ	Δ	ν: 0	.1	.2	.4	.6	.8	.9
.4	.8	.38	.36	.34	.31	.29	.273	.266
1.5	.5	.19	.17	.15	.12	.096	.082	.077
1.5	1.0	-.0018	-.015	-.027	-.044	-.054	-.059	-.059
1.5	1.5	-.108	-.114	-.118	-.123	-.124	-.121	-.119
1.5	2.0	-.1426	-.1428	-.1421	-.139	-.134	-.128	-.125
1.5	3.0	-.097	-.093	-.089	-.083	-.077	-.072	-.070

X_{max}) explains the decrease in the upward bias of X_{max} as ν increases in the configuration of (2.24). For small δ values, X^* will behave similarly to X_{max}, as Table 1 shows for $\delta = 0.4$, $\Delta = 0.8$.

For larger δ values, Figure 18.1 shows a negative bias when Δ is moderately large. This phenomenon reflects the fact that although $X_{max} = X_1$ with very high probability, X^* will with high probability be an average of X_1 and one or both of X_2, X_3 even when $X_1 > 0$. (For $\delta = 1.6$, $\Delta = 1.5$, simulations show that X^* includes X_1 97% of the time, and X_2, X_3 each appear about 50% of the time.) Thus, X^* overcompensates for the upward bias of X_{max}. For these same combinations of δ and Δ, Table 1 shows that as ν increases from zero, the bias can become more negative. This can be explained if one assumes that for small values of ν, the chance of including X_3 decreases only slightly while the contribution of X_3 to X^* decreases by ν when it is included. The simulations bear out this hypothesis, since for $\delta = 1.6$, $\Delta = 1.5$, the inclusion percentage for X_3 drops only from 55% to 52% when ν rises from 0 to 0.1 and falls only to 49% for $\nu = 0.2$.

Regardless of δ, for very large Δ, the bias becomes small because the chance of including either X_2 or X_3 in X^* is very small. Similarly for fixed Δ, when ν gets sufficiently large the percentage of time that X_3 is included in X^* drops off enough so that the bias becomes less negative. (In the $\delta = 1.6$, $\Delta = 1.5$ case, for $\nu = 1$, X_3 is included only about 20% of the time.)

Since an underestimate is just as undesirable as an overestimate, we propose to choose δ to minimize the maximum absolute bias for the configuration (1.5). To distinguish δ which achieves this goal from δ which minimizes maximum bias over all configurations with $\theta_1 - \theta_2 \geq \Delta$, we shall refer to the δ we calculate as "(1.5)-minimax." Since the bias at

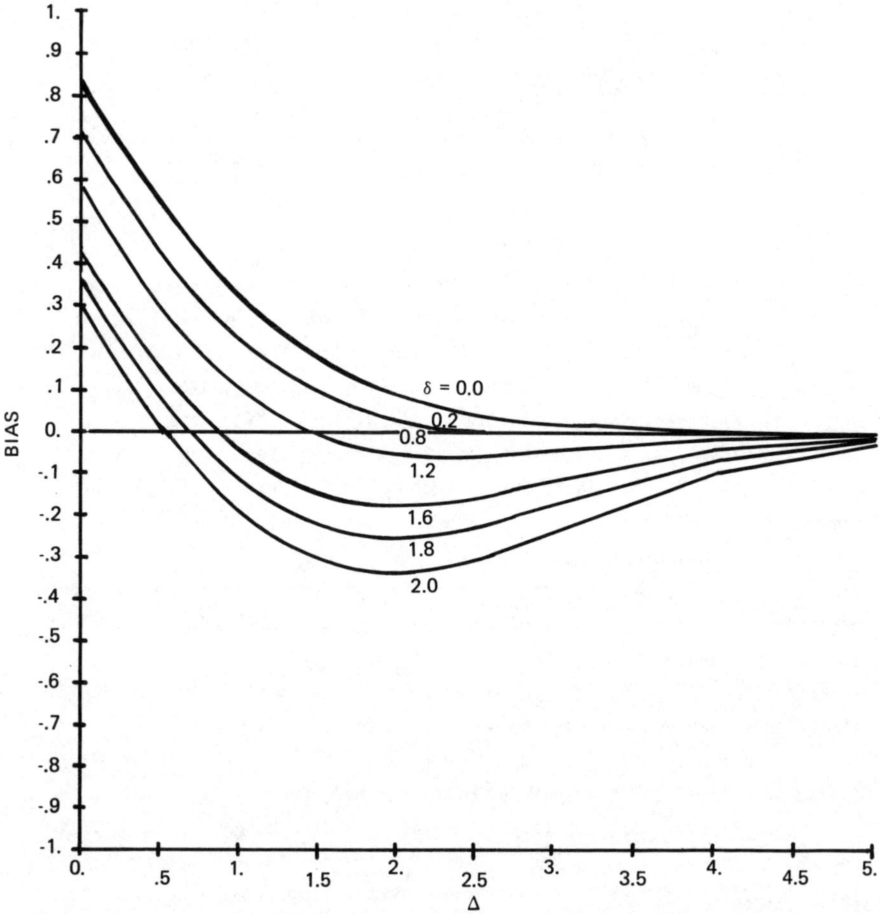

FIGURE 18.1 Bias of X^* as a function of Δ for selected values of δ.

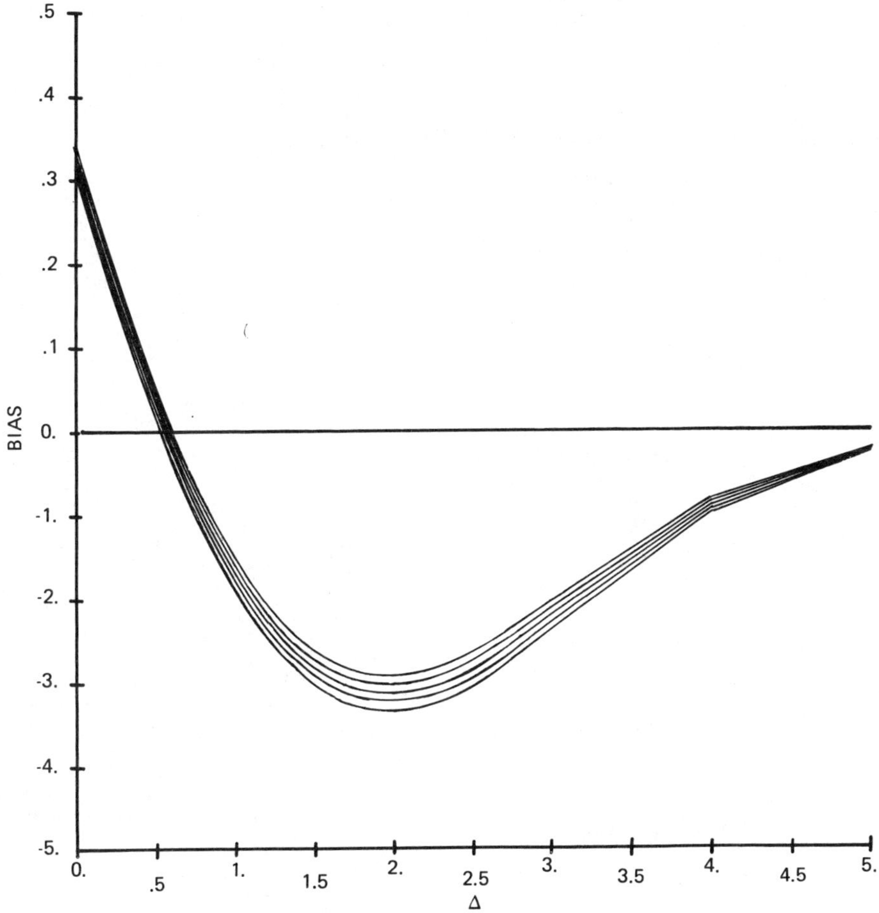

FIGURE 18.2 Expanded plot of bias of X^* against Δ for $\delta = 1.90(.025)2.00$.

$\Delta = 0$ decreases monotonically with δ while the absolute bias around $\Delta \cong 2$ increases, the minimax goal will be achieved when the two biases are equalized for some δ. From Figure 18.1, we see that this will be achieved for $1.9 < \delta < 2.0$. In Figure 18.2, we plot bias for this range of δ in (.025) increments and it appears that the (1.5)-minimax value is $\delta = 1.98$. From Table II of [2], it appears that for $k = 2$, the true minimax δ value lies between $(1/\sqrt{2})$ and 1.

3. MEAN SQUARED ERROR

In the location parameter case,

$$
\begin{aligned}
E(X^*-\theta)^2 &= EX^{*2} \\
&= \sum_{m=1}^{k} \sum_{j=1}^{k} \frac{1}{m^2} \sum_{C(j)} \int_{x_j=-\infty}^{\infty} \int_{x_{i_1}=x_j-\delta}^{x_j} \cdots \int_{x_{i_{m-1}}=x_j-\delta}^{x_j} \\
&\quad \cdot (x_j+x_{i_1}+\dots+x_{i_{m-1}})^2 g(x_j+\Delta_j)g(x_{i_1}+\Delta_{i_1})\dots g(x_{i_{m-1}}+\Delta_{i_{m-1}}) \\
&\quad \cdot \prod_{\substack{r\neq j \\ r\neq(i_1,\dots,i_{m-1})}} G(x_j-\delta+\Delta_r)dx_1\dots dx_k,
\end{aligned}
\tag{3.1}
$$

where $\Delta_1,\dots,\Delta_k$ are as in Lemma 2.1. For the configuration (1.5), we can write (3.1) as

$$
\begin{aligned}
EX^{*2} = \sum_{m=1}^{k} \frac{1}{m^2} \{&\tbinom{k-1}{m-1}[J_1+(m-1)J_2+(m-1)(m-2)J_3]+\tbinom{k-2}{m-1}[J_4+(m-1)J_5+(m-1)(m-2)J_6] \\
&+ \tbinom{k-2}{m-2}[J_7+J_8+(m-2)J_9+2(m-2)J_{10}+(m-2)(m-3)J_{11}]\}
\end{aligned}
\tag{3.2}
$$

where

$$J_1 = \int x^2 f(x)U^{m-1}(x;\delta;\Delta)G^{k-m}(x+\Delta-\delta)dx$$

$$J_2 = \int \int_{y=x-\delta}^{x} (y^2+2xy)g(x)g(y+\Delta)U^{m-2}(x;\delta;\Delta)G^{k-m}(x+\Delta-\delta)dxdy$$

$$J_3 = \int \int_{x-\delta}^{x} \int_{x-\delta}^{x} yzg(x)g(y+\Delta)g(z+\Delta)U^{m-3}(x;\delta;\Delta)G^{k-m}(x+\Delta-\delta)dxdydz$$

$$J_4 = \int x^2 g(x+\Delta)U^{m-1}(x;\delta;\Delta)G(x-\delta)G^{k-m-1}(x+\Delta-\delta)dx$$

$$J_5 = \int \int_{x-\delta}^{x} (y^2+2xy)g(x+\Delta)g(y+\Delta)U^{m-2}(x;\delta;\Delta)G(x-\delta)G^{k-m-1}(x+\Delta-\delta)dxdy$$

$$J_6 = \int \int_{x-\delta}^{x} \int_{x-\delta}^{x} yzg(x+\Delta)g(y+\Delta)g(z+\Delta)U^{m-3}(x;\delta;\Delta)G(x-\delta)G^{k-m-1}(x+\Delta-\delta)\,dxdydz$$

$$J_7 = \int x^2 g(x+\Delta)U(x;\delta)U^{m-2}(x;\delta;\Delta)G^{k-m}(x+\Delta-\delta)\,dx$$

$$J_8 = \int \int_{x-\delta}^{x} (y^2+2xy)g(x+\Delta)g(y)U^{m-2}(x;\delta;\Delta)G^{k-m}(x+\Delta-\delta)\,dxdy$$

$$J_9 = \int \int_{x-\delta}^{x} (y^2+2xy)g(x+\Delta)g(y+\Delta)U(x;\delta)U^{m-3}(x;\delta;\Delta)G^{k-m}(x+\Delta-\delta)\,dxdy$$

$$J_{10} = \int \int_{x-\delta}^{x} \int_{x-\delta}^{x} yzg(y)g(z+\Delta)g(x+\Delta)U^{m-3}(x;\delta;\Delta)G^{k-m}(x+\Delta-\delta)\,dxdydz$$

$$J_{11} = \int \int_{x-\delta}^{x} \int_{x-\delta}^{x} yzg(x+\Delta)g(y+\Delta)g(z+\Delta)U(x;\delta)U^{m-4}(x;\delta;\Delta)G^{k-m}(x+\Delta-\delta)\,dxdydz.$$

For the normal distribution, using the fact that

$$\int_a^b x^2 g(x+\Delta)dx = (a-\Delta)g(a+\Delta) - (b-\Delta)g(b+\Delta) + (1+\Delta^2)(G(b+\Delta) - G(a+\Delta)) \quad (3.3)$$

along with extensive algebraic rearrangement leads to

$$EX^{*2} = A_1 + (k-1)B_1 + ((k-1)/4)(A_2 + (k-2)B_2 + C_2) + \sum_{m=3}^{k} (1/m^2)\{\binom{k-1}{m-1}A_m + (m-1)[\binom{k-2}{m-1}B_m + \binom{k-2}{m-2}C_m]\} \quad (3.4)$$

where

$$A_1 = \int x^2 g(x)G^{k-1}(x+\Delta-\delta)\,dx$$

$$B_1 = \int x^2 g(x+\Delta)G(x-\delta)G^{k-2}(x+\Delta-\delta)\,dx$$

$$A_2 = \int g(x)G^{k-2}(x+\Delta-\delta)\{U(x;\delta;\Delta)[x^2-2\Delta x+1+\Delta^2] - [(3x-\Delta)U'(x;\delta;\Delta) + \delta g(x+\Delta-\delta)]\}dx$$

$$B_2 = \int g(x+\Delta)G(x-\delta)G^{k-3}(x+\Delta-\delta)[(x^2-2\Delta x+1+\Delta^2)U(x;\delta;\Delta) + 2(\Delta-x)U'(x;\delta;\Delta) + U''(x;\delta;\Delta)]dx$$

$$C_2 = \int g(x+\Delta)G^{k-2}(x+\Delta-\delta)\{U''(x;\delta) - 2xU'(x;\delta) + (1+x^2)U(x;\delta)\}dx$$

and for $m \geq 3$,

$$A_m = \int g(x)G^{k-m}(x+\Delta-\delta)U^{m-3}(x;\delta;\Delta)\{U^2(x;\delta;\Delta)[x^2- 2(m-1)x\Delta + (m-1)(1+\Delta^2(m-1))] - (m-1)U(x;\delta;\Delta)[(3x - (2m-3)\Delta)U'(x;\delta;\Delta) + \delta g(x+\Delta-\delta)] + (m-1)(m-2)(U'(x;\delta;\Delta))^2\}dx$$

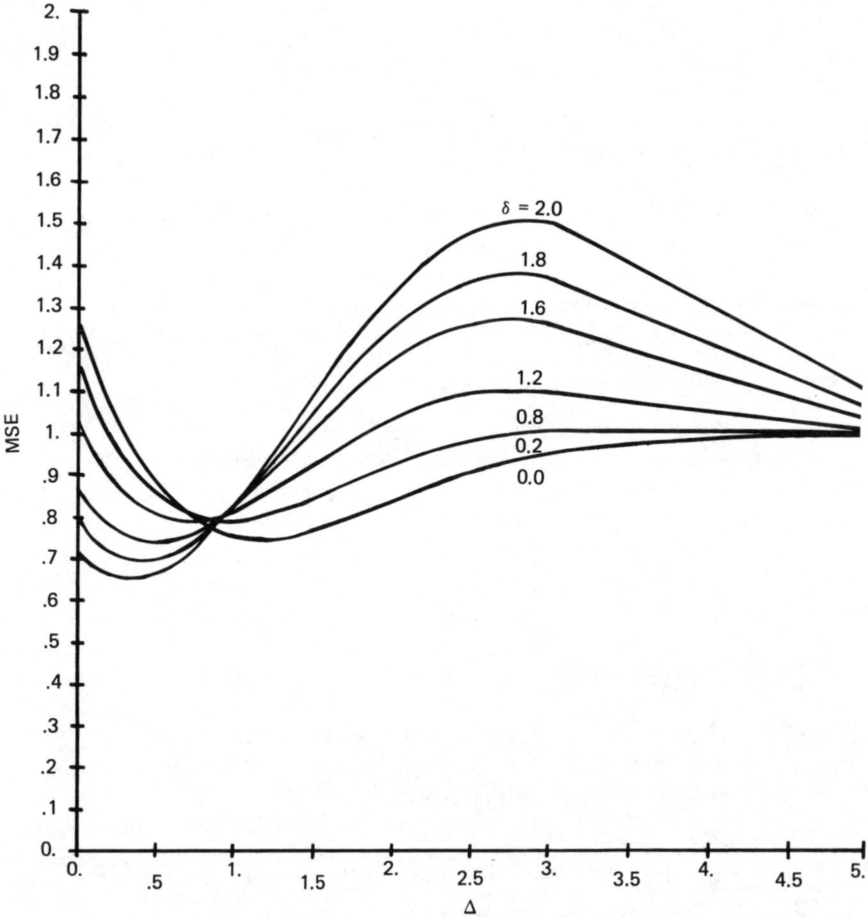

FIGURE 18.3 MSE of X^* as a function of Δ for selected values of δ.

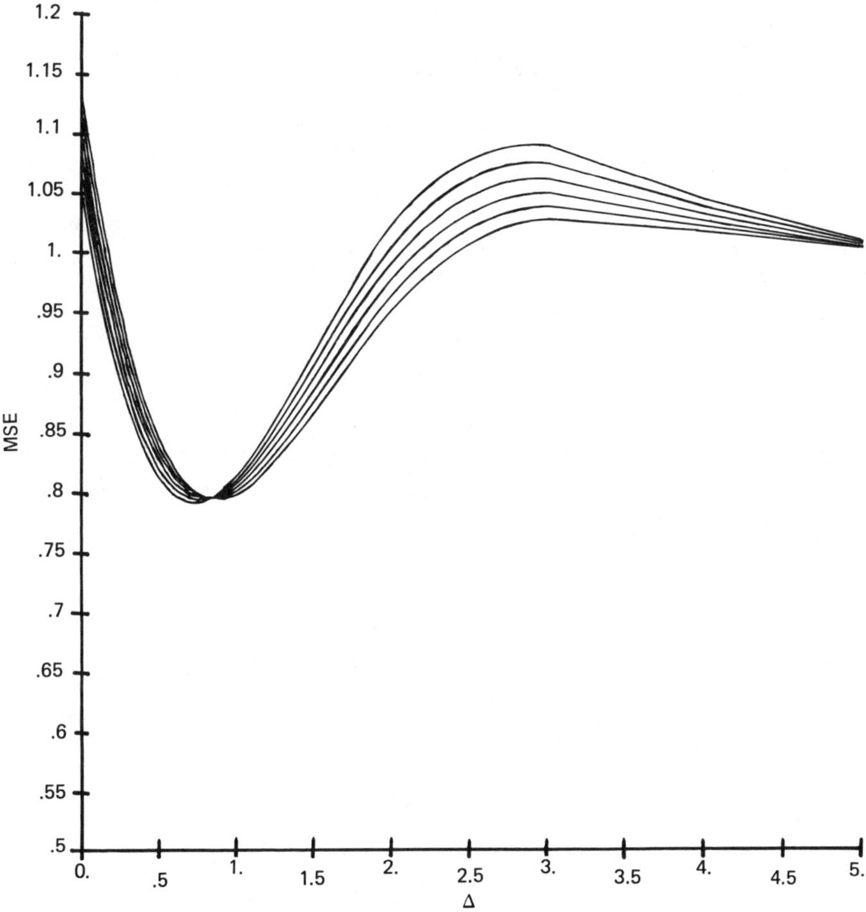

FIGURE 18.4 Expanded plot of MSE of X^* against Δ for $\delta = 0.90(.05)1.15$.

$$B_m = \int g(x+\Delta)G^{k-m-1}(x+\Delta-\delta)G(x-\delta)U^{m-3}(x;\delta;\Delta)\{x^2U^2(x;\delta;\Delta) + (m-1)U(x;\delta;\Delta)$$
$$\cdot [U''(x;\delta;\Delta) + 2(\Delta-x)U'(x;\delta;\Delta) + (1+\Delta^2-2\Delta x)U(x;\delta;\Delta)]$$
$$+ (m-1)(m-2)[U'(x;\delta;\Delta) + \Delta U(x;\delta;\Delta)]^2\}dx$$

$$C_m = \int g(x+\Delta)G^{k-m}(x+\Delta-\delta)U^{m-4}(x;\delta;\Delta)\{U^2(x;\delta;\Delta)[U''(x;\delta) + 2((m-2)\Delta-x)U'(x;\delta)$$
$$+ (m-1+x^2-2(m-2)\Delta x + (m-2)^2\Delta^2)U(x;\delta)]$$
$$+ (m-2)U(x;\delta;\Delta)[2U'(x;\delta)U'(x;\delta;\Delta) + U(x;\delta)(U''(x;\delta;\Delta)$$
$$+ 2((m-2)\Delta-x)U'(x;\delta;\Delta))]$$
$$+ (m-2)(m-3)U(x;\delta)(U'(x;\delta;\Delta))^2\}dx$$

where $U'(x;\delta;\Delta) = \frac{d}{dx} U(x;\delta;\Delta)$, $U''(x;\delta;\Delta) = \frac{d^2}{dx^2} U(x;\delta;\Delta)$, etc.

When $\delta = 0$, we obtain the special case

$$EX_{max}^2 = A_1 + (k-1)B_1 = \int x^2G^{k-2}(x+\Delta)[g(x)G(x+\Delta) + (k-1)g(x+\Delta)G(x)]dx$$
$$= \int x^2 d(G(x)G^{k-1}(x+\Delta)) \qquad (3.5)$$

which is valid for the general location parameter case.

Numerical integration of (3.4) and (3.5) gives the M.S.E. curves of Figure 18.3. Increasing δ decreases the MSE near $\Delta = 0$, but increases MSE for larger values of Δ (the peak occurs around $\Delta = 2.5$, whereas max|Bias| for these estimators occurs around $\Delta = 2.0$). Taking a (1.5)-minimax approach, any of the estimators with $\delta \leq 1.6$ has a smaller maximum M.S.E. than X_{max} and would be preferred to it. The (1.5)-minimax estimator whose MSE at $\Delta = 0$ equals its peak at $\Delta \approx 2.5$ seems from our computations to be at $\delta = 1.07$ with max MSE = 1.07. (From Figure 18.4)

4. CONCLUDING REMARKS

Since our original goal was to reduce bias, we would suggest using $\delta = 1.6$ This gives a considerable improvement over X_{max} in bias while maintaining the same value of max MSE. The (1.5)-minimax bias estimator with $\delta = 1.98$ beats $\delta = 1.6$ in terms of maximum bias but sacrifices in MSE, and the (1.5)-minimax MSE estimator with $\delta = 1$ sacrifices potential improvement in bias (although it does improve on X_{max} on (1.5)-minimax bias). Also, the fact that (1.5) is not least favorable militates for a δ smaller than 1.98. Thus we view $\delta = 1.6$ as a compromise which emphasizes bias reduction but does not ignore MSE. It remains for future work to determine how much a (1.5)-minimax δ differs from a true minimax δ.

Additional computation will be necessary to determine recommended δ values for $k \geq 4$. Programs to compute bias and MSE for any k now exist, but the calculations have yet to be carried out. Since the bias of X_{max} at $\Delta = 0$ goes up rapidly with k, the potential for improvement is considerable with larger k. Whether some formula could be deduced for the minimax δ as a function of k remains open.

Although this study was aimed at comparing X^* with X_{max}, comparisons with the maximum likelihood or Pitman-type estimators (as was done for $k = 2$ in [2] and [3]) would be interesting. For $k \geq 3$, these comparisons are complicated because formulas for the alternative estimators have never been derived, and it is suspected that they are so complex that numerical integration of bias and MSE will not be feasible. This will leave Monte Carlo as the only possibility, and that can be expected to be expensive because of the complex computations needed to evaluate the estimators (m.l.e. or Pitman) at each iteration. Nonetheless, the author plans to carry out these comparisons in the near future.

SUMMARY

To estimate the largest of k location parameters based on one observation from each population, a computationally simple alternative to using the largest observation (X_{max}) is proposed. Namely, average all of the observations which exceed $(X_{max}-\delta)$, where δ is chosen by the experimenter. Numerical integration of the bias and mean squared error for a range of δ values when the distributions are normal and $k = 3$ leads to recommendations for δ values which improve on X_{max} with respect to either bias, mean squared error, or both.

ACKNOWLEDGMENT

The author wishes to express his appreciation to Ms. Shiela O'Leary for programming and running the computations which were so vital to this research. The author also thanks the referee for supplying the streamlined proof of Lemma 2.1.

REFERENCES

[1] Blumenthal, S. and Cohen, A. (1968). Estimation of the larger translation parameter. Ann. Math. Statist. 39, 501-516.

[2] Blumenthal, S. and Cohen, A. (1968). Estimation of the larger of two normal means. J. Amer. Statist. Assoc. 63, 861-876.

[3] Dhariyal, I.D., Dudewicz, E.J. and Blumenthal, S. (1982). Estimation of the larger mean. SIAM J. Appl. Math. 42, 1265-1275.

[4] Dudewicz, E.J. (1971). Maximum likelihood estimators for ranked means. Z. Wahrsch. Verw. Gebiete 19, 29-42.

[5] Dudewicz, E.J. (1973). Maximum probability estimators for ranked means. Ann. Inst. Statist. Math. 25, 467-477.

[6] Dudewicz, E.J. (1976). Generalized maximum likelihood estimators for ranked means. Z. Wahrsch. Verw. Gebiete 35, 283-297.

AN ASYMPTOTIC COMPARISON BETWEEN MAXIMUM LIKELIHOOD AND METHOD OF MOMENTS IN A PARTICULAR ERRORS-IN-VARIABLES REGRESSION MODEL

Raymond J. Carroll*

Department of Statistics
University of North Carolina
Chapel Hill, North Carolina

Paul P. Gallo[†]

Statistical Design
and Analysis Department
Lederle Laboratories
Pearl River, New York

1. INTRODUCTION

The problem of estimating linear regression parameters when the variables are subject to measurement or observation error has a long history and has recently been the focus of considerable attention. Reilly and Patino-Leal (1981) list a number of recent publications concerning situations in which the problem arises; see Wu, Ware and Feinlieb (1980) for a simple but particularly interesting example in a biomedical context. Blomquist (1977), Nussbaum (1980), Fuller (1980) and Gleser (1981) have recently addressed various theoretical aspects of the problem.

The purpose of this paper is to compare maximum likelihood and method of moments in a very particular functional errors-in-variables regression model. In a different model, for simple linear regression Dorff and Gurland (1961) showed that if there is no equation error as in a "pure" functional model, and if the predictor and the response are replicated, then the maximum likelihood estimator is asymptotically more efficient than method of moments. We obtain the same result for multiple linear regression in a model which differs from theirs.

*This research was supported by the Air Force Office of Scientific Research under Grant AFOSR F49620 82 C 0009.

[†]This research was supported in part by a National Science Foundation Graduate Fellowship and by the Air Force Office of Scientific Research under Grant AFOSR F49620 82 C 0009.

The concept of "equation error" was introduced by Fuller (1980) and is natural to the typical user of regression. The pure functional model has no equation error and the maximum likelihood estimate takes advantage of this fact. On the other hand, the definition of the moments estimator is not affected by the introduction of equation error. It thus makes sense to conjecture that the moments estimator will be more robust than maximum likelihood to possible substantial but unexpected equation error. While we do not verify this conjecture theoretically, we do lend some credence to it through a Monte-Carlo study.

2. NOTATION AND ESTIMATES

We consider a particular errors-in-variables (EIV) regression model in which some subset of the variables is subject to error, while some are observed exactly; the response is replicated R times and the predictor variables subject to measurement error are also replicated R times. Specifically,

$$\begin{aligned} Y &= X_1\beta_1 + X_2\beta_2 + \delta && && (2.1)\\ Y_i &= y + V_i && i = 1,\ldots,R\ (\geq 2), \\ C_j &= X_2 + U_j && j = 1,\ldots,R. \end{aligned}$$

Here, β_1 is a $(p_1\times 1)$ vector and β_2 is $(p_2\times 1)$, $p = p_1+p_2$. The vectors Y_i, δ, and V_i are of dimension $(N\times 1)$, where N is the sample size in the study. Because of measurement error, we can only observe Y_i, not y. X_1 is observable, however, because of measurement error U_j, X_2 is not observable but rather the $(N\times P_2)$ matrices C_j are observed. The $(N\times 1)$ random vector δ is called the equation error, while the $\{V_i\}$ are the measurement errors in the response. The assumption that X_1 and X_2 are constant puts us in a functional EIV model.

Throughout the main part of this paper we will assume that $\delta = 0$, i.e., there is no equation error so that we are in the "pure" functional case.

We assume that the vectors (U_i, V_i) are independent, with independent rows having the covariance matrix

$$\Sigma = \begin{pmatrix} \Sigma_u & \Sigma_{vu} \\ \Sigma_{vu}' & \sigma^2 \end{pmatrix}.$$

We also define

$$X = [X_1 : X_2]$$

and assume that X is of full column rank such that

$$\Delta = \lim_{N\to\infty} N^{-1}X'X \quad \text{is positive definite.} \tag{2.2}$$

We further define

$$U_{j*} = [\underset{N\times p_1}{0} : U_j] \quad \beta' = [\beta_1' \; \beta_2'] \quad C_{j*} = [X_1 : C_j]$$

$$\Sigma_{u*} = \begin{bmatrix} 0 & 0 \\ 0 & \Sigma_u \end{bmatrix} \quad \Delta = \begin{bmatrix} \Delta_{11} & \Delta_{12} \\ \Delta_{21} & \Delta_{22} \end{bmatrix} \quad \Sigma_{vu*} = \binom{0}{\Sigma_{vu}}$$

where Δ_{11} and the upper left-hand submatrix of zeroes in Σ_{u*} are $(p_1 \times p_1)$.

3. THE ESTIMATORS AND THEIR LIMIT DISTRIBUTIONS

In defining the maximum likelihood estimate for our functional relationship, we assume that all the random variables of the previous section are normally distributed. However, this assumption is used only to define the estimators; once the definitions are made, asymptotic distributions will be calculated for a general distribution.

Let θ_R be the smallest eigenvalue of $T_1^{-1}T_2$, where

$$T_1 = \sum_{i=1}^{R}\sum_{j=1}^{R} [C_i:Y_i]' \quad (\gamma_{ij} - R^{-1})I_N[C_j:Y_j]$$

$$T_2 = T_1 + R^{-1}(\sum_{i=1}^{R} [C_i:Y_i]')G(\sum_{i=1}^{R} [C_i:Y_i]),$$

$$G = I_N - X_1(X_1'X_1)^{-1}X_1',$$

where γ_{ij} is the Kronecker delta, the indicator of $i = j$. Then with

$$m_{ij} = (R(\theta_R - 1)\gamma_{ij} - \theta_R)$$

we define the maximum likelihood estimate

$$\hat{\beta}_{MR} = (\sum_{j=1}^{R}\sum_{k=1}^{R} m_{jk}C_{j*}'C_{k*})^{-1} \sum_{j=1}^{R}\sum_{k=1}^{R} m_{jk}C_{j*}'Y_k. \tag{3.1}$$

We shall call this the normality - MLE, which has been derived by Anderson (1951) and Healy (1980).

A method of moments estimator motivated by the work of Fuller (1980) is

$$\hat{\beta}_R = (\sum_{\substack{j=1 \\ j\neq k}}^{R}\sum_{k=1}^{R} C_{j*}'C_{k*})^{-1} \sum_{\substack{j=1 \\ j\neq k}}^{R}\sum_{k=1}^{R} C_{j*}'Y_k. \tag{3.2}$$

Under various conditions which we do not state here, but which concern the existence of high enough order moments, the estimators are consistent and asymptotically normally distributed; Gallo (1982a, 1982b). Here we merely state the results.

Theorem 1. (Moments) $N^{1/2}(\hat{\beta}_R-\beta)$ is asymptotically normally distributed with mean zero and covariance matrix

$$\mathrm{Cov}(\hat{\beta}_R) = R^{-1}d\Delta^{-1} + R^{-1}(R-1)^{-1}\Delta^{-1}(d\Sigma_{u*}+D_R)\Delta^{-1}. \qquad (3.3)$$

where

$$D_R = (\Sigma_{u*}\beta - \Sigma_{vu*})(\Sigma_{u*}\beta - \Sigma_{vu*})',$$

$$d = [\beta_2' \ -1]\Sigma[\beta_2' \ -1]'.$$

(Maximum Likelihood) $N^{1/2}(\hat{\beta}_{MR}-\beta)$ is asymptotically normally distributed with mean zero and covariance matrix

$$\mathrm{Cov}(\hat{\beta}_{MR}) = (dR^{-1})\{\Delta^{-1} + (R-1)^{-1}\Delta^{-1}\begin{bmatrix}0 & 0\\ 0 & Q\end{bmatrix}\Delta^{-1}\}, \qquad (3.4)$$

$$Q = ([I_{p_2}\beta_2]\Sigma^{-1}[I_{p_2}\beta_2]')^{-1}.$$

Remark: Although in constructing the maximum likelihood estimate we assumed that all random variables were normally distributed, Theorem 1 holds without change even without normality, as long as the elements of all random vectors have finite fourth moments.

4. COMPARISON

The main result of this paper shows that the moments estimator can never be more efficient than the maximum likelihood estimator in the pure functional model. Note that the following result does not assume that any of the underlying random variables are normally distributed.

At this point it is worth reminding oneself that, while one might intuitively believe that the maximum likelihood estimator should dominate a moments estimator at the normal model, even such a special subcase of our result does not follow from the usual asymptotic optimality theory for maximum likelihood. The reason is that, in the functional regression model, the number of parameters grows proportionally with the sample size, including as it does not only β and Σ but also the $(N\times p_2)$ unknown vector X_2. It should be noted that our model is a special case of Healy's, who shows that the maximum likelihood estimate of the covariance matrix Σ is inconsistent.

Theorem 2. The normality-MLE (3.1) is asymptotically no worse than the moment estimator (3.2), i.e.,

$$\text{Cov}(\hat{\beta}_R) - \text{Cov}(\hat{\beta}_{MR})$$

is positive semi-definite.

Proof of Theorem 2. The limiting covariance matrix of the moments estimator $\hat{\beta}_R$ is from (3.3)

$$\Delta^{-1}\{R^{-1}d\Delta + R^{-1}(R-1)^{-1}(d\Sigma_{u*}+D_R)\}\Delta^{-1}.$$

Comparing this with (3.4), since D_R is positive semi-definite, it will suffice to show that for all $(p_2\times 1)$ vectors γ,

$$\gamma'([I_{p_2}\beta]\Sigma^{-1}[I_{p_2}\beta_2]')^{-1}\gamma' \leq \gamma'\Sigma_u\gamma. \tag{4.1}$$

Now since Σ^{-1} can be expressed as

$$\begin{bmatrix} \Sigma_u^{-1} & 0 \\ 0 & 0 \end{bmatrix} + \begin{bmatrix} \Sigma_u^{-1}\Sigma_{vu} \\ -1 \end{bmatrix} (\sigma^2 - \Sigma_{vu}'\Sigma_u^{-1}\Sigma_{vu})^{-1}[\Sigma_{vu}'\Sigma_u^{-1} \quad -1]$$

we have

$$[I_{p_2}:\beta_2]\Sigma^{-1}[I_{p_2}:\beta_2]' = \Sigma_u^{-1} + (\sigma^2-\Sigma_{vu}'\Sigma_u^{-1}\Sigma_{vu})^{-1}(\Sigma_u^{-1}\Sigma_{vu}-\beta_2)(\Sigma_u^{-1}\Sigma_{vu}-\beta_2)'$$

so that

$$\gamma'[I_{p_2}:\beta_2]\Sigma^{-1}[I_{p_2}:\beta_2]'\gamma \geq \gamma'\Sigma_u^{-1}\gamma. \tag{4.2}$$

Equation (4.1) now follows immediately from (4.2) and Graybill (1969), Theorem 12.2.14 (5).

5. MONTE-CARLO

We performed a small Monte-Carlo study to illustrate the main result Theorem 2, as well as to explore the role of equation error. Random numbers were generated using the IMSL routines GGNPM and GGUBS. There were 500 Monte-Carlo replications. The true model was simple linear regression following the format of Section 2 with $R = 2$ replications. The intercept was 10 and the slope was -4. In the notation of Section 2, X_1 is a column vector of $N = 40$ ones, $\beta_1 = 10$, $\beta_2 = -4$ and X_2 is a column vector obtained as the values of X_2 from Table 1 of Jobson and Fuller (1980).

Although the estimates were calculated in the forms which do not assume $\Sigma_{vu} = 0$, we fixed $\Sigma_{vu} = 0$ and performed the Monte-Carlo study. The Monte-Carlo is meant to be illustrative, so we made the choice $\Sigma_{vu} = 0$ for convenience only. Naturally, the maximum likelihood estimate has been computed when Σ_{vu} is unknown, so our simulations are most directly relevant for the case that we do not know that $\Sigma_{vu} = 0$ when indeed it is. The rows of the error terms δ, V_i and U_i were thus generated independently; all three were either normally distributed or had a contaminated normal distribution. In general, any random variable was either $N(0,\sigma^2)$ (Normal) or $N(0,\sigma^2)$ with probability 0.9 and $N(0,c^2\sigma^2)$ with probability 0.1 (Contaminated Normal), with

<u>Equation Error</u> (δ): $\sigma_\delta^2 = 0$ or 4, $c = 5$;

<u>Y Measurement Error</u> (V): $\sigma_v^2 = 4$, $c = 5$;

<u>X Measurement Error</u> (U): $\sigma_u^2 = 1,2$ or 4, $c = 3$.

The normality-MLE is calculated assuming $\sigma_\delta^2 = 0$, so the results for $\sigma_\delta^2 = 4$ give some idea of its robustness against this assumption. The normality-MLE was computed by (3.1). The moments estimator was (3.2), with exception that the term in (3.2)

$$H = C_{1*}'C_{2*} + C_{2*}'C_{1*} \tag{5.1}$$

was replaced by a type of Fuller's modification (Fuller (1980), page 414-415) using Fuller's $\alpha = 1$. Specifically we replaced H in (5.1) by

$$H_* = (W_1 + N^{-1}W_2),$$

where

$$\begin{aligned}
W_1 &= W_3 - W_2 \quad \text{if} \quad \gamma \geq 1 + 1/N \\
&= W_3 - (\gamma - 1/N)W_2 \quad \text{if} \quad \gamma < 1 + 1/N; \\
W_2 &= (C_{1*}-C_{2*})'(C_{1*}-C_{2*})/2; \\
W_3 &= (C_{1*}+C_{2*})'(C_{1*}+C_{2*})/2; \\
\gamma &= \text{smallest root of } |M - \gamma W_2|; \\
M &= \begin{bmatrix} M_{YY} & M_{YX} \\ M_{YX} & W_3 \end{bmatrix}; \\
M_{YY} &= (Y_1+Y_2)'(Y_1+Y_2)/2; \\
M_{YX} &= (Y_1+Y_2)'(C_{1*}+C_{2*})/2.
\end{aligned}$$

This modification is crucial to get the best performance of the moment's estimator, although other modifications may give even better answers. Although this modification occurs with negligible probability asymptotically, we found in our study that it occurred often.

The least squares estimator is given by $W_3^{-1}M_{YX}'$.

Mean square error (MSE) efficiencies of the moments estimator and the least squares estimator LSE relative to the normality-MLE are given in Table 1, along with the percentage of times Fuller's modification was used. Efficiencies are also given in Table 2 for the 95th percentile of the absolute errors for the different estimators. The first twelve lines of each table are for the situation of no equation error $(\delta = \sigma_\delta^2 = 0)$, assumed in calculating the normality MLE. Note as suggested by Theorem 2 that the normality-MLE generally dominates the moments estimator (but not vastly so), even at non-normal distributions. The last twelve lines of Tables 1 and 2 reflect the situation $\sigma_\delta^2 = 4$, i.e., there is substantial

Table 1

δ and V error distribution	σ_δ^2	U error distribution	σ_U^2	MSE efficiencies for slope: LSE	MSE efficiencies for slope: Moments	% Fuller modification	MSE efficiencies for intercept: LSE	MSE efficiencies for intercept: Moments
N	0	N	1	.90	.98	2.8	.89	.98
N	0	N	2	.52	.98	31.2	.52	.98
N	0	N	4	.23	1.02	50.0	.23	1.03
N	0	CN	1	.68	.95	13.4	.69	.95
N	0	CN	2	.31	.92	45.0	.31	.91
N	0	CN	4	.16	.88	57.8	.16	.89
CN	0	N	1	.96	.98	00.0	.97	.98
CN	0	N	2	.63	.94	06.4	.62	.94
CN	0	N	4	.27	.96	36.2	.26	.96
CN	0	CN	1	.81	.96	01.0	.82	.96
CN	0	CN	2	.37	.84	19.4	.37	.84
CN	0	CN	4	.16	.80	47.6	.17	.80
N	4	N	1	1.15	1.19	00.0	1.17	1.21
N	4	N	2	.79	1.24	04.6	.79	1.25
N	4	N	4	.30	1.20	34.8	.30	1.21
N	4	CN	1	1.02	1.17	00.4	1.01	1.18
N	4	CN	2	.46	1.07	22.4	.45	1.06
N	4	CN	4	.19	1.00	49.4	.19	.99
CN	4	N	1	1.28	1.27	00.0	1.34	1.32
CN	4	N	2	1.57	1.92	00.2	.69	2.01
CN	4	N	4	.71	1.79	12.2	.74	1.85
CN	4	CN	1	1.48	1.51	09.0	1.52	1.55
CN	4	CN	2	1.14	1.76	03.2	1.16	1.82
CN	4	CN	4	.40	1.27	29.0	.40	1.29

Table 2

δ and V error distribution	σ_δ^2	U error distribution	σ_U^2	95th percentile of absolute errors-efficiencies-intercept LSE	Moments	95th percentile of absolute errors - efficienceies - slope LSE	Moments
N	0	N	1	1.00	.96	.99	.99
N	0	N	2	.81	1.01	.82	1.00
N	0	N	4	.68	1.08	.67	1.09
N	0	CN	1	.83	1.01	.79	.97
N	0	CN	2	.64	.98	.60	.94
N	0	CN	4	.53	.96	.52	.90
CN	0	N	1	.97	1.00	1.02	.96
CN	0	N	2	.87	.96	.86	.97
CN	0	N	4	.69	.97	.68	.97
CN	0	CN	1	.92	1.00	.92	1.02
CN	0	CN	2	.65	.94	.68	.93
CN	0	CN	4	.58	.92	.55	.89
N	4	N	1	1.08	1.15	1.09	1.11
N	4	N	2	.95	1.11	.97	1.14
N	4	N	4	.78	1.14	.79	1.17
N	4	CN	1	1.04	1.12	1.06	1.14
N	4	CN	2	.73	1.07	.76	1.06
N	4	CN	4	.58	1.00	.62	1.06
CN	4	N	1	1.09	1.09	1.17	1.14
CN	4	N	2	1.34	1.37	1.34	1.39
CN	4	N	4	1.06	1.32	1.09	1.34
CN	4	CN	1	1.20	1.19	1.25	1.20
CN	4	CN	2	1.13	1.34	1.13	1.33
CN	4	CN	4	.82	1.19	.83	1.12

equation error. Here the normality-MLE calculated assuming $\sigma_\delta^2 = 0$ does particularly poorly. Clearly, the normality-MLE is not robust against violations of the linear functional relationship (no equation error).

Note that the Fuller modification actually came into play a considerable percentage of the time. It turned out that the moments estimate calculated without correction was considerably less efficient than the modified version reported in Tables 1 and 2.

6. CONCLUSION

We have shown that in a particular pure functional errors-in-variables regression model, the normal-based maximum likelihood estimate will be preferable asymptotically to moments estimates. We also indicated that the moments estimators greatly benefitted from Fuller's finite sample correction. Finally, we have indicated the maximum likelihood estimate for the pure functional model is not robust against possible equation error.

SUMMARY

We study a particular functional errors-in-variables regression model. In the case of no equation error (all randomness due to measurement errors), we show that the maximum likelihood estimator computed assuming normality is asymptotically better than the usual moments estimator, even if the errors are not normally distributed.

REFERENCES

Anderson, T.W. (1951). Estimating Linear restrictions on regression coefficient for multivariate normal distributions. Ann. Math. Statist. 22, 327-351.

Blomquist, N. (1977). On the relation between change and initial value. J. Amer. Statist. Assoc. 72, 746-750.

Dorff, M. and Gurland, J. (1961). Estimation of the parameters in a linear functional relation. J. Roy. Statist. Soc. Ser. B. 23, 160-170.

Fuller, W.A. (1975). Regression analysis for sample survey. Sankhya, Series C. 37, 117-132.

Fuller, W.A. (1980). Properties of some estimators for the errors-in-variables model. Ann. Statist. 8, 407-422.

Gallo, P.P. (1982a). Properties of estimators in errors-in-variables regression models. Ph.D. dissertation, University of North Carolina at Chapel Hill.

Gallo, P.P. (1982b). Consistency of regression estimates when some variables are subject to error. Comm. in Statist., Series A. 11, 973-983.

Gleser, L.J. (1981). Estimation in multivariate "errors-in-variables" regression models: large sample results. Ann. Statist. 9, 24-44.

Graybill, F.A. (1969). Introduction to Matrices with Applications in Statistics. Wadsworth Publishing Co., Belmont, CA.

Healy, J.D. (1975). Estimation and tests for unknown linear restrictions in multivariate linear models. Ph.D. dissertation, Purdue University.

Healy, J.D. (1980). Maximum likelihood estimation of a multivariate linear functional relationship. J. Multivariate Anal. 10, 243-251.

Jobson, J.D. and Fuller, W.A. (1980). Least squares estimation when the covariance matrix and parameter vector are functionally related. _J. Amer. Statist. Assoc._ _75_, 176-181.

Nussbaum, M. (1978). Asymptotic efficiency of estimators of a multivariate linear functional relation. _Akademie der Wissenschaften der DDR_, Berlin.

Reilly, P.M. and Patino-Leal, H. (1981). A Bayesian study of the errors-in-variables model. _Technometrics_ 23, 221-231.

Wu, M., Ware, J.H. and Feinlieb, M. (19800. On the relation between blood pressure change and initial value. _Jour. of Chronic Diseases_ 33, 637-644.

ESTIMATING A CONSTANT OF PROPORTIONALITY FOR EXCHANGEABLE RANDOM VARIABLES*

Irwin Guttman

Department of Statistics
University of Toronto
Toronto, Canada

Ingram Olkin

Department of Statistics
Stanford University
Stanford, California

1. INTRODUCTION

This study is generated by an experiment in which use of, say, Method 1, generates $x = (x_1,\ldots,x_p)$, whereas use of another Method 2, generates $y = (y_1,\ldots,y_p)$ so that x and y each comprise p measurements. An example is furnished by a study of the effectiveness of drugs 1 and 2, with scores on p diagnostic tests when drug 1 and drug 2 are used.

The models under consideration in this paper assume the exchangeability and normality of x and y. The exchangeability condition is equivalent to

$$\mu_1^{(1)} = \cdots = \mu_p^{(1)} \equiv \mu_1^*, \quad \mu_1^{(2)} = \cdots = \mu_p^{(2)} \equiv \mu_2^*, \tag{1.1}$$

$$\Sigma_1 = \sigma_1^2 \begin{pmatrix} 1 & \rho_1 \cdots \rho_1 \\ & 1 \cdots \rho_1 \\ & \cdots \\ & \cdots \\ & 1 \end{pmatrix}, \qquad \Sigma_2 = \sigma_2^2 \begin{pmatrix} 1 & \rho_2 \cdots \rho_2 \\ & 1 \cdots \rho_2 \\ & \cdots \\ & \cdots \\ & 1 \end{pmatrix}.$$

Of concern is the hypothesis that x and y differ by a constant of proportionality, that is

*This work was supported in part by the National Science Foundation, and by NSERC (Canada).

$$H_1: \mu_2^* = \gamma\mu_1^*, \quad \sigma_2^2 = \sigma_1^2, \quad \rho_2 = \rho_1, \tag{1.2}$$

or

$$H_2: \mu_2^* = \gamma\mu_1^*, \quad \sigma_2^2 = \gamma^2\sigma_1^2, \quad \rho_2 = \rho_1. \tag{1.3}$$

For each of the models H_1 and H_2 we obtain the maximum likelihood estimator (MLE) and Bayesian estimator of the constant of proportionality γ, or equivalently, of $\delta \equiv 1/\gamma$.

These are prototype models for determining whether two drugs are "equivalent" when the tests given to the patient are homogeneous. It also arises in determining whether one training program is more difficult than another, where achievement tests given to each student are homogeneous.

As a consequence of determining the MLE under (1.2) or (1.3) it is straightforward to also obtain the likelihood ratio test for testing one model against another.

Suppose that we have samples of size N_1 and N_2 on x and y, respectively. Then the sufficient statistics under the hypothesis of exchangeability are

$$\begin{aligned} w_1 &= \bar{x}e'/\sqrt{p} \sim \mathcal{N}(\theta_1, \alpha_1/N_1), \\ w_2 &= \bar{y}e'/\sqrt{p} \sim \mathcal{N}(\theta_2, \alpha_2/N_2), \\ u_1/\alpha_1 &\sim \chi^2_{n_1}, \quad u_2/\alpha_2 \sim \chi^2_{n_2}, \\ v_1/\beta_1 &\sim \chi^2_{(p-1)n_1}, \quad v_2/\beta_2 \sim \chi^2_{(p-1)n_2}, \end{aligned} \tag{1.4}$$

where $\bar{x} = (\bar{x}_1, \ldots, \bar{x}_p)$ and $\bar{y} = (\bar{y}_1, \ldots, \bar{y}_p)$ are the vectors of sample means, $e = (1, \ldots, 1)$, $n_1 = N_1 - 1$, $n_2 = N_2 - 1$, $u_i = eS_ie'/p$, $v_i =$ tr $S_i - eS_ie'/p$, $i = 1,2$, and

$$S_1 = \sum_{i=1}^{N_1} (x_i - \bar{x})'(x_i - \bar{x}), \quad S_2 = \sum_{i=1}^{N_2} (y_i - \bar{y})'(y_i - \bar{y}). \tag{1.5}$$

The parameters in the model are

$$\begin{aligned} \theta_1 &= \mu^*\sqrt{p}, & \theta_2 &= \nu^*\sqrt{p}, \\ \alpha_1 &= \sigma_1^2[1 + (p-1)\rho_1], & \alpha_2 &= \sigma_2^2[1 + (p-1)\rho_2], \\ \beta_1 &= \sigma_1^2(1-\rho_1), & \beta_2 &= \sigma_2^2(1-\rho_2). \end{aligned} \tag{1.6}$$

In (1.4), we have used the notation $\mathcal{N}(\theta_1, \alpha_1/N_1)$, for example, to mean a random variable whose distribution is normal with expectation θ_1 and variance α_1/N_1. We note that because of the assumptions, the six statistics w_i, u_i, v_i, $i = 1,2$ are all independent random variables.

For references to other multivariate models in which the equivalence of drugs (treatments) is considered see Guttman, Kim and Olkin (1983).

2. MAXIMUM LIKELIHOOD ESTIMATES OF THE PARAMETERS

The results for the model under H_1 reduce to a standard two-sample t-test. From (1.4) we have, under H_1, that

$$w_1 \sim N(\theta_1, \alpha/N_1), \quad w_2 \sim N(\gamma\theta_1, \alpha/N_2), \tag{2.1}$$
$$(u_1+u_2)/\alpha \sim \chi^2_n, \quad (v_1+v_2)/\beta \sim \chi^2_{(p-1)n},$$

where $n = n_1+n_2$ and $N = N_1+N_2$. By relabeling $\gamma\theta_1 = \theta_2$, the model is familiar and well-documented.

For model H_2 we have

$$\begin{aligned} &w_1 \sim N(\theta_1, \alpha_1/N_1), && w_2 \sim N(\gamma\theta_1, \gamma^2\alpha_1/N_2), \\ &u_1/\alpha_1 \sim \chi^2_{n_1}, && u_2/\gamma^2\alpha_1 \sim \chi^2_{n_2}, \\ &v_1/\beta_1 \sim \chi^2_{(p-1)n_1}, && v_2/\gamma^2\beta_1 \sim \chi^2_{(p-1)n_2}, \end{aligned} \tag{2.2}$$

so that the loglikelihood L, ignoring constants is

$$\begin{aligned} &-\frac{N_1(w_1-\theta_1)^2}{2\alpha_1} - \frac{N_2(w_2\delta-\theta_1)^2}{2\alpha_1} - \frac{u_1}{2\alpha_1} - \frac{\delta^2 u_2}{2\alpha_1} - \frac{v_1}{2\beta_1} - \frac{\delta^2 v_2}{2\beta_1} \\ &\quad - \frac{N}{2}\log\alpha_1 - \frac{(p-1)n}{2}\log\beta_1 + (pn_2+1)\log\delta. \end{aligned} \tag{2.3}$$

From (2.3) we readily obtain that

$$\begin{aligned} \hat{\theta}_1 &= (N_1 w_1 + N_2\delta w_2)/N, \\ \hat{\alpha}_1 &= \left[\frac{N_1 N_2}{N}(w_1-\delta w_2)^2 + u_1 + \delta^2 u_2\right]/N, \\ \hat{\beta}_1 &= (v_1 + \delta^2 v_2)/(p-1)n. \end{aligned} \tag{2.4}$$

Inserting (2.4) in (2.3) and taking the derivative of L with respect to δ yields a fourth degree equation, so that we have to obtain $\hat{\delta}$ numerically. However, we can obtain the asymptotic distribution of $\hat{\delta}$.

Theorem. If N_1 and N_2 grow at the same rate so that $t = N_1/N$ and $\bar{t} = N_2/N$ are fixed, then

$$\sqrt{N}(\hat{\delta}-\delta) \sim N(0,\sigma_\infty^2),$$

where

$$\sigma_\infty^2 = \delta^2/t\bar{t}[2p+\theta^2/\alpha].$$

Proof. The proof is based on employing the delta method for which we require the matrix of partial derivatives of the second degree. From (2.3) we obtain (dropping subscripts on θ_1, α_1, β_1)

$$E\left(-\frac{\partial^2 L}{\partial\delta^2}\right) = \gamma^2 N_2\left[\frac{\theta^2}{\alpha} + 2\left(\frac{pn_2+1}{N_2}\right)\right], \quad E\left(-\frac{\partial^2 L}{\partial\delta\,\partial\theta}\right) = \frac{-N_2\gamma\theta}{\alpha},$$

$$E\left(-\frac{\partial^2 L}{\partial\delta\,\partial\alpha}\right) = \frac{-\gamma N_2}{\alpha}, \quad E\left(-\frac{\partial^2 L}{\partial\delta\,\partial\beta}\right) = \frac{-\gamma(p-1)n_2}{\beta},$$

$$E\left(-\frac{\partial^2 L}{\partial\theta^2}\right) = \frac{N}{\alpha}, \quad E\left(-\frac{\partial^2 L}{\partial\theta\,\partial\alpha}\right) = 0 = E\left(-\frac{\partial^2 L}{\partial\theta\,\partial\beta}\right),$$

$$E\left(-\frac{\partial^2 L}{\partial\alpha^2}\right) = \frac{N}{2\alpha^2}, \quad E\left(-\frac{\partial^2 L}{\partial\alpha\,\partial\beta}\right) = 0, \quad E\left(-\frac{\partial^2 L}{\partial\beta^2}\right) = \frac{(p-1)n}{2\beta^2}.$$

We use the above to construct the information matrix and we obtain the asymptotic variance of $\sqrt{N}\,\delta$ as the limit of the reciprocal of

$$\frac{1}{N}\left\{\gamma^2 N_2\left[\frac{\theta^2}{\alpha} + 2\left(\frac{pn_2+1}{N_2}\right)\right] - \frac{(N_2\gamma\theta)^2}{\alpha^2}\frac{\alpha}{N} - \left(\frac{\gamma N_2}{\alpha}\right)^2\frac{2\alpha^2}{N} - \frac{\gamma^2(p-1)^2 n_2^2}{\beta^2}\,\frac{(2\beta^2)}{(p-1)n}\right\}, \tag{2.5}$$

where the limit is taken as $N \to \infty$ with N_1/N and N_2/N tending to t and $\bar{t}$ respectively. After taking the limit and simplifying (2.5) tends to

$$\gamma^2 t\bar{t}[\theta^2/\alpha + 2p] = t\bar{t}[\theta^2/\alpha + 2p]/\delta^2, \tag{2.6}$$

which on taking the reciprocal completes the proof.

3. A BAYESIAN APPROACH

We first briefly note that under H_1 the variances of $\sqrt{N_1}\,w_1$ and $\sqrt{N_2}\,w_2$ are the same, namely α, and that information about α exists using the fact that $(u_1+u_2)/\alpha \sim \chi_n^2$, so that indeed, after relabeling $\gamma\theta$, by θ_2 (see (2.1)) we are in a well-known and fully documented situation. Hence we turn to the model specified by H_2 stemming from (1.4). As indicated by (2.3), the likelihood ℓ is proportional to

$$\ell(\delta,\alpha,\beta,\theta|\text{data}) \propto \delta^{pn_2+1}\alpha^{-N/2}\beta^{-(p-1)n/2}\exp -\frac{1}{2}Q, \tag{3.1}$$

where, after simplification,

$$Q = [N(\theta-\tilde{\theta})^2 + \frac{N_1N_2}{N}(w_1-\delta w_2)^2 + u_1 + \delta^2 u_2]/\alpha + (v_1+\delta^2 v_2)/\beta \tag{3.2}$$

with

$$\tilde{\theta} = (N_1w_1 + N_2\delta w_2)/N. \tag{3.3}$$

Assuming a vague prior about δ,α,β and θ:

$$p(\delta,\alpha,\beta,\theta) \propto (\delta\alpha\beta)^{-1}, \tag{3.4}$$

and combining (3.1) with (3.4) we obtain the posterior

$$p(\delta,\alpha,\beta,\theta|\text{data}) \propto \delta^{pn_2}\alpha^{-(N+2)/2}\beta^{-((p-1)n+2)/2}\exp -\frac{1}{2}Q. \tag{3.5}$$

We may use (3.5) immediately to find $(\delta_m,\alpha_m,\beta_m,\theta_m)$, the joint posterior modes of the posterior (3.5). Apart from a constant,

$$\log p(\delta,\alpha,\beta,\theta|\text{data}) = pn_2\log\delta - \frac{1}{2}(N+2)\log\alpha - \frac{1}{2}((p-1)n+2)\log\beta - \frac{1}{2}Q. \tag{3.6}$$

Inspection of (3.6)--see also (3.2)--yields the maximization over θ at $\theta = \theta_m = \tilde{\theta} = \hat{\theta}$ (see (2.4)). Further, differentiation of $\log p(\delta,\alpha,\beta,\theta|$ data) with respect to β yields

$$\beta_m = (v_1+\delta^2v_2)/((p-1)n+2) = \hat{\beta}(p-1)n/[(p-1)n+2], \tag{3.7}$$

where $\hat{\beta}$ is given in (2.4).

Similarly, we find easily that

$$\alpha_m = [\frac{N_1N_2}{N}(w_1-\delta w_2)^2 + u_1 + \delta^2u_2]/(N+2) = \hat{\alpha}N/(N+2). \tag{3.8}$$

To find δ_m, differentiate $\log p(\delta,\alpha_m,\beta_m,\theta_m|\text{data})$, which, except for constants, is

$$\log p(\delta,\alpha_m,\beta_m,\theta_m|\text{data}) = pn_2\log\delta - \frac{1}{2}(N+2)\log[\frac{N_1N_2}{N}(w_1-\delta w_2)^2 + u_1 + \delta^2u_2]$$
$$- \frac{1}{2}((p-1)n+2)\log(v_1+\delta^2v_2). \tag{3.9}$$

(C.f. (3.9) with the expression we would obtain if (2.4) were inserted in (2.3).) Differentiation of (3.9) is easily seen to lead to a fourth degree equation in δ, similar (except for minor changes in the coefficients) to the equation alluded to in Section 2 (see after (2.4)).

Of course, we may use (3.5) to find the marginal posterior of δ, and integration of (3.5) with respect to α,β,θ yields

$$p(\delta|\text{data}) \propto \delta^{pn_2}\left[\frac{N_1N_2}{N}(w_1-\delta w_2)^2 + u_1 + \delta^2 u_2\right]^{-(N-1)/2} \times [v_1+\delta^2 v_2]^{-(p-1)n/2}. \tag{3.10}$$

Numerical integration techniques can be used to find the normalizing constant, the posterior mean, variance and confidence interval of Box and Tiao (1973). In tabulating $p(\delta|\text{data})$, necessary for these operations, we may graph (3.10) and determine the mode. Analytically, apart from constants,

$$\log p(\delta|\text{data}) = pn_2 \log \delta - \frac{1}{2}(p-1)n \log(v_1+\delta^2 v_2) - \frac{1}{2}(N-1)\log\left[\frac{N_1N_2}{N}(w_1-\delta w_2)^2 + u_1 + \delta^2 u_2\right]. \tag{3.11}$$

Differentiation of (3.11), with respect to δ, leads to a fourth degree equation, but comparing this to (3.9) we see that the marginal mode obtained from (3.11) does not equal δ_m. However, as N gets large, the marginal mode will be close to δ_m. The marginal mode does not equal $\hat{\delta}$ of Section 2, primarily due to the difference in the coefficients pn_2 appearing in (3.11) and pn_2+1 appearing in (2.3), but again the results will be equivalent asymptotically.

SUMMARY

The effectiveness of a drug is made on the basis of p diagnostic tests assumed to be homogeneous, i.e. the means are equal, the variances are equal, and the correlations are equal. To compare two drugs A and B that may differ in potency we obtain maximum likelihood and Bayesian estimates of the potency factor, γ, that relates the two means by $\mu_B = \gamma\mu_A$ and the two variances by $\sigma_B^2 = \gamma^2\sigma_A^2$. The correlations ρ_A and ρ_B are assumed equal.

REFERENCES

Box, G.E.P. and Tiao, G.C. (1973). Bayesian inference in statistical analysis. Addison-Wesley Pub. Co., Reading, Mass.

Guttman, I., Kim, D.Y., and Olkin, I. (1983). Statistical inference for constants of proportionality, Technical Report No. 195, Stanford University, Stanford, California.

AN ESTIMATION PROBLEM RELATING TO SUBSET SELECTION FOR NORMAL POPULATIONS

S. Jeyaratnam

Department of Mathematics
Southern Illinois University
Carbondale, Illinois

S. Panchapakesan[†]

Department of Mathematics
Southern Illinois University
Carbondale, Illinois

1. INTRODUCTION

Let $\pi_1, \pi_2, \ldots, \pi_k$ be k independent normal populations with unknown means $\theta_1, \theta_2, \ldots, \theta_k$, respectively, and a common known variance τ^2. The population associated with the largest θ_i is called the <u>best</u> population. In the subset selection approach, we want to select a nonempty subset of the k populations so that it includes the best population with a minimum guaranteed probability $P^*(1/k < P^* < 1)$. The basic idea of the subset selection approach is that the number of populations to be selected should depend upon the evidence supplied by the data. The size of the selected subset depends on the sample size and the confidence level P^* associated with the claim that a <u>correct selection</u> (i.e. selection of any subset that contains the best) occurs. It can be said that subject to the P^*-value and the sample size, we cannot make finer distinction among the populations that are selected in seeking the best. In this case, one may decide to use the selected populations in equal proportions in the future. In this sense, the average worth of the selected subset is given by

$$M = \sum_{i \in S} \theta_i \Big/ \sum_{i \in S} I_i \tag{1.1}$$

[†] The research of this author was supported by the Office of Naval Research contract N00014-75-C-0455 at Purdue University and reproduction for any purpose of the United States Government is permitted.

where S denotes the set of indices of the selected populations, and $I_i = 1$ or 0 according as π_i is or is not included in the selected subset. Our interest is to estimate M, which we call the <u>mean of the selected subset</u>. It is important to note that M is a random variable.

In this paper, we consider the subset selection rule of Gupta (1956; 1965), which has known optimality properties; see Gupta and Panchapakesan (1979), and Berger and Gupta (1980). Let $Y_1, Y_2, \ldots, Y_k$ be the sample means based on n independent observations from each population. The rule R of Gupta (1965) is:

$$\text{"Select } \pi_i \text{ if and only if } Y_i \geq \max_{1 \leq j \leq k} Y_j - d\tau/\sqrt{n}\text{ "}$$

where $d > 0$ is to be determined such that the minimum probability of a correct selection (PCS) is P*. This value of d is shown to be given by

$$\int_{-\infty}^{\infty} \Phi^{k-1}(t+d)\, \phi(t)\, dt = P^* , \tag{1.2}$$

where Φ and ϕ denote the standard normal cdf and density function respectively.

Our present investigations relate to only the case of $k = 2$. The following notations hold for the entire paper:

$$\left.\begin{array}{llll} \sigma^2 = 2\tau^2/n & ; c = d/\sqrt{2} & ; \delta = (\theta_1-\theta_2)/\sigma ; \\ \theta = \theta_1-\theta_2 & ; \theta^* = \theta_1+\theta_2 & ; \\ Y = Y_1-Y_2 & ; Y^* = Y_1+Y_2 & ; \\ I_1 = I_{\{Y>c\sigma\}} & ; I_2 = I_{\{Y < -c\sigma\}} & ; \\ \text{where } I_A \text{ denotes the indicator function of the set A.} \end{array}\right\} \tag{1.3}$$

Now, for $k = 2$, we get

$$\begin{aligned} c &= \Phi^{-1}(P^*) , \\ M &= \theta_1 I_1 + \theta_2 I_2 + \frac{\theta_1+\theta_2}{2}(1-I_1-I_2) . \end{aligned} \tag{1.4}$$

When $c = 0$, the rule R selects the population that yields the largest sample mean and M is the mean of the selected population. Of course, in this case, the minimum PCS cannot be guaranteed for $P^* > 1/k$ unless additional modifications are made in the formulation of the selected problem. This is the aspect not considered by those who discussed the estimation of M in this case; these are Sarkadi (1967) and Dahiya (1974) for $k = 2$ and known τ; Hsieh (1981) for $k = 2$ and unknown τ; and Cohen and Sackrowitz (1982) for $k \geq 2$ and known τ.

For any estimator $\hat{M}$ of M, the bias $B(\hat{M}) = E(\hat{M}-M)$ and the mean squared error $MSE(\hat{M}) = E(\hat{M}-M)^2$. It can be shown (Theorem 2.1) that no unbiased estimator of M, having a finite MSE, exists. In Section 2, we define the 'natural' estimator T and three classes of estimators $T_{1\lambda}$, $T_{2\lambda}$ and $T_{3\lambda}$, obtained by making adjustments for the bias of T. The biases and the MSEs of $T_{1\lambda}$, $T_{2\lambda}$ and $T_{3\lambda}$ are discussed in Sections 2 and 3 respectively. Numerical comparisons of the performances of these estimators are made in Section 4.

2. THE ESTIMATORS AND THEIR BIASES

Since $Y \sim N(\theta,\sigma^2)$, it is easy to see that

$$\begin{aligned} E(M) &= \theta_1\{1-\Phi(c-\delta)\} + \theta_2\{1-\Phi(c+\delta)\} \\ &\qquad + \frac{\theta_1+\theta_2}{2}\{\Phi(c+\delta) + \Phi(c-\delta) - 1\} \\ &= \frac{1}{2}[\theta^* + \delta\sigma\{\Phi(c+\delta) - \Phi(c-\delta)\}] . \end{aligned} \tag{2.1}$$

Theorem 2.1. No unbiased estimator of M, with finite MSE, exists.

Proof. See Appendix A1. □

Let us now consider the 'natural' estimator

$$T = \begin{cases} \max(Y_1,Y_2) & \text{if } |Y| > c\sigma, \\ \dfrac{Y^*}{2} & \text{otherwise.} \end{cases} \tag{2.2}$$

Since $\max(Y_1,Y_2) = \frac{1}{2}\{Y^* + |Y|\}$, we obtain

$$E(T) = \frac{\theta^*}{2} + \frac{1}{2}\int_{-\infty}^{-c\sigma} |y| \frac{1}{\sigma}\phi\left(\frac{y-\theta}{\sigma}\right) dy + \frac{1}{2}\int_{c\sigma}^{\infty} |y| \frac{1}{\sigma}\phi\left(\frac{y-\theta}{\sigma}\right) dy .$$

By changing the variable of integration by setting $t = (y-\theta)/\sigma$ and using (A2.1) in Appendix 2, E(T) simplifies to

$$E(T) = \frac{\theta^*}{2} + \frac{\delta\sigma}{2}\{\Phi(c+\delta) - \Phi(c-\delta)\} + \frac{\sigma}{2}\{\phi(c+\delta) + \phi(c-\delta)\}$$

$$= E(M) + \frac{\sigma}{2}\{\phi(c+\delta) + \phi(c-\delta)\} . \tag{2.3}$$

Thus $B(T) = \frac{\sigma}{2}\{\phi(c+\delta) + \phi(c-\delta)\}$.

2.1 Estimators $T_{1\lambda}$

Since the bias of T is positive, we define $T_{1\lambda}$ by

$$T_{1\lambda} = T - \frac{\lambda\sigma}{2}\{\phi(c + \frac{Y}{\sigma}) + \phi(c - \frac{Y}{\sigma})\}, \quad \lambda \geq 0 . \tag{2.4}$$

The bias of $T_{1\lambda}$ is

$$B(T_{1\lambda}) = E[T - M - \frac{\lambda\sigma}{2}\{\phi(c + \frac{Y}{\sigma}) + \phi(c - \frac{Y}{\sigma})\}]$$

$$= \frac{\sigma}{2}\{\phi(c+\delta) + \phi(c-\delta)\} - \frac{\lambda\sigma}{2\sqrt{2}}\{\phi(\frac{c+\delta}{\sqrt{2}}) + \phi(\frac{c-\delta}{\sqrt{2}}) , \tag{2.5}$$

by using (2.3) and (A2.7). It should be noted that $T_{1\lambda}$ becomes T when $\lambda = 0$, and that it reduces to $\hat{M}_\lambda$ of Dahiya (1974) when c = 0.

2.2 Estimators $T_{2\lambda}$

To motivate the definition of $T_{2\lambda}$, $\lambda \geq 0$, consider the following estimator

$$U = \frac{Y^*}{2} + \frac{Y}{2}[\Phi\{\lambda(c + \frac{Y}{\sigma})\} - \Phi\{\lambda(c - \frac{Y}{\sigma})\}]. \tag{2.6}$$

For $\lambda = 1$, U is the maximum likelihood estimator (MLE) of E(M).

$$E(U) = \frac{\theta^*}{2} - \frac{\theta}{2} + \frac{1}{2}E[Y\Phi\{\lambda(c + \frac{Y}{\sigma})\}] + \frac{1}{2}E[Y\Phi\{\lambda(\frac{Y}{\sigma} - c)\}]$$

and using (A2.8) this is simplified to

$$E(U) = \frac{\theta^*}{2} + \frac{\theta}{2}\left[\Phi\left\{\frac{\lambda(\delta+c)}{\sqrt{1+\lambda^2}}\right\} - \Phi\left\{\frac{\lambda(c-\delta)}{\sqrt{1+\lambda^2}}\right\}\right]$$

$$+ \frac{\sigma\lambda}{2\sqrt{1+\lambda^2}}\left[\phi\left\{\frac{\lambda(\delta+c)}{\sqrt{1+\lambda^2}}\right\} + \phi\left\{\frac{\lambda(\delta-c)}{\sqrt{1+\lambda^2}}\right\}\right]. \tag{2.7}$$

We note that the sum of the first two terms in (2.7) tends to E(M) as $\lambda \to \infty$. Also, it is seen clearly from (A2.7) that the last term in (2.7) is unbiasedly estimated by $\frac{\sigma\lambda}{2}[\phi\{\lambda(\frac{Y}{\sigma}+c)\}+\phi\{\lambda(\frac{Y}{\sigma}-c)\}]$. By subtracting this unbiased estimator from U, we define

$$T_{2\lambda} = \frac{Y^*}{2} + \frac{Y}{2}[\Phi\{\lambda(c+\frac{Y}{\sigma})\} - \Phi\{\lambda(c-\frac{Y}{\sigma})\}] - \frac{\sigma\lambda}{2}[\phi\{\lambda(c+\frac{Y}{\sigma})\} + \phi\{\lambda(c-\frac{Y}{\sigma})\}]. \quad (2.8)$$

The bias of $T_{2\lambda}$ is

$$B(T_{2\lambda}) = \frac{\theta}{2}\left[\Phi\left\{\frac{\lambda(\delta+c)}{\sqrt{1+\lambda^2}}\right\} + \Phi\left\{\frac{\lambda(\delta-c)}{\sqrt{1+\lambda^2}}\right\} - \Phi(\delta+c) - \Phi(\delta-c)\right], \quad (2.9)$$

which tends to zero as λ tends to infinity. Finally, we note that $T_{2\lambda}$ corresponds to t_c of Dahiya (1974) when c = 0; Dahiya's c corresponds to our λ.

2.3 Estimators $T_{3\lambda}$

Let us first consider T_3, the MLE of E(M); this is the same as U in (2.6) with $\lambda = 1$. Thus, from (2.7) with $\lambda = 1$,

$$E(T_3) = \frac{\theta^*}{2} + \frac{\theta}{2}\left[\Phi(\frac{\delta+c}{\sqrt{2}}) - \Phi(\frac{c-\delta}{\sqrt{2}})\right] + \frac{\sigma}{2\sqrt{2}}\left[\phi(\frac{\delta+c}{\sqrt{2}}) + \phi(\frac{\delta-c}{\sqrt{2}})\right]. \quad (2.10)$$

Hence the bias of T_3 is

$$B(T_3) = \frac{\theta}{2}\left[\Phi(\frac{\delta+c}{\sqrt{2}}) + \Phi(\frac{\delta-c}{\sqrt{2}}) - \Phi(\delta+c) - \Phi(\delta-c)\right] + \frac{\sigma}{2\sqrt{2}}\left[\phi(\frac{\delta+c}{\sqrt{2}}) + \phi(\frac{\delta-c}{\sqrt{2}})\right]. \quad (2.11)$$

Noting that the last term in (2.11) is unbiasedly estimated by $\frac{\sigma}{2}[\phi(\frac{Y}{\sigma}+c)+\phi(\frac{Y}{\sigma}-c)]$, we define

$$T_{3\lambda} = T_3 - \lambda\left[\frac{Y}{2}\left\{\Phi(\frac{c}{\sqrt{2}}+\frac{Y}{\sqrt{2}\sigma}) + \Phi(\frac{Y}{\sqrt{2}\sigma}-\frac{c}{\sqrt{2}}) - \Phi(\frac{Y}{\sigma}+c) - \Phi(\frac{Y}{\sigma}-c)\right\} + \frac{\sigma}{2}\left\{\phi(\frac{Y}{\sigma}+c) + \phi(\frac{Y}{\sigma}-c)\right\}\right]. \quad (2.12)$$

The bias of $T_{3\lambda}$ is

$$B(T_{3\lambda}) = B(T_3) - \frac{\lambda\sigma}{2\sqrt{2}}\left[\phi(\frac{\delta+c}{\sqrt{2}}) + \phi(\frac{\delta-c}{\sqrt{2}})\right]$$
$$- \frac{\lambda}{2} E\left[Y\left\{\Phi(\frac{c}{\sqrt{2}} + \frac{Y}{\sqrt{2}\sigma}) + \Phi(\frac{Y}{\sqrt{2}\sigma} - \frac{c}{\sqrt{2}}) - \Phi(\frac{Y}{\sigma} + c) - \Phi(\frac{Y}{\sigma} - c)\right\}\right]. \quad (2.13)$$

Now, using (A2.8) to evaluate the expectation in (2.13) and carrying out routine manipulations, we obtain

$$B(T_{3\lambda}) = \frac{\theta}{2}\left[(1+\lambda)\left\{\Phi(\frac{\delta+c}{\sqrt{2}}) + \Phi(\frac{\delta-c}{\sqrt{2}})\right\} - \lambda\left\{\Phi(\frac{\delta+c}{\sqrt{3}}) + \Phi(\frac{\delta-c}{\sqrt{3}})\right\}\right.$$
$$\left. - \{\Phi(\delta+c) + \Phi(\delta-c)\}\right]$$
$$+ \frac{\sigma}{2\sqrt{2}}\left[\left\{\phi(\frac{\delta+c}{\sqrt{2}}) + \phi(\frac{\delta-c}{\sqrt{2}})\right\} - \lambda\sqrt{2/3}\left\{\phi(\frac{\delta+c}{\sqrt{3}}) + \phi(\frac{\delta-c}{\sqrt{3}})\right\}\right]. \quad (2.14)$$

It should be noted that $T_{3\lambda}$ is slightly different from the estimator T_λ of Dahiya (1974) when c = 0 because he does not estimate part of $B(T_3)$ unbiasedly as we do in (2.12).

3. MEAN SQUARED ERRORS

We give here the expressions for the MSEs of T and $T_{1\lambda}$. For $T_{2\lambda}$ and $T_{3\lambda}$ the derivations become more tedious. For the numerical comparisons of the MSEs we use numerical integration as will be explained later in this section.

3.1 Mean Squared Error of T

We note that M and T can be written as

$$\left.\begin{aligned} M &= \frac{1}{2}\{\theta^* + \theta(I_1-I_2)\}, \\ T &= \frac{1}{2}\{Y^* + Y(I_1-I_2)\}. \end{aligned}\right\} \quad (3.1)$$

Now, $T-M = \frac{1}{2}\{(Y^*-\theta^*) + (Y-\theta)(I_1-I_2)\}$. Since Y and Y* are independent and

$E(Y^*-\theta^*) = 0$, it is easy to see by direct evaluation that

$$\text{MSE}(T) = \frac{\sigma^2}{4} + \frac{\sigma^2}{4} \int_{c-\delta}^{\infty} t^2\phi(t)dt + \frac{\sigma^2}{4} \int_{-\infty}^{-c-\delta} t^2\phi(t)dt$$

$$= \frac{\sigma^2}{4} \{2 + (c-\delta)\phi(c-\delta) + (c+\delta)\phi(c+\delta) + \Phi(\delta-c) - \Phi(\delta+c)\}, \tag{3.2}$$

using (A2.3).

3.2 Mean Squared Error of $T_{1\lambda}$

Letting $V = \phi(c + \frac{Y}{\sigma}) + \phi(c - \frac{Y}{\sigma})$, we have

$$\text{MSE}(T_{1\lambda}) = E[(T-M) - \frac{\lambda\sigma}{2} V]^2$$

$$= \text{MSE}(T) + (\lambda^2\sigma^2/4)E(V^2) - \lambda\sigma E[(T-M)V]. \tag{3.3}$$

By repeated applications of (A2.6), $E(V^2)$ can be evaluated in a straightforward manner to yield

$$E(V^2) = \frac{1}{\sqrt{6\pi}} \left[\phi\left\{\sqrt{2/3}\,(\delta+c)\right\} + \phi\left\{\sqrt{2/3}\,(\delta-c)\right\} + 2\phi\left\{\sqrt{\frac{2(\delta^2+3c^2)}{3}}\right\}\right]. \tag{3.4}$$

Again, noting that V (which is a function of Y) and Y^* are independent, it is easy to see that

$$E[(T-M)V] = \frac{1}{2} E[(Y-\theta)I_1V] - \frac{1}{2} E[(Y-\theta)I_2V]. \tag{3.5}$$

The right-hand side of (3.5) can be written as a sum of four integrals each of which is either of the form in (A2.4) or in (A2.5). Thus we get $4\,E[(T-M)V] = B_1 + B_2 + B_3 + B_4$, where

$$\left.\begin{aligned}
B_1 &= \sigma\phi(\tfrac{c+\delta}{\sqrt{2}})[\phi(\tfrac{3c-\delta}{\sqrt{2}}) + \tfrac{c+\delta}{\sqrt{2}}\Phi(\tfrac{3c-\delta}{\sqrt{2}}) - \tfrac{c+\delta}{\sqrt{2}}],\\
B_2 &= \sigma\phi(\tfrac{\delta-c}{\sqrt{2}})[\phi(\tfrac{\delta-c}{\sqrt{2}}) - \tfrac{\delta-c}{\sqrt{2}}\Phi(\tfrac{\delta-c}{\sqrt{2}})],\\
B_3 &= \sigma\phi(\tfrac{c+\delta}{\sqrt{2}})[\phi(\tfrac{c+\delta}{\sqrt{2}}) + \tfrac{c+\delta}{\sqrt{2}}\Phi(\tfrac{-c-\delta}{\sqrt{2}})],\\
B_4 &= \sigma\phi(\tfrac{\delta-c}{\sqrt{2}})[\phi(\tfrac{3c+\delta}{\sqrt{2}}) + \tfrac{\delta-c}{\sqrt{2}}\Phi(\tfrac{-3c-\delta}{\sqrt{2}})].
\end{aligned}\right\} \quad (3.6)$$

Now we can obtain MSE $(T_{1\lambda})$ by using (3.2), (3.4), and (3.6) in (3.3).

Remark 3.1. For i = 1,2,3, $T_{i\lambda}$ - M is of the form $\frac{1}{2}$ $(Y^*-\theta^*) + g(Y)$, where g is some known function. Hence $MSE(T_{i\lambda}) = \sigma^2/4 + E[g^2(Y)]$; the expectation of the product term is zero. After suitable change of variable, $E[g^2(Y)]$ can be numerically evaluated using the Gauss-Hermite quadrature formula. Our results in the next section were obtained by using the 20-point formula (see Abramowitz and Stegun, 1965, p. 924). It should be noted that $MSE(T_{1\lambda})$ is also computed using the same quadrature formula so that the comparisons are made on the same footing.

4. COMPARISON OF THE ESTIMATORS

In this section we make some comparisons of the performances of $T_{1\lambda}$, $T_{2\lambda}$ and $T_{3\lambda}$ based on the values of their biases and the MSEs computed in units of σ and σ^2, respectively, for δ = 0 (0.5) 4 (1) 7; c = 1.645, 2.576; $\lambda = 0, \sqrt{2}\exp\{-c^2/4\}, 1, \sqrt{2}$. These biases and MSEs are given in Tables 1A, 1B and Tables 2A, 2B, respectively. For convenience, $\sqrt{2}\exp\{-c^2/4\}$ is denoted by λ_c in the tables.

Remark 4.1. The choices for values of λ are based on the following considerations. For $\lambda = 0$, $T_{1\lambda}$ becomes T. The value $\lambda = 1$ corresponds to using MLEs in constructing the estimators. Further, $B(T_{1\lambda})$ decreases in λ and if we let λ_δ to be the value for which $B(T_{1\lambda}) = 0$, $\lambda_\delta \leq \sqrt{2}$. Also, λ_δ tends to 0 or $\sqrt{2}\exp\{-c^2/4\}$ according as δ tends to infinity or zero.

Remark 4.2. The c-values chosen here correspond to P* = 0.90, 0.99, usual values of interest in selection problems. The value of c = 0 (P* = 0.50) is not of interest in our selection problems. This is the case considered by Dahiya (1974). As pointed out earlier, $T_{1\lambda}$ and $T_{2\lambda}$ in this case coin-

Table 1A

Biases of $T_{1\lambda}$ (top entry), $T_{2\lambda}$ (middle entry), and $T_{3\lambda}$ (bottom entry) Expressed in Units of σ. P*=0.90; c=1.645; λ_c=0.7190.

δ \ λ	0	λ_c	1	$\sqrt{2}$
0.0	.1031 0 .1434	-.0000 0 .0379	-.0403 0 -.0033	-.0997 0 -.0641
0.5	.1236 -.0275 .1549	.0184 .0091 .0495	-.0227 .0086 .0083	-.0833 .0062 -.0524
1.0	.1680 -.1277 .1707	.0590 .0183 .0683	.0164 .0190 .0284	-.0464 .0142 -.0306
1.5	.1988 -.3311 .1556	.0894 -.0063 .0648	.0466 .0034 .0293	-.0164 .0047 -.0230
2.0	.1876 -.6386 .0973	.0856 -.0732 .0292	.0458 -.0444 .0026	-.0129 -.0260 -.0367
2.5	.1384 -1.0046 .0218	.0526 -.1504 -.0161	.0190 -.0977 -.0309	-.0304 -.0583 -.0527
3.0	.0797 -1.3684 -.0329	.0151 -.1952 -.0410	-.0101 -.1227 -.0441	-.0473 -.0700 -.0488
3.5	.0357 -1.6944 -.0507	-.0073 -.1907 -.0371	-.0242 -.1105 -.0319	-.0490 -.0580 -.0241
4.0	.0125 -1.9815 -.0421	-.0129 -.1517 -.0185	-.0228 -.0774 -.0092	-.0375 -.0360 .0044
5.0	.0007 -2.4990 -.0126	-.0054 -.0619 .0062	-.0077 -.0211 .0136	-.0112 -.0067 .0245
6.0	.0000 -3.0000 -.0019	-.0009 -.0165 .0053	-.0012 -.0031 .0080	-.0017 -.0005 .0121
7.0	.0000 -3.5000 -.0002	-.0001 -.0031 .0015	-.0001 -.0003 .0021	-.0002 -.0000 .0030

Table 1B

Biases of $T_{1\lambda}$ (top entry), $T_{2\lambda}$ (middle entry), and $T_{3\lambda}$ (bottom entry) Expressed in Units of σ. P*=0.99; c=2.576; λ_c=0.2692

δ \ λ	0	λ_c	1	$\sqrt{2}$
0.0	.0145 0 .0537	.0000 0 .0332	-.0392 0 -.0225	-.0615 0 -.0541
0.5	.0249 -.0045 .0709	.0084 .0162 .0479	-.0364 .0096 -.0144	-.0618 .0053 -.0497
1.0	.0579 -.0287 .1163	.0360 .0537 .0874	-.0236 .0347 .0090	-.0574 .0200 -.0355
1.5	.1119 -.1057 .1682	.0828 .0782 .1333	.0040 .0603 .0386	-.0406 .0363 -.0151
2.0	.1690 -.2823 .1896	.1338 .0411 .1527	.0384 .0590 .0527	-.0157 .0367 -.0039
2.5	.1989 -.5871 .1519	.0578 .0109 .0323	.1609 -.0889 .1197	-.0006 .0069 -.0172
3.0	.1823 -.9963 .0653	.1460 -.2909 .0438	.0474 -.0696 -.0145	-.0085 -.0432 -.0476
3.5	.1302 -1.4389 -.0243	.0995 -.4978 -.0323	.0162 -.1383 -.0539	-.0310 -.0832 -.0662
4.0	.0724 -1.8456 -.0746	.0495 -.6442 -.0706	-.0126 -.1595 -.0596	-.0478 -.0905 -.0534
5.0	.0106 -2.4808 -.0565	.0018 -.7028 -.0429	-.0219 -.0890 -.0058	-.0353 -.0406 .0152
6.0	.0006 -2.9991 -.0148	-.0015 -.5980 -.0060	-.0070 -.0223 .0178	-.0101 -.0068 .0313
7.0	.0000 -3.5000 -.0020	-.0003 -.4602 .0010	-.0010 -.0031 .0091	-.0015 -.0005 .0138

Table 2A

MSEs of $T_{1\lambda}$ (top entry), $T_{2\lambda}$ (middle entry), and $T_{3\lambda}$ (bottom entry) Expressed in Units of σ^2. P*=0.90; c=1.645; λ_c=0.7190.

δ \ λ	0	λ_c	1	$\sqrt{2}$
0.0	.3825 .2500 .3195	.3605 .2932 .3704	.3580 .2925 .3962	.3606 .2935 .4403
0.5	.3841 .2614 .3196	.3596 .2917 .3726	.3564 .2942 .3999	.3582 .2997 .4466
1.0	.3836 .3288 .3256	.3566 .2992 .3795	.3529 .3100 .4085	.3545 .3273 .4594
1.5	.3778 .5318 .3525	.3575 .3418 .3952	.3566 .3622 .4214	.3625 .3916 .4699
2.0	.3783 .9402 .4107	.3764 .4332 .4265	.3822 .4570 .4434	.3973 .4882 .4791
2.5	.3983 1.5628 .4878	.4188 .5508 .4697	.4320 .5652 .4732	.4568 .5818 .4891
3.0	.4337 2.3499 .5516	.4691 .6464 .5061	.4865 .6398 .4974	.5157 .6310 .4941
3.5	.4676 3.2456 .5780	.5051 .6857 .5198	.5218 .6552 .5039	.5485 .6244 .4877
4.0	.4885 4.2284 .5697	.5184 .6710 .5128	.5311 .6243 .4952	.5509 .5854 .4738
5.0	.4996 6.4994 .5247	.5100 .5842 .4909	.5142 .5412 .4789	.5206 .5187 .4627
6.0	.5000 9.2500 .5042	.5019 .5264 .4916	.5027 .5071 .4868	.5038 .5017 .4800
7.0	.5000 12.5002 .5004	.5002 .5058 .4973	.5003 .5007 .4961	.5004 .5001 .4943

Table 2B

MSEs of $T_{1\lambda}$ (top entry), $T_{2\lambda}$ (middle entry), and $T_{3\lambda}$ (bottom entry) Expressed in Units of σ^2. P*=0.99; c=2.576; λ_c=0.2692.

δ \ λ	0	λ_c	1	$\sqrt{2}$
0.0	.2729 .2500 .2687	.2717 .2637 .2721	.2723 .2587 .2830	.2750 .2574 .2902
0.5	.2852 .2514 .2753	.2829 .2658 .2802	.2816 .2612 .2960	.2839 .2600 .3065
1.0	.3167 .2667 .2922	.3114 .2707 .3011	.3045 .2685 .3305	.3055 .2688 .3505
1.5	.3518 .3399 .3153	.3423 .2821 .3287	.3281 .2853 .3751	.3275 .2904 .4079
2.0	.3722 .5600 .3482	.3602 .3210 .3626	.3426 .3272 .4189	.3424 .3405 .4618
2.5	.3750 1.0313 .4030	.3650 .4211 .4117	.3546 .4124 .4599	.3597 .4312 .5033
3.0	.3779 1.8026 .4836	.3748 .5957 .4800	.3825 .5353 .5007	.3970 .5474 .5322
3.5	.3983 2.8230 .5685	.4039 .8087 .5509	.4317 .6536 .5352	.4556 .6440 .5472
4.0	.4337 3.9832 .6227	.4452 .9916 .5951	.4848 .7167 .5492	.5128 .6801 .5421
5.0	.4885 6.4663 .5989	.4986 1.1131 .5711	.5284 .6534 .5107	.5468 .5989 .4862
6.0	.4996 9.2491 .5302	.5030 1.0119 .5151	.5128 .5456 .4783	.5186 .5197 .4602
7.0	.5000 12.5001 .5047	.5006 .8689 .4994	.5023 .5072 .4856	.5033 .5016 .4783

cide with $\hat{M}_\lambda$ and t_c of Dahiya. For $T_{1\lambda}$, our choices for values of λ are included in Dahiya's tables. For $T_{2\lambda}$, $\lambda = 1$ is the only common choice. However, we do not report our values in this case, as it is not of main interest here. Finally, Dahiya defines a hybrid estimator H_c for his problem; however, H_c is really our T. But his c-values are chosen arbitrarily and they do not correspond to P*-values of common interest.

Remark 4.3. Though the biases and MSEs are tabulated here only for P* = 0.90 and 0.99, the comparisons made below include also the values for P* = 0.95 available in a technical report by the authors (1983).

Now, considering the biases, we see that $T_{2\lambda}$ performs better than $T_{1\lambda}$ and $T_{2\lambda}$ for small values of δ ($\delta \leq 1$ for P* = 0.90 and $\delta \leq 0.5$ for P* = 0.95, 0.99). As δ increases, $T_{2\lambda}$ becomes increasingly bad without adjustment for bias (i.e., $\lambda = 0$); however, with λ increasing, the bias of $T_{2\lambda}$ is very much reduced. For large δ ($\delta \geq 5$ for P* = 0.90, $\delta \geq 6$ for P* = 0.95, 0.99), $T_{1\lambda}$ performs better than $T_{3\lambda}$ and is generally better than $T_{2\lambda}$ as well. For small δ ($\delta \leq 1$ for P* = 0.90, 0.95 and $\delta \leq 2$ for P* = 0.99) $T_{1\lambda}$ performs better than $T_{3\lambda}$ for $\lambda \leq \lambda_c$. For moderate δ, $T_{3\lambda}$ is better than $T_{1\lambda}$ for $\lambda \leq \lambda_c$.

However, from the more important point of view of MSE, an overall picture emerges as follows: For $\delta \leq 1$, $T_{2\lambda}$ is the best. For $1.5 \leq \delta \leq 3$, $T_{1\lambda}$ is the best. When $3.5 \leq \delta \leq 4$, $T_{1\lambda}$ with small λ or $T_{3\lambda}$ with large λ is the best. Finally, for $\delta \geq 5$, $T_{3\lambda}$ with λ away from zero is the best.

Taking the bias and MSE together into account, the relative performances of these estimators can be largely summarized as follows: For $\delta \leq 2$, $T_{2\lambda}$ is the best. For $2 < \delta < 5$, one of $T_{1\lambda}$ and $T_{3\lambda}$ is always better than or as good as $T_{2\lambda}$. When $\delta \geq 5$, $T_{2\lambda}$ is never the best. However, for $\delta > 2$, there is no clear choice between $T_{1\lambda}$ and $T_{3\lambda}$ because the bias and the MSE pull each in the opposite directions.

SUMMARY

Let $\pi_1,\ldots,\pi_k$ be k independent normal populations with unknown means $\theta_1,\ldots,\theta_k$, respectively, and a common known variance τ^2. Let $Y_1,\ldots,Y_k$ denote the means of independent samples of size n from $\pi_1,\ldots,\pi_k$. For selecting a nonempty subset containing the best population (the one associated with the largest θ_i) we consider the rule of Gupta which selects π_i if and only if $Y_i \geq \max_{1\leq j\leq k} Y_j - d\tau/\sqrt{n}$, where $d > 0$ is determined such that the probability of including the best population in the selected subset is at least equal to a preassigned level P^* $(1/k < P^* < 1)$. We are interested in estimating M, the mean of all the θ_i's corresponding to the populations that are selected. In this paper we consider the case of k = 2. In Section 2, we define the 'natural' estimator T and three classes of estimators $T_{1\lambda}$, $T_{2\lambda}$ and $T_{3\lambda}$, obtained by making adjustments for the bias of T. The biases and the MSEs of $T_{1\lambda}$, $T_{2\lambda}$ and $T_{3\lambda}$ are discussed in Sections 2 and 3 respectively. Numerical comparisons of the performances of these estimators are made in Section 4.

APPENDIX 1

Proof of Theorem 2.1. Suppose that $\hat{M}$ is an unbiased estimator of M with finite MSE($\hat{M}$). Then it follows that $\hat{M}$ is an unbiased estimator of E(M) with finite variance, $V(\hat{M}) = E(\hat{M} - E(M))^2$. Thus it is enough to show that no unbiased estimator of E(M) with finite variance exists. Since Y^* and Y are jointly sufficient for E(M), we restrict our attention to unbiased estimators that are functions of Y^* and Y. Let $\eta(Y^*,Y)$ be an unbiased estimator of E(M) such that $V(\eta) < \infty$. Let Z be such that it is independent of Y^* and Y, and $Z \sim N(\delta/b^2, 1/b^2)$, where $b > 0$ is known. Applying the Neyman Factorization Theorem to the joint distribution of Y^*, Y and Z, we see that Y^* and $W = Z + Y/\sigma$ are sufficient for θ^* and δ. Thus $\eta_1 = E[\eta|(Y^*,W)]$ is a function of Y^* and W only. Also, $E(\eta_1) = E(\eta)$ and $V(\eta_1) \leq V(\eta)$ by Rao-Blackwell Theorem. Now, let

$$\eta_2 = [\frac{Y^*}{2} + \frac{b^2W}{1+b^2}\{\Phi(bc^* + DW) - \Phi(bc^* - DW)\}]$$

$$- [D\phi\{bc^* + DW\} + D\phi\{bc^* - DW\}]$$

$$= R_1 - R_2, \text{ say,}$$

where $bc^* = c\sqrt{1+b^2}$ and $D = b^2/\sqrt{1+b^2}$. Using (A2.7) and (A2.8), it can be seen that $E(\eta_2) = E(M)$. Since $E(\eta_1 - \eta_2) = 0$, it follows that $\eta_1 = \eta_2$ with probability 1 by the completeness of (Y^*,W). Thus $V(\eta_2) = V(\eta_1) \leq V(\eta) < \infty$.

To complete the proof, we obtain a contradiction by showing that $V(\eta_2)$ can be increased indefinitely by letting $b \to \infty$. To see this, we first note that $V(R_1) < \infty$ which implies that $V(R_2) < \infty$. It is also easy to see that $0 < E(R_2) < \infty$. Hence, $E(R_2^2) < \infty$ which implies that $E[D^2\phi^2\{bc^* + DW\}]$

$< \infty$. On the other hand, using (A2.7) and the fact that $\phi^2(a) = \phi(\sqrt{2}a)/\sqrt{2\pi}$, we can see that

$$E[D^2\phi^2\{bc^* + DW\}] = \frac{1}{\sqrt{2\pi}} \frac{b^4}{1+b^2} \frac{1}{\sqrt{1+2b^2}} \phi\left\{\frac{\sqrt{2}b(\delta+c^*)}{\sqrt{1+2b^2}}\right\},$$

a quantity which tends to infinity as $b \to \infty$. □

APPENDIX 2

We state a few results relating to normal distribution. We omit the proofs, which are straightforward.

$$\int_a^b t\phi(t)\,dt = \phi(a) - \phi(b) . \tag{A2.1}$$

$$\int_{-\infty}^{\infty} \phi\{\alpha(t+\beta)\}\phi(t)\,dt = \frac{1}{\sqrt{1+\alpha^2}} \phi\left(\frac{\alpha\beta}{\sqrt{1+\alpha^2}}\right) . \tag{A2.2}$$

$$\int_a^b t^2\phi(t)\,dt = a\phi(a) - b\phi(b) + \Phi(b) - \Phi(a) . \tag{A2.3}$$

$$\int_a^{\infty} t\phi(t)\phi(t+\alpha)\,dt = \frac{1}{2}\,\phi\left(\frac{\alpha}{\sqrt{2}}\right)\left[\phi\left(\sqrt{2}a + \frac{\alpha}{\sqrt{2}}\right) + \frac{\alpha}{\sqrt{2}}\,\Phi\left(\sqrt{2}a + \frac{\alpha}{\sqrt{2}}\right) - \frac{\alpha}{\sqrt{2}}\right] . \tag{A2.4}$$

$$\int_{-\infty}^{-b} t\phi(t)\phi(t+\alpha)\,dt = -\frac{1}{2}\,\phi\left(\frac{\alpha}{\sqrt{2}}\right)\left[\phi\left(\sqrt{2}b - \frac{\alpha}{\sqrt{2}}\right) + \frac{\alpha}{\sqrt{2}}\,\Phi\left(\frac{\alpha}{\sqrt{2}} - \sqrt{2}b\right)\right] . \tag{A2.5}$$

$$\int_{-\infty}^{\infty} \phi(t+\alpha)\phi(t+\beta)\phi(t)\,dt = \frac{1}{\sqrt{6\pi}}\,\phi\left\{\sqrt{\frac{2(\alpha^2+\beta^2-\alpha\beta)}{3}}\right\} . \tag{A2.6}$$

Let $Y \sim N(\theta,\sigma^2)$ and $\delta = \theta/\sigma$. Then, using (A2.2) on each expression,

$$E\left[\phi\left\{\alpha\left(\frac{Y}{\sigma} + c\right)\right\}\right] = \frac{1}{\sqrt{1+\alpha^2}}\,\phi\left\{\frac{\alpha(\delta+c)}{\sqrt{1+\alpha^2}}\right\} , \tag{A2.7}$$

$$E\left[Y\Phi\left\{\alpha\left(\frac{Y}{\sigma} + c\right)\right\}\right] = \theta\Phi\left\{\frac{\alpha(\delta+c)}{\sqrt{1+\alpha^2}}\right\} + \frac{\sigma\alpha}{\sqrt{1+\alpha^2}}\,\phi\left\{\frac{\alpha(\delta+c)}{\sqrt{1+\alpha^2}}\right\} . \tag{A2.8}$$

REFERENCES

Abramowitz, M. and Stegun, I. A. (1965). Handbook of Mathematical Functions. Dover, New York.

Berger, R. L. and Gupta, S. S. (1980). Minimax subset selection rules with applications to unequal variance (unequal sample size) problems. Scand. J. Statist. 7, 21-26.

Cohen, A. and Sackrowtiz, H. B. (1982). Estimating the mean of the selected population. Statistical Decision Theory and Related Topics III, Vol. 1, S. S. Gupta and J. O. Berger (eds.). Academic Press, New York.

Dahiya, R. C. (1974). Estimation of the mean of the selected population. J. Amer. Statist. Assoc. 69, 226-230.

Gupta, S. S. (1956). On a decision rule for a problem in ranking means. Ph.D. Thesis (Mimeo. Ser. No. 150), Institute of Statistics, University of North Carolina, Chapel Hill.

Gupta, S. S. (1965). On some multiple decision (selection and ranking) rules. Technometrics 7, 225-245.

Gupta, S. S. and Panchapakesan, S. (1979). Multiple Decision Procedures: Theory and Methodology of Selecting and Ranking Populations. John Wiley, New York.

Hsieh, H.-K. (1981). On estimating the mean of the selected population with unknown variance. Comm. Statist. A - Theory Methods 10, 1869-1878.

Jeyaratnam, S. and Panchapakesan, S. (1983). An estimation problem relating to subset selection for normal populations. Tech. Report No. 83-38, Department of Statistics, Purdue University, West Lafayette, Indiana.

Sarkadi, K. (1967). Estimation after selection. Studia Sci. Math. Hungar. 2, 341-350.